DEC 20 2007

12/08 - 9x

East Meadow Public Library
East Meadow, New York
516-794-2570
www.eastmeadow.info

Oxford Graduate Texts in Mathematics

Series Editors

S. K. Donaldson   S. Hildebrandt
M. J. Taylor   R. Cohen

OXFORD GRADUATE TEXTS IN MATHEMATICS

---

1. Keith Hannabuss: *An introduction to quantum theory*
2. Reinhold Meise and Dietmar Vogt: *Introduction to functional analysis*
3. James G. Oxley: *Matroid Theory*
4. N.J. Hitchin, G.B. Segal, and R.S. Ward: *Integrable systems: twistors, loop groups, and Riemann surfaces*
5. Wulf Rossmann: *Lie groups: an introduction through linear groups*
6. Q. Liu: *Introduction to algebraic curves over Dedekind domains*

Lie Groups

An Introduction Through Linear Groups

Wulf Rossmann

Department of Mathematics and Statistics
University of Ottawa

OXFORD
UNIVERSITY PRESS
Great Clarendon Street, Oxford OX2 6DP
Oxford University Press is a department of the University of Oxford.
It furthers the University's objective of excellence in research, scholarship,
and education by publishing worldwide in
Oxford New York
Auckland Cape Town Dar es Salaam Hong Kong Karachi
Kuala Lumpur Madrid Melbourne Mexico City Nairobi
New Delhi Shanghai Taipei Toronto
With offices in
Argentina Austria Brazil Chile Czech Republic France Greece
Guatemala Hungary Italy Japan South Korea Poland Portugal
Singapore Switzerland Thailand Turkey Ukraine Vietnam
Published in the United States
by Oxford University Press Inc., New York
© W. Rossmann, 2002
The moral rights of the author have been asserted
Database right Oxford University Press (maker)
First published 2002
Reprinted 2003 (with corrections), 2004, 2005 (with corrections)
All rights reserved. No part of this publication may be reproduced,
stored in a retrieval system, or transmitted, in any form or by any means,
without the prior permission in writing of Oxford University Press,
or as expressly permitted by law, or under terms agreed with the appropriate
reprographics rights organizations. Enquiries concerning reproduction
outside the scope of the above should be sent to the Rights Department,
Oxford University Press, at the address above
You must not circulate this book in any other binding or cover
and you must impose this same condition on any acquirer
British Library Cataloguing in Publication Data
Data available
Library of Congress Cataloging in Publication Data
Rossmann, Wulf, 1948–
Lie groups: an introduction through linear groups/Wulf Rossmann.
p. cm. — (Oxford graduate texts in mathematics; 5)
Includes bibliographical references and index.
1. Lie groups. I. Title. II. Series.
QA387 R68 2002 512′.55—dc21 2001050008
ISBN 0 19 859683 9 (Hbk)
10 9 8 7 6 5 4
Typeset by
Newgen Imaging Systems (P) Ltd., Chennai, India
Printed in Great Britain
on acid-free paper by
Biddles Ltd., King's Lynn, Norfolk

Preface

This book was started many years ago, in the enthusiasm of expounding the theory of Lie groups for the first time to a class of advanced undergraduate and beginning graduate students. As it turned out, most of them did not have the equipment required for the customary path to Lie groups and representation theory, lacking things like differential geometry, functional analysis, or general topology. So I decided to look for an alternative route, starting straight from calculus and linear algebra. Somewhat to my surprise, this lowly starting point turned out to be perfectly satisfactory, so much so that I have come to prefer it myself: it now seems natural and appropriate for an introduction. Over the years, when I taught Lie groups again, I might take shortcuts and add detours, but the basic course remained the same. The trail of lecture notes left on the way was gathered into a set of notes, which eventually became this book.

The route is through linear groups. By this I mean any group of matrices, a notion which is both narrower and broader than is customary in Lie theory: it is narrower in that the groups are required to be linear and it is broader in that they are not required to be manifolds at the outset. (Nor indeed are manifolds themselves required at that point.) That it is possible to base a satisfactory theory on such simple foundations is not new: in the customary development it usually transpires at some point that presupposing a manifold structure is unnecessary in so far as every connected subgroup of a Lie group is of its own a manifold in a natural way. Yet, as far as I am aware, this circumstance has never been taken as basis for the theory. Apart from questions of taste and outlook, one explanation may be that the proof of the Lie correspondence between connected subgroups and Lie subalgebras customarily proceeds by Frobenius's theory of involutive differential systems, where the manifolds are appropriate and linearity is useless. Here an elementary argument, based essentially on the inverse function theorem, fills in what is needed. Be that as it may, a desire to shorten the list of prerequisites is not the only reason for the point of view taken: the restriction to linear groups seems desirable to me, even if the prerequisites are available; for it puts into focus from the beginning the essential aspects of the theory, free of technicalities. The help with matters of notation is also welcome, be it only writing matrix products instead of differentials of left and right translation. As further benefit, the presence of the matrix exponential allows a strikingly simple formulation of the basic fact of Lie theory: there is a natural one-to-one correspondence between connected linear groups and linear Lie algebras. Some care needs to be taken with the definition of the topology for these groups, but that is easily done, and quite appropriately in calculus style. Additional restrictions, such as closedness or even compactness, would not simplify the exposition, but would rather have the opposite effect.

Thus the prerequisites are modest. What is required is an understanding of linear algebra and multivariable calculus, such as might be gained from standard undergraduate courses, together with the rudiments of abstract group theory. This material can be found in most textbooks on these subjects, for example in Hoffman–Kunze (1961), Marsden (1974), and Herstein (1964) and is reviewed in the Appendix on page 250. The simplicity of the prerequisites is not an awkward constraint, but is entirely appropriate for the subject as presented here. Auxiliary material beyond these basics is developed as needed, less an interruption than an opportunity of acquiring new concepts in a meaningful setting, where they can be put to immediate use. In spite of its elementary nature, this book covers the essentials of the theory of Lie groups, so that it can serve not only as introduction to the subject for its own sake, but also as preparation for an advanced study of Lie groups and their representations.

Among the essentials of the theory of Lie groups I include the Cartan–Weyl theory of finite dimensional representations of semisimple (or reductive) groups. I present this theory 'by example', discussing first $\mathrm{U}(n)$ and $\mathrm{GL}(n, \mathbb{C})$, then the modifications needed for the other classical groups. I did not want to go into the general structure theory of semisimple Lie algebras, since there are already several expositions of the algebraic theory that could complement what is presented here. Nor did I want to get involved in the questions of functional analysis needed to prove the Peter–Weyl Theorem for arbitrary compact groups.

While the structure theory of semisimple groups has not been developed in general, the classical groups are discussed in considerable detail. Structural paraphernalia, such as roots, are introduced as auxiliary notions for the study of the classical groups. My hope is that this discussion will complement the abstract treatments found elsewhere.

Chapters 1 and 2 contain what is known as 'Lie theory' in the context of linear groups. Together with parts of chapters 3 or 4, these chapters could serve for a one-semester course at an entirely elementary level. At a somewhat more advanced level, some of the material chapters 1–4 could be omitted or left to the student and the emphasis placed on representation theory (chapter 6) instead. I have taught in this way one-semester courses based on chapter 6, covering only the bare essentials needed from previous chapters.

A course on Lie groups, as proposed here, could well be combined with a course on Lie algebras, based for example on Humphreys (1972), Jacobson (1962), or Serre (1966). The two courses could be given in sequence (in either order), or concurrently. There would be no duplication; the two courses would rather complement each other.

The numerous problems should not intimidate: they are generally not required within the text, but are intended to supplement the theory. They are not research projects, but exercises for the weekly problem set.

Here and there I have added some historical tidbits, not with scholarly intentions, but only because I found them remarkable or curious. The references, historical or otherwise, are kept to a minimum, certainly not intended as a guide to the literature.

Theorems, equations, figures, etc. are numbered independently in each section. A parenthetical (QED) asks that a proof be supplied.

W. R.

To Michin and Erik

Contents

1

The exponential map

1.1 Vector fields and one-parameter groups of linear transformations[1]

A linear transformation X of the n-dimensional real or complex vector space E can be thought of as a vector field: X associates to each point p in E, the vector $X(p)$. As an aid to one's imagination one may think of a fluid in motion, so that the velocity of the fluid particles passing through the point p is always $X(p)$; the vector field is then the *current* of the flow and the paths of the fluid particles are the *trajectories*. Of course, this type of flow is very special, in that the current $X(p)$ at p is independent of time (*stationary*) and depends linearly on p. Figure 1 shows the trajectories of such flows in the plane. (We shall see later, that in appropriate linear coordinates every plane flow corresponds to exactly one of these pictures.)

Consider the trajectory of the fluid particle that passes a certain point p_0 at time $\tau = \tau_0$. If $p(\tau)$ is its position at time τ, then its velocity at this time is $p'(\tau) = dp(\tau)/d\tau$. Since the current at $p(\tau)$ is supposed to be $X(p(\tau))$, one finds that $p(\tau)$ satisfies the differential equation

$$p'(\tau) = X(p(\tau)) \tag{1}$$

together with the initial condition

$$p(0) = p_0. \tag{2}$$

The solution of (1) and (2) may be obtained as follows. Try for $p(\tau)$ a power series in τ:

$$p(\tau) = \sum_{k=0}^{\infty} \tau^k p_k$$

[1] Refer to the appendix to §1.1 for an explanation of the notation as necessary.

with coefficients $p_k \in E$ to be determined. Ignoring convergence question for the moment, eqn (1) becomes

$$\sum_{k=1}^{\infty} k\tau^{k-1} p_k = \sum_{k=0}^{\infty} \tau^k X^k(p_k),$$

which leads to

$$p_{k+1} = \frac{1}{k+1} X(p_k).$$

These equations determine p_k inductively in terms of p_0:

$$p_k = \frac{1}{k!} X^k(p_0).$$

We write the solution $p(\tau)$ thus found in the form

$$p(\tau) = \exp(\tau X) p_0, \tag{3}$$

where the *exponential* of a matrix X is defined by

$$\exp X = \sum_{k=0}^{\infty} \frac{1}{k!} X^k.$$

This exponential function is a fundamental concept, which underlies just about everything to follow. First of all, the exponential series converges in norm for all matrices X, and (3) represents a genuine solution of (1) and (2). (See the appendix to this section.) We shall see momentarily that it is in fact the *only* differentiable solution, but first we shall prove some basic properties of the matrix exponential:

Proposition 1.
(a) For any matrix X,

$$\frac{d}{d\tau} \exp \tau X = X \exp(\tau X) = \exp(\tau X) X,$$

and $a(\tau) = \exp \tau X$ is the unique differentiable solution of

$$a'(\tau) = X a(\tau), \quad a(0) = 1$$

(and also of $a'(\tau) = a(\tau)X, a(0) = 1$).
(b) For any two commuting matrices X, Y,

$$\exp X \exp Y = \exp(X + Y).$$

(c) $\exp X$ is invertible for any matrix X and

$$(\exp X)^{-1} = \exp(-X).$$

(d) For any matrix X and scalars σ, τ

$$\exp(\sigma + \tau) X = \exp(\sigma X) \exp(\tau X),$$

and $a(\tau) = \exp \tau X$ is the unique differentiable solution of $a(\sigma + \tau) = a(\sigma) a(\tau)$, $a(0) = 1$, $a'(0) = X$.

Proof.

(a) Because of the norm-convergence of the power series for $\exp \tau X$ (appendix), we can differentiate it term-by-term:

$$\frac{d}{d\tau}(\exp \tau X) = \frac{d}{d\tau}\sum_{j=0}^{\infty}\frac{\tau^j}{j!}X^j = \sum_{j=1}^{\infty}\frac{\tau^{j-1}}{(j-1)!}X^j,$$

and this is $X\exp(\tau X) = \exp(\tau X)X$, as one sees by factoring out X from this series, either to the left or to the right.

To prove the second assertion of (a), assume $a(\tau)$ satisfies

$$a'(\tau) = X(a(\tau)), \quad a(0) = 1.$$

Differentiating according to the rule $(ab)' = a'b + ab'$ (appendix) we get

$$\begin{aligned}\frac{d}{d\tau}(\exp(-\tau X)a(\tau)) &= \left(\frac{d}{d\tau}\exp -\tau X\right)a(\tau) + \exp(-\tau X)\left(\frac{d}{d\tau}a(\tau)\right)\\ &= \exp(-\tau X)(-X)a(\tau) + \exp(-\tau X)Xa(\tau) = 0.\end{aligned}$$

So $\exp(-\tau X)a(\tau)$ is independent of τ, and equals 1 for $\tau = 0$, hence it equals 1 identically. The assertion will now follow from (c).

(b) As for scalar series, the product of two norm-convergent matrix series can be computed by forming all possible products of terms of the first series with terms of the second and then summing in any order. If we apply this recipe to $\exp X \exp Y$, we get,

$$\exp X \exp Y = \left(\sum_{j=0}^{\infty}\frac{X^j}{j!}\right)\left(\sum_{k=0}^{\infty}\frac{Y^k}{k!}\right) = \sum_{j,k=0}^{\infty}\frac{X^jY^k}{j!k!}. \tag{4}$$

On the other hand, *assuming* X and Y *commute*, we can rearrange and collect terms while expanding $(1/m!)(X+Y)^m$ to get

$$\exp(X+Y) = \sum_{m=0}^{\infty}\frac{1}{m!}(X+Y)^m = \sum_{m=0}^{\infty}\frac{1}{m!}\left(\sum_{j+k=m}\frac{m!}{j!k!}X^jY^k\right) = \sum_{j,k=0}^{\infty}\frac{X^jY^k}{j!k!}. \tag{5}$$

Comparison of (4) and (5) gives (b).

(c) This follows from (b) with $Y = -X$.

(d) The first assertion comes from (b) with X replaced by σX and Y by τX. To prove the second assertion, assume $a(\tau)$ has the indicated property. Then

$$a'(\tau) = \frac{d}{d\sigma}a(\tau+\sigma)\Big|_{\sigma=0} = a(\tau)\,\frac{d}{d\sigma}a(\sigma)\Big|_{\sigma=0} = a(\tau)X, \tag{6}$$

and the assertion follows from the second part of (a). QED

Remark 2. For (b) it is essential that X and Y commute. In fact, the following statements are equivalent:

(a) X and Y commute.

(b) $\exp \sigma X$ and $\exp \tau Y$ commute for all scalars σ, τ.

(c) $\exp(\sigma X + \tau Y) = \exp(\sigma X)\exp(\tau Y)$ for all scalars σ, τ. (QED)

It is now easy to see that $p'(\tau) = X(p(\tau))$ and $p(0) = p_0$ implies $p(\tau) = \exp(\tau X)p_0$: that $p(\tau) = \exp(\tau X)p_0$ satisfies $p'(\tau) = X(p(\tau))$ and $p(0) = p_0$ is clear from part (a) of the proposition; and the uniqueness of this solution is seen by differentiating $\exp(-\tau X)p(\tau)$ as in the proof of the uniqueness of (a).

The properties of the matrix exponential summarized in the proposition are of basic importance. Parts (c) and (d) may be rephrased by saying that, for fixed X, the map $\tau \to \exp \tau X$ is a *homomorphism* of the group of scalars under addition ($\mathbb{R}$ or $\mathbb{C}$) into the *general linear group* $\mathrm{GL}(E)$ of all invertible linear transformations of E. This map $\tau \to \exp \tau X$ is called the *one-parameter group generated by* X, even though it is a group homomorphism rather than a group.

The group property $\exp(\sigma X)\exp(\tau X) = \exp(\sigma + \tau)X$ of exp is intimately related to the stationary property of flow in the physical picture: it may be interpreted as saying that two fluid particles passing through the same point at different times will travel along the same path, passing through the same points, after equal time intervals.

There is another simple property of exp that is frequently used:

Proposition 3. *For any matrix X and any invertible matrix a,*

$$a(\exp X)a^{-1} = \exp(aXa^{-1}).$$

Proof.

$$a\exp(X)a^{-1} = a\left(\sum_{k=0}^{\infty} \frac{X^k}{k!}\right)a^{-1} = \sum_{k=0}^{\infty} \frac{(aXa^{-1})^k}{k!} = \exp(aXa^{-1}).$$

QED

Incidentally, the property of exp expressed by the proposition is shared by any convergent matrix-power series. It only relies on the fact that the *conjugation* operation $X \to aXa^{-1}$ in the matrix space $M = L(E)$ is linear and preserves product of matrices:

$$a(\alpha X + \beta Y)a^{-1} = \alpha(aXa^{-1}) + \beta(aYa^{-1}),$$
$$a(XY)a^{-1} = (aXa^{-1})(aYa^{-1}).$$

Proposition 3 is often useful for computing matrix exponentials. For example, if X is diagonalizable, then $X = aYa^{-1}$, where Y is the diagonal, and $\exp X = a(\exp Y)a^{-1}$. The exponential of a diagonal matrix is easy to compute:

$$\exp\begin{bmatrix}\lambda_1 & & \\ & \ddots & \\ & & \lambda_n\end{bmatrix} = \sum_{k=0}^{\infty}\frac{1}{k!}\begin{bmatrix}\lambda_1^k & & \\ & \ddots & \\ & & \lambda_n^k\end{bmatrix} = \begin{bmatrix}e^{\lambda_1} & & \\ & \ddots & \\ & & e^{\lambda_n}\end{bmatrix}.$$

Example 4 (Exponential of 2×2 matrices).

(i) Every real 2×2 matrix is conjugate to exactly one of the following types with $\alpha, \beta \in \mathbb{R}$, $\beta \neq 0$.

(a) *Elliptic:* $\alpha \begin{bmatrix} 1 & 0 \\ 0 & 1 \end{bmatrix} + \beta \begin{bmatrix} 0 & -1 \\ 1 & 0 \end{bmatrix}$. (b) *Hyperbolic:* $\alpha \begin{bmatrix} 1 & 0 \\ 0 & 1 \end{bmatrix} + \beta \begin{bmatrix} 0 & 1 \\ 1 & 0 \end{bmatrix}$.

(c) *Parabolic:* $\alpha \begin{bmatrix} 1 & 0 \\ 0 & 1 \end{bmatrix} + \beta \begin{bmatrix} 0 & 0 \\ 1 & 0 \end{bmatrix}$. (d) *Scalar:* $\alpha \begin{bmatrix} 1 & 0 \\ 0 & 1 \end{bmatrix}$.

(ii) The matrices X given above in (a)–(d) generate the following one-parameter groups $\exp \tau X$.

(a) *Elliptic:* $e^{\alpha\tau} \begin{bmatrix} \cos \beta\tau & -\sin \beta\tau \\ \sin \beta\tau & \cos \beta\tau \end{bmatrix}$. (b) *Hyperbolic:* $e^{\alpha\tau} \begin{bmatrix} \cosh \beta\tau & \sinh \beta\tau \\ \sinh \beta\tau & \cosh \beta\tau \end{bmatrix}$.

(c) *Parabolic:* $e^{\alpha\tau} \begin{bmatrix} 1 & 0 \\ \beta\tau & 1 \end{bmatrix}$. (d) *Scalar:* $e^{\alpha\tau} \begin{bmatrix} 1 & 0 \\ 0 & 1 \end{bmatrix}$.

The normal forms in (i) may be derived by manipulating with eigenvalues and eigenvectors, which we omit. As a sample calculation with exp, we verify the formulas in (ii).

(a)

$$\exp\left(\alpha\tau \begin{bmatrix} 1 & 0 \\ 0 & 1 \end{bmatrix} + \beta\tau \begin{bmatrix} 0 & -1 \\ 1 & 0 \end{bmatrix}\right) = \exp\left(\alpha\tau \begin{bmatrix} 1 & 0 \\ 0 & 1 \end{bmatrix}\right) \exp\left(\beta\tau \begin{bmatrix} 0 & -1 \\ 1 & 0 \end{bmatrix}\right)$$

because the matrices commute. The first factor is $e^{\alpha\tau} \begin{bmatrix} 1 & 0 \\ 0 & 1 \end{bmatrix}$.

To evaluate the second one, note that

$$\begin{bmatrix} 0 & -1 \\ 1 & 0 \end{bmatrix}^{2k} = (-1)^k \begin{bmatrix} 1 & 0 \\ 0 & 1 \end{bmatrix}, \quad \begin{bmatrix} 0 & -1 \\ 1 & 0 \end{bmatrix}^{2k+1} = (-1)^k \begin{bmatrix} 0 & -1 \\ 1 & 0 \end{bmatrix}.$$

Therefore,

$$\exp\left(\tau\beta \begin{bmatrix} 0 & -1 \\ 1 & 0 \end{bmatrix}\right) = \sum_{k=0}^{\infty} \frac{(-1)^k (\tau\beta)^{2k}}{(2k)!} \begin{bmatrix} 1 & 0 \\ 0 & 1 \end{bmatrix} + \sum_{k=0}^{\infty} \frac{(-1)^k (\tau\beta)^{2k+1}}{(2k+1!)} \begin{bmatrix} 0 & -1 \\ 1 & 0 \end{bmatrix}.$$

Recognizing the sin and cos series one gets (a).

Comment. The exponential just calculated may also be found by diagonalization:

$$\begin{bmatrix} 0 & -1 \\ 1 & 0 \end{bmatrix} = \begin{bmatrix} 1 & -i \\ -i & 1 \end{bmatrix}^{-1} \begin{bmatrix} -i & 0 \\ 0 & i \end{bmatrix} \begin{bmatrix} 1 & -i \\ -i & 1 \end{bmatrix}.$$

(b) This is similar.

(c) In this case, one actually gets a finite sum for exp of the second (non-scalar) summand in part (c) of (i):

$$\exp \tau \begin{bmatrix} 0 & 0 \\ 1 & 0 \end{bmatrix} = \begin{bmatrix} 1 & 0 \\ 0 & 1 \end{bmatrix} + \tau \begin{bmatrix} 0 & 0 \\ 1 & 0 \end{bmatrix}.$$

This kind of thing will evidently, always happen if one takes the exponential of a nilpotent matrix, i.e. a matrix X for which $X^k = 0$ for some k.

Using the formulas in (ii) one may verify that the pictures in Figure 1 correspond to the normal forms of the vector fields in (i).

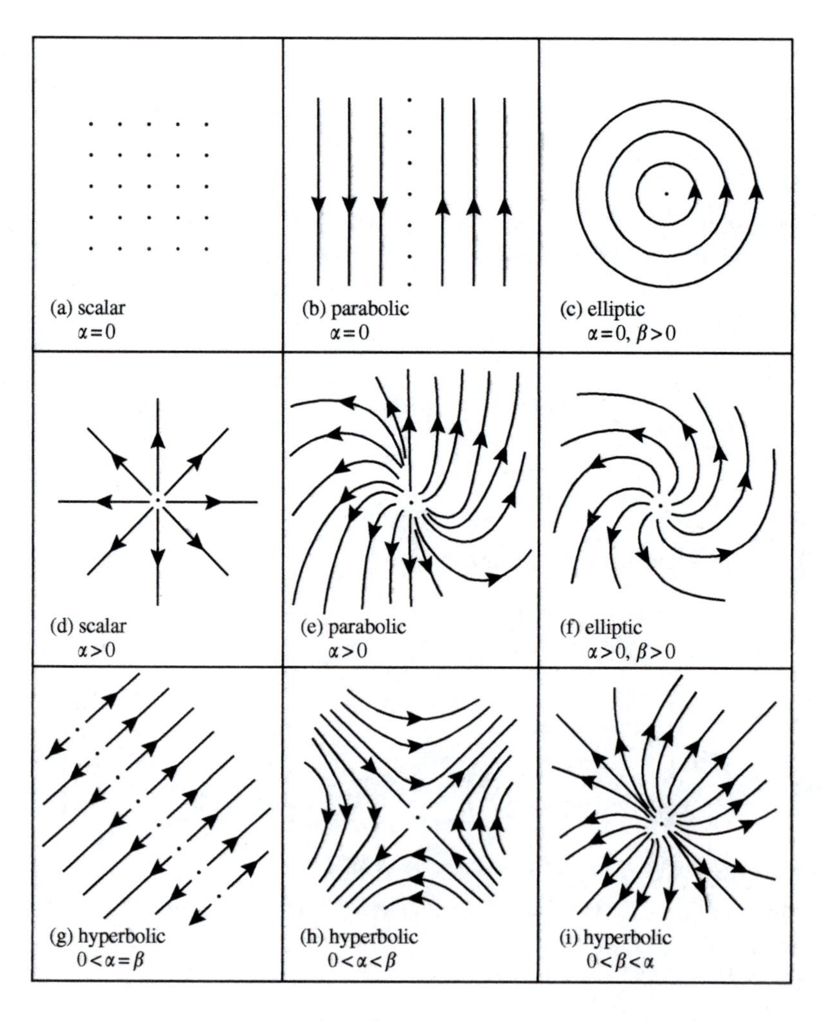

Fig. 1

Problems for §1.1

1. Prove Remark 2.

2. Verify the formula for exponential in the hyperbolic case given in Example 4: (a) by power series calculation, (b) by diagonalization.

3. Show that $\exp \tau X$ is orthogonal (unitary) for all $\tau \in \mathbb{R}$ if and only if X is skew-symmetric (skew-Hermitian).

4. Let $E = \mathbb{R}^3$, $0 \neq x \in E$, X be the linear transformation of $p \to (x \times p)$ of E (cross-product).

   (a) Choose a right-handed orthonormal basis (e_1, e_2, e_3) for E *with* e_3 a unit vector parallel to x. Show that

   $$\begin{aligned} \exp(X)e_1 &= \cos\|x\|e_1 + \sin\|x\|e_2, \\ \exp(X)e_2 &= -\sin\|x\|e_1 + \cos\|x\|e_2, \\ \exp(X)e_3 &= e_3. \end{aligned}$$

   [For the purpose of this problem, an ordered orthonormal basis (e_1, e_2, e_3) for E may be defined to be *right-handed*, if it satisfies the usual cross-product relation given by the 'right-hand rule', i.e.

   $$e_1 \times e_2 = e_3, \quad e_2 \times e_3 = e_1, \quad e_3 \times e_1 = e_2.$$

   Any orthonormal basis can be ordered so that it becomes right–handed. The above formula means that $\exp(X)$ is the right-handed rotation around x with an angle $\|x\|$.]

   (b) Show that

   $$\exp(X) = 1 + \frac{\sin\|x\|}{\|x\|}(X) + \frac{1 - \cos\|x\|}{\|x\|^2}(X)^2.$$

5. Show:

$$\exp\begin{bmatrix} \lambda & 1 & 0\cdots & 0 \\ 0 & \lambda & 1\cdots & 0 \\ \vdots & \vdots & \vdots & \vdots \\ 0 & 0 & 0\cdots & \lambda \end{bmatrix} = \begin{bmatrix} e^\lambda & e^\lambda & e^\lambda/2!\cdots & e^\lambda/(n-1)! \\ 0 & e^\lambda & e^\lambda\cdots & e^\lambda/(n-2)! \\ \vdots & \vdots & \vdots & \vdots \\ 0 & 0 & 0\cdots & e^\lambda \end{bmatrix}.$$

6. Let E be the space of all polynomials $f(\xi) = c_0 + c_1\xi + \cdots + c_{n-1}\xi^{n-1}$ of degree $\leq(n-1)$ (some fixed n; the coefficients c_j can be taken either real or complex). The derivative $Df(\xi) = f'(\xi)$ defines a linear transformation D of E. Show:

   $$\exp(\tau D)f(\xi) = f(\xi + \tau).$$

   This problem is related to the previous one. Explain how.

7. Show that
$$\begin{bmatrix} -1 & 1 \\ 0 & -1 \end{bmatrix}$$
is not the exponential of any real 2×2 matrix. [Suggestion: consider eigenvalues and eigenvectors.]

8. An *affine transformation* of E is a transformation of the form $p \to ap + b$ where $a \in M$ is linear and $b \in E$ operates as a translation $p \to p + b$. An affine transformation may again be interpreted as a vector field on E, called *affine vector field*, typically denoted by $p \to Xp + v$ where $X \in M$ and $v \in E$.

   Think of E as embedded on the hyperplane $E^+ = \{(p, 1)\}$ in a space $E \times \mathbb{R}$ (or $E \times \mathbb{C}$) of one higher dimension by identifying $p \in E$ with $p^+ = (p, 1) \in E^+$. The vectors of $E \times \mathbb{R}$ (or $E \times \mathbb{C}$) tangential to E^+ are those of the form $(y, 0)$ [*not* $(y, 1)$] and a vector field on E^+ associates to each point $(p, 1)$ such a vector $(y, 0)$.

   (a) Show: the affine transformation $p \to ap + b$ of E corresponds to the restriction of E^+ to the linear transformation
   $$\begin{bmatrix} a & b \\ 0 & 1 \end{bmatrix}$$
   and the affine vector field $p \to Xp + v$ on E corresponds to the restriction to E^+ of the linear vector field
   $$\begin{bmatrix} X & v \\ 0 & 0 \end{bmatrix}.$$

   (b) Show: the affine vector field $A : p \to Xp + v$ generates the one-parameter group of affine transformations
   $$\exp \tau A : p \to \exp(\tau X)p + \frac{\exp \tau X - 1}{X} v.$$
   [Start by explaining what this means.]

9. Let $X \in M$. Show:

   (a) There is exactly one trajectory for the vector field X through each point of E.

   (b) If $a \in \mathrm{GL}(E)$ commutes with X, then the transformation $E \to E$, $x \to ax$, permutes with the trajectories of X.

10. Let $X \in M$. Show:

    (a) For $p \in E$, $\exp(\tau X)p = p$ for all $\tau \in \mathbb{R}$ if and only if $X(p) = 0$. [In words: p is a stationary point of the flow, if and only if there is no current at p.]

    (b) If $\lim_{\tau \to \infty} \exp(\tau X)p_0 = p_\infty$ for some $p_0, p_\infty \in E$, then $Xp_\infty = 0$.

11. Two matrices P, Q are said to satisfy *Heisenberg's Commutation Relation* if

$$PQ - QP = k1$$

for some scalar k. Show that this is the case, if and only if

$$\exp \sigma P \exp \tau Q = e^{\sigma\tau k} \exp \tau Q \exp \sigma P$$

for all real σ, τ.

[Suggestion. Write $\exp\{\exp(\sigma P)\tau Q \exp(-\sigma P)\} = \exp\{\sigma\tau k + \tau Q\}$ and differentiate the exponent $\{\cdots\}$ with respect to σ. Comment. The equation $\exp(\sigma P)\exp(\tau Q) = e^{\sigma\tau k}\exp(\tau Q)\exp(\sigma P)$ is Weyl's group theoretic formulation of the Canonical Commutation Relation $PQ - QP = k1$ in quantum mechanics. (Weyl, 1950, §14). As he emphasizes on p. 95, such matrices P and Q exist only in infinite dimensions (if $k \neq 0$); the formal equivalence of Weyl's equation and CCR, however, is simply a consequence of the properties of the matrix exponential listed in Proposition 1.]

Appendix to §1.1: Notation and background

E denotes a n-dimensional real or complex vector space. Whenever necessary, we shall assume that E comes equipped with a positive definite inner product, written (x, y). We shall frequently have to express elements of E in terms of their components with respect to some basis and linear transformations of E in terms of their matrix coefficients. We could of course take E to be $\mathbb{R}^n$ or $\mathbb{C}^n$ from the outset, but this is of no help when one needs to choose a basis adapted to a particular situation rather than the standard basis. We therefore use the following expedient. Having fixed a basis (sometimes without mentioning this explicitly), we identify elements of E with (column) n-vectors and linear transformations of E with $n \times n$ matrices without change of notation. In this spirit, and in the interest of brevity, linear transformations of E are frequently referred to as *matrices*. The space of all linear transformations of E is denoted by $M = M(E)$ (*matrix space*). Elements of M are typically denoted X, Y, or Z when thought of as vector fields on E. Invertible linear transformations of E are usually denoted a, b, or c; they form the *general linear group* $\mathrm{GL}(E) = \{a \in M(E) | \det a \neq 0\}$ of all invertible linear transformations of E. If E is $\mathbb{R}^n$ one writes $M_n(\mathbb{R})$ and $\mathrm{GL}(n, \mathbb{R})$ for $M(E)$ and $\mathrm{GL}(E)$; similarly if E is $\mathbb{C}^n$.

A basis $\{e_k\}$ for E gives a basis $\{E_{ij}\}$ for the matrix space M consisting of the matrices E_{ij} with ij-entry 1 and 0 elsewhere. If e_k is orthonormal for the inner product (x, y) on E, then E_{ij} is orthonormal for the inner product (X, Y) on M given by

$$(X, Y) = \mathrm{tr}(Y^* X) = \sum_{ij} X_{ij} \bar{Y}_{ij}.$$

The asterisk denotes the adjoint with respect to the inner product (x, y) : $(Xx, y) = (x, X^* y)$. If X is represented as a matrix relative to an orthonormal basis, as above, then $X^* = \bar{X}^t$ (conjugate transpose).

The *matrix norm* in M is given by

$$\|X\|^2 = (X, X).$$

In terms of the orthonormal basis $\{E_{ij}\}$, $\|X\|^2$ is the usual sum of the squares of the absolute values entries of matrix X:

$$\|X\|^2 = \sum_{ij} |X_{ij}|^2.$$

It satisfies the inequalities

$$\|X + Y\| \leq \|X\| + \|Y\|, \tag{A.1}$$
$$\|XY\| \leq \|X\|\|Y\|. \tag{A.2}$$

The first of these is the familiar *triangle inequality*, which holds for any inner product space. The second one follows from the *Schwarz inequality* as follows: the ij-entry of XY, satisfies

$$\left|\sum_k X_{ik}Y_{kj}\right|^2 \leq \left(\sum_l |X_{il}|^2\right)\left(\sum_l |Y_{lj}|^2\right)$$

the Schwarz Inequality. Summing over ij and taking the square root one gets (A.2).

A real vector space may always be thought of as the real subspace of a complex vector space of the same dimension, consisting of those vectors whose components with respect to a fixed basis are real. Similarly for real matrices. More formally, a real vector space E may be embedded in a complex vector space $E \oplus iE$ whose elements are formal sums $x + iy$, $x, y \in E$, added and multiplied by complex scalars in the obvious way. $E \oplus iE$ is called the *complexification* of E, denoted $E_{\mathbb{C}}$.

On the other hand, a complex vector space E may be considered as a real vector space (of twice the dimension) by forgetting about multiplication by i. The real and imaginary parts of the components of a vector in E relative to a complex basis $\{e_k\}$ are the components of the vector with respect to the real basis $\{e_k, ie_k\}$.

A linear transformation of a real vector space E extends uniquely to a linear transformation of its complexification $E_{\mathbb{C}}$. Both are represented by the same matrix with respect to a real basis for E considered also as a complex basis for $E_{\mathbb{C}}$. By the *eigenvalues* of a linear transformation of a real vector space E, we always mean the eigenvalues of the corresponding transformation of $E_{\mathbb{C}}$ (i.e. we always allow complex eigenvalues, as is customary).

As real vector space, M may be identified with $M_n(\mathbb{R}) = \mathbb{R}^{n \times n}$ or, if E is complex, with $M_n(\mathbb{C}) \approx \mathbb{R}^{2n \times 2n}$ (with the real and imaginary parts of the entries of complex matrices as real coordinates). Thus all notions from analysis on $\mathbb{R}^N$ apply to M without further comment. A special feature of analysis on the matrix space M are M-valued functions of a matrix variable $X \in M$ defined by matrix-power series:

$$\sum_{k=0}^{\infty} \alpha_k X^k. \tag{A.3}$$

A tail segment of such a series may be estimated by

$$\left\| \sum_{k=N}^{M} \alpha_k X^k \right\| \leq \sum_{k=N}^{M} |\alpha_k| \, \|X^k\| \leq \sum_{k=N}^{M} |\alpha_k| \, \|X\|^k, \tag{A.4}$$

obtained from the inequalities (A.1) and (A.2). From this, one sees that the matrix series (A.3) converges whenever $\sum_{k=0}^{\infty} |\alpha_k| \|X^k\|$ converges. One then says that the matrix series *converges in norm* or is *norm-convergent.* Such is in particular the case when the power series of norms $\sum_{k=0}^{\infty} |\alpha_k| \|X\|^k$ converges. Substituting τX for X into the series (A.3) one gets a power series in the scalar variable τ with coefficients depending on X:

$$\sum_{k=0}^{\infty} \alpha_k \tau^k X^k. \tag{A.5}$$

Generally, a power series with matrix coefficients, $\sum \tau^k A_k (A_k \in M)$, has a radius of (norm)-convergence $R \geq 0$ and may be treated according to the usual rules for scalar power series for $|\tau| < R$. For example, the series can be differentiated or integrated term-by-term with respect to τ within its radius of norm-convergence ($|\tau| < R$):

$$\frac{d}{d\tau} \left(\sum \tau^k A_k \right) = \sum k \tau^{k-1} A_k,$$

$$\int \left(\sum \tau^k A_k \right) d\tau = \sum \frac{\tau^{k+1}}{k+1} A_k + C.$$

For any matrix valued function $a(\tau)$ of a real variable τ (defined on some interval), the derivative $a'(\tau) = da/d\tau$ is defined by

$$a'(\tau) = \lim_{\epsilon \to 0} \frac{1}{\epsilon} (a(\tau + \epsilon) - a(\tau)), \tag{A.6}$$

whenever the limit exists. In addition to the usual rules for the derivative of vector valued functions one has the product rule

$$(ab)' = a'b + ab'. \tag{A.7}$$

This follows directly from the definition (A.6), since

$$(ab)(\tau + \epsilon) - (ab)(\tau) = (a(\tau + \epsilon) - a(\tau)) b(\tau + \epsilon) + a(\tau)(b(\tau + \epsilon) - b(\tau)).$$

Note, incidentally, that the product rule (A.7) depends only on the fact that the matrix product ab is bilinear in a and b. The formula (A.7) therefore holds for any bilinear function $ab = f(a, b)$. The variables a and b and the value $f(a, b)$ may come from any (possibly different) vector spaces.

Where $\det a(\tau) \neq 0$, the inverse $a(\tau)^{-1}$ is differentiable and its derivative may be found by differentiating the relation $aa^{-1} = 1$ according to the rule (A.7), which gives:

$$\frac{da}{d\tau} a^{-1} + a \frac{da^{-1}}{d\tau} = 0,$$

hence

$$\frac{da^{-1}}{d\tau} = -a^{-1}\frac{da}{d\tau}a^{-1}. \qquad \text{(A.8)}$$

Consult the appendix on Analytic Functions and the Inverse Function Theorem on page 250 for a review of these topics.

1.2 Ad, ad, and $d\exp$

We continue with exp, considered as a mapping $X \to \exp X$ of the matrix space M onto itself. First item:

Proposition 1. *The map* $\exp : M \to M$ *carries a neighbourhood of 0 one-to-one onto a neighbourhood of 1.*

More precisely, in a neighbourhood of 0, $\exp : M \to M$ has a local inverse[2] $\log : M \cdots \to M$, defined in a neighbourhood of 1 by the series

$$\log a = \sum_{k=1}^{\infty} \frac{(-1)^{k-1}}{k}(a-1)^k. \qquad (1)$$

This series converges in norm for $\|a-1\| < 1$, and

$$\log(\exp X) = X \quad \text{for } \|X\| < \log 2, \qquad (2)$$

$$\exp(\log a) = a \quad \text{for } \|a-1\| < 1. \qquad (3)$$

Proof. The series (1) converges in norm for $\|a-1\| < 1$ because

$$\sum_{k=1}^{\infty} \frac{1}{k}\|a-1\|^k,$$

converges for $\|a-1\| < 1$.

Working towards a proof of (2), we first try to calculate $\log(\exp X)$ naively by substituting the log series into the exp series:

$$\begin{aligned}\log(\exp X) &= \left(X + \frac{1}{2!}X^2 + \cdots\right) - \frac{1}{2}\left(X + \frac{1}{2!}X^2 + \cdots\right)^2 \\ &\quad + \frac{1}{3}\left(X + \frac{1}{2!}X^2 + \cdots\right)^3 + \cdots \\ &= X + \left(\frac{1}{2!}X^2 - \frac{1}{2}X^2\right) + \left(\frac{1}{3!}X^3 - \frac{1}{2}X^3 + \frac{1}{3}X^3\right) + \cdots \\ &= X + 0 + 0 + \cdots\end{aligned}$$

except that it is not immediately clear that the remaining dots are really all equal to 0. But one can argue as follows. First of all, since $\|\exp X - 1\| \le e^{\|X\|} - 1$, the double series for $\log(\exp X)$ converges in norm provided $e^{\|X\|} - 1 < 1$, i.e. $\|X\| < \log 2$. Assuming this, the terms of the double series for $\log(\exp X)$ may

[2] The dots indicate a partially defined map, here and elsewhere.

be rearranged at will. But then the coefficients of X^k must add up to 1 for $k = 1$ and to 0 otherwise, because this is the case when X is a scalar variable, and the operations on the coefficients are evidently the same whether X is thought of as a scalar or as a matrix. This proves (2); (3) is seen in the same way. QED

The trick just used, though obvious, is worth recording:
Substitution Principle 2. Any equation involving power series in a variable X, which holds as an identity of absolutely convergent scalar series, also holds as an identity of norm-convergent matrix series. If the scalar series converge for $|X| < \rho$, then the matrix series converge for $\|X\| < \rho$.

Remark 2 (Substitution principle).
(a) A more detailed version of the Substitution Principle may be formulated as follows:

Let $F(z)$ and $G(z)$ be power series in a real or complex variable z with real or complex coefficients. Let $\sigma > 0$ be the radius of convergence of $F(z)$, $\rho > 0$ that of $G(z)$. Let $X \in M$ be a real or complex matrix. Then

(i) $(F+G)(X) = F(X) + G(X)$, if $\|X\| < \min(\sigma, \rho)$

(ii) $(FG)(X) = F(X)G(X)$, if $\|X\| < \min(\sigma, \rho)$

(iii) $(F \circ G)(X) = F(G(X))$, if $\|X\| < \rho$, $\|G(X)\| < \sigma$, and $G(0) = 0$.

The series $F + G$, FG, $F \circ G$ is defined by formal calculation the coefficients of this series. The condition $G(0) = 0$ in (iii) ensures that only finitely many coefficients of F and G contribute to a given coefficient of $F \circ G$.

The idea behind the Substitution Principle may apply even if the statement itself does not. For example, if X is a nilpotent matrix ($X^k = 0$ for some k), then $F(X)$ exists for any power series F, even though $\|X\|$ may lie outside its radius of convergence: consider $F(\tau X)$ as a power series in scalar variable τ. Some caution with the Substitution Principle is nevertheless advised, as one sees already from the equation $\log \exp z = z$ when one tries to substitute $z = 2\pi i$.

(b) The Substitution Principle extends in an obvious way to power series in several *commuting* matrices. All of the formulas of Proposition 1 of §1.1 could have been derived from the scalar case by the Substitution Principle in this form.

(c) The result of Proposition 1 is not the best possible. In fact, $\exp : M \to M, X \to a = \exp X$, maps the region of matrices X whose eigenvalues λ satisfy $|\mathrm{Im}\,\lambda| < \pi$ one-to-one onto the region of matrices whose eigenvalues α are *not* real and ≤ 0. These λ and α are in bijection under the complex exponential function $\alpha = \exp \lambda$. The assertion concerning matrices ultimately comes down to this. (See problem 3(c).) The notation log is also used for the inverse of exp on this larger domain, where series (1) need not converge.

There are two operations with matrices which are of importance in connection with exp. First, for any invertible $a \in M$ it is customary to denote by $\mathrm{Ad}(a)$ the conjugation operation by a as linear transformation of M:

$$\mathrm{Ad}(a)Y = aYa^{-1}.$$

Note that

$$\mathrm{Ad}(ab) = \mathrm{Ad}(a)\mathrm{Ad}(b), \quad \mathrm{Ad}(a^{-1}) = \mathrm{Ad}(a)^{-1},$$

so that Ad is a homomorphism from the group of invertible linear transformation of E to that of M:

$$\mathrm{Ad} : \mathrm{GL}(E) \to \mathrm{GL}(M).$$

This homomorphism is called the *adjoint representation* of $\mathrm{GL}(E)$. Second, in addition to big Ad, we introduce little ad by:

$$\mathrm{ad}(X)Y = XY - YX.$$

$\mathrm{ad}(X) : M \to M$ is a linear transformation of M defined for all $X \in M$ (invertible or not), and certainly not a group homomorphism. When thought of as an operation on matrices, $\mathrm{ad}(X)Y$ is also written as a *bracket*:

$$[X, Y] = XY - YX.$$

This bracket operation is evidently bilinear and skew-symmetric in X and Y; in addition it satisfies the *Jacobi Identity*

$$[[X, Y], Z] + [[Z, X], Y] + [[Y, Z], X] = 0.$$

(XYZ are permuted cyclically to form the terms of this sum.) It is equivalent to

$$(\mathrm{ad}\, Z)[X, Y] = [(\mathrm{ad}\, Z)X, Y] + [X, (\mathrm{ad}\, Z)Y],$$

which is called the *derivation property* of $\mathrm{ad}\, Z$. (The Jacobi Identity for matrices is verified by a simple calculation, which we omit. It derives its name from Jacobi's investigations in mechanics (1836).)

Big Ad and little ad are intertwined by exp in the following sense:

Proposition 4. *For any $X \in M$,*

$$\mathrm{Ad}(\exp X) = \exp(\mathrm{ad}\, X).$$

Explanation. There are two different exp's in this formula: on the left is exp of the linear transformation X of E, on the right, exp of the linear transformation $\mathrm{ad}\, X$ of M. Written out explicitly the formula says that for all $X, Y \in M$,

$$\exp(X)Y \exp(-X) = \sum_{k=0}^{\infty} \frac{1}{k!} (\mathrm{ad}\, X)^k Y.$$

Proof. Fix $X \in M$ and let $A(\tau) = \mathrm{Ad}(\exp \tau X)$ for $\tau \in \mathbb{R}$. Calculate its derivative:

$$\begin{aligned} A'(\tau)Y &= \frac{d}{d\tau}(\exp(\tau X)Y \exp(-\tau X)) \\ &= X \exp(\tau X)Y \exp(-\tau X) + \exp(\tau X)Y \exp(-\tau X)(-X) \\ &= (\mathrm{ad}\, X)\mathrm{Ad}(\exp \tau X)Y. \end{aligned}$$

Thus

$$A'(\tau) = UA(\tau)$$

with $U = \mathrm{ad}\, X$, and

$$A(0) = 1.$$

From Proposition 1 of §1.1 we know that the only solution of these equations is $A(\tau) = \exp(\tau U)$, which gives the desired result. QED

The next item is a differentiation formula for exp:

Theorem 5.
$$\frac{d}{d\tau}\exp X = \exp(X)\frac{1-\exp(-\operatorname{ad} X)}{\operatorname{ad} X}\frac{dX}{d\tau}.$$

Explanation. In this formula $X = X(\tau)$ is any matrix-valued differentiable function of a scalar variable τ. The fraction of linear transformations of M is defined by its (everywhere convergent) power series

$$\frac{1-\exp(-\operatorname{ad} X)}{\operatorname{ad} X} = \sum_{k=0}^{\infty}\frac{(-1)^k}{(k+1)!}(\operatorname{ad} X)^k.$$

The proposition may also be read as saying that the differential of $\exp : M \to M$ at any $X \in M$ is the linear transformation $d\exp_X : M \to M$ of M given by the formula
$$d\exp_X Y = \exp(X)\frac{1-\exp(-\operatorname{ad} X)}{\operatorname{ad} X}Y.$$

Historical Comment. The formula goes back to the beginnings of Lie theory. It was first proved by F. Schur (1891) (not to be confused with the better known I. Schur), and was taken up later from a different point of view by Poincaré (1899).

Proof. Let $X = X(\tau)$ be a differentiable curve in M and set

$$Y(\sigma,\tau) = (\exp -\sigma X(\tau))\frac{\partial}{\partial\tau}\exp\sigma X(\tau)$$

for $\sigma, \tau \in \mathbb{R}$. Differentiate with respect to σ:

$$\begin{aligned}
\frac{\partial Y}{\partial\sigma} &= (\exp -\sigma X)(-X)\frac{\partial}{\partial\tau}\exp(\sigma X) + \exp(-\sigma X)\frac{\partial}{\partial\tau}(X\exp\sigma X)\\
&= (\exp -\sigma X)(-X)\frac{\partial}{\partial\tau}\exp(\sigma X) + \exp(-\sigma X)\frac{dX}{d\tau}\exp(\sigma X)\\
&\quad + (\exp -\sigma X)X\frac{\partial}{\partial\tau}\exp(\sigma X)\\
&= (\exp -\sigma X)\frac{dX}{d\tau}\exp(\sigma X)\\
&= \operatorname{Ad}(\exp -\sigma X)\frac{dX}{d\tau}\\
&= \exp(\operatorname{ad} - \sigma X)\frac{dX}{d\tau}.
\end{aligned}$$

Now

$$\exp(-X)\frac{d}{d\tau}\exp X = Y(1,\tau) = \int_0^1 \frac{\partial}{\partial\sigma}Y(\sigma,\tau)\,d\sigma \quad [\text{since } Y(0,\tau)=0]$$

and

$$\frac{\partial Y}{\partial \sigma} = (\exp -\sigma \operatorname{ad} X)\frac{dX}{d\tau} = \sum_{k=0}^{\infty} \frac{(-1)^k \sigma^k}{k!} (\operatorname{ad} X)^k \frac{dX}{d\tau}.$$

Integrate this series term-by-term from $\sigma = 0$ to $\sigma = 1$ to get

$$\exp(-X)\frac{d}{d\tau} \exp X = \sum_{k=0}^{\infty} \frac{(-1)^k}{(k+1)!} (\operatorname{ad} X)^k \frac{dX}{d\tau},$$

which is the desired formula. QED

The theorem, together with the Inverse Function Theorem (Appendix), gives information on the local behaviour of the exponential map: the Inverse Function Theorem says that exp has a local inverse around a point $X \in M$ at which its differential $d\exp_X$ is invertible, and the theorem says that this is the case precisely when $(1 - \exp(-\operatorname{ad} X))/\operatorname{ad} X$ is invertible, i.e. when zero is not an eigenvalue of this linear transformation of M. To find these eigenvalues we use a general result:

Lemma 6. *Let $f(z) = \sum_{k=0}^{\infty} \alpha_k z^k$ be a power series with real or complex coefficients. Suppose U is a linear transformation so that the series $f(U) = \sum_{k=0}^{\infty} \alpha_k U^k$ converges. If $\lambda_1, \lambda_2, \ldots, \lambda_N$ are the eigenvalues of U, listed with multiplicities, then $f(\lambda_1), f(\lambda_2), \ldots, f(\lambda_N)$ are the eigenvalues of $f(U)$, listed with multiplicities.*

Proof. Choose a basis so that U is triangular with diagonal entries $\lambda_1, \lambda_2, \ldots, \lambda_N$. (This may require passing to complex scalars, even if U starts out as a real linear transformation.) For each $k = 0, 1, \ldots, U^k$ is then also triangular with diagonal entries $\lambda_1^k, \lambda_2^k, \ldots, \lambda_N^k$. Thus, $f(U)$ itself is a triangular matrix with diagonal entries given by the power series $f(\lambda_1), f(\lambda_2), \ldots, f(\lambda_N)$; these power series $f(\lambda_j)$ converge, because $f(U)$ is assumed to converge. QED

¿From the lemma and the remarks preceding it one obtains now:

Proposition 7. *If none of the eigenvalues of* $\operatorname{ad} X$ *are of the form $\lambda = 2\pi i k$, $k = \pm 1, \pm 2, \ldots$, then* $\exp : M \to M$ *has a local inverse near X.*

Proof. By the lemma, the eigenvalues of $(1 - \exp -U)/U$ are of the form $(1 - e^{-\lambda})/\lambda$, λ an eigenvalue of U. The given values of λ are precisely the solutions of the equation $(1 - e^{-\lambda})/\lambda = 0$; this gives the conclusion when one takes $U = \operatorname{ad} X$. QED

It remains to determine the eigenvalues of $\operatorname{ad} X$:

Lemma 8. *If X has n eigenvalues $\{\lambda_j | j = 1, 2, \ldots, n\}$, then* $\operatorname{ad} X$ *has n^2 eigenvalues $\{\lambda_j - \lambda_k | j, k = 1, 2, \ldots, n\}$.*

Proof. Let $\{e_j\}$ be a basis so that X is triangular, say

$$Xe_j = \lambda_j e_j + \cdots ,$$

where the dots indicate a linear combination of e_k's with $k > j$. Let E_{jk} be the corresponding basis for M (E_{jk} has jk-entry 1 and all other entries 0). Order the basis E_{jk} of M so that $j - k < j' - k'$ implies E_{jk} precedes $E_{j'k'}$. One checks by matrix computation that

$$\mathrm{ad}(X)E_{jk} = (\lambda_j - \lambda_k)E_{jk} + \cdots ,$$

where the dots indicate a linear combination of $E_{j'k'}$'s that come after E_{jk} in the chosen order. This means that $\mathrm{ad}(X)$ is triangular with diagonal entries $(\lambda_j - \lambda_k)$. QED

In view of Lemma 8, Proposition 7 can be rephrased as

Proposition 7′. If no two eigenvalues of $X \in M$ have a difference of the form $2\pi ik$, $k = 1, 2, \ldots$, then $\exp : M \to M$ has a local inverse near X.

Example 9 (exp **for real** 2×2 **matrices).** We start with the observation that for any matrix X

$$\det(\exp X) = e^{\mathrm{tr}\, X};$$

this is immediate from the formulas for det and tr in terms of eigenvalues. This formula implies first of all that any exponential of a real matrix must have a positive determinant. Furthermore, since $\exp(\alpha 1 + X) = e^{\alpha} \exp X$, it suffices to consider matrices with $\mathrm{tr}\, X = 0$, as far as the behaviour of the real matrix exponential is concerned.

We now specialize to the case of real 2×2 matrices. In view of the preceding remarks, we specialize further by *assuming throughout that* $\mathrm{tr}(X) = 0$. The characteristic polynomial of a 2×2 matrix X is

$$\det(\lambda 1 - X) = \lambda^2 - (\mathrm{tr}\, X)\lambda + (\det X)1.$$

According to Cayley–Hamilton,

$$X^2 - (\mathrm{tr}\, X)X + (\det X)1 = 0.$$

The assumption that $\mathrm{tr}\, X = 0$ implies $X^2 = -(\det X)1$. Use this fact to compute:

$$\begin{aligned}
\exp X &= \sum_{k=0}^{\infty} \frac{X^k}{k!} \\
&= \sum_{k=0}^{\infty} \frac{X^{2k}}{(2k)!} + \sum_{k=0}^{\infty} \frac{X^{2k+1}}{(2k+1)!} \\
&= \sum_{k=0}^{\infty} \frac{(-1)^k}{(2k)!}(\det X)^k 1 + \sum_{k=0}^{\infty} \frac{(-1)^k}{(2k+1)!}(\det X)^k X.
\end{aligned}$$

Recognizing these series one finds

$$\exp X = (\cos\sqrt{\det X})1 + \frac{\sin\sqrt{\det X}}{\sqrt{\det X}}X. \tag{4}$$

(The functions in (4) are independent of the choice of the sign of the root. The root may be imaginary; $\sin\theta$ and $\cos\theta$ are defined for complex arguments by:

$$\sin\theta = \tfrac{1}{2i}(e^{i\theta} - e^{-i\theta}), \quad \cos\theta = \tfrac{1}{2}(e^{i\theta} + e^{-i\theta}).$$

Looking at formula (4) more closely one sees:

For a 2×2 matrix a with $\det a = 1$, the equation $\exp X = a$ has a solution X if and only if $\frac{1}{2}\operatorname{tr} a > -1$ or $a = -1$. If so, the solutions are given as follows:

$$\text{(a) For } -1 < \tfrac{1}{2}\operatorname{tr} a < 1\text{:}\quad X = \frac{\xi}{\sin\xi}\left(a - \tfrac{1}{2}\operatorname{tr} a1\right)$$

with $\xi > 0$ satisfying $\cos\xi = \frac{1}{2}\operatorname{tr} a$. There are countably many solutions.

$$\text{(b) For } \tfrac{1}{2}\operatorname{tr} a > 1\text{:}\quad X = \frac{\xi}{\sinh\xi}\left(a - \tfrac{1}{2}\operatorname{tr} a1\right)$$

with $\xi > 0$ satisfying $\cosh\xi = \frac{1}{2}\operatorname{tr} a$. There is a unique solution.

$$\text{(c) For } \tfrac{1}{2}\operatorname{tr} a = 1,\ a \neq 1\text{:}\quad X = a - 1.$$

(d) $a = \pm 1$: X = any matrix on one of the family of surfaces in the space of matrices 2×2 of trace zero is given by the equations

$$\det X = \begin{cases} (\pi 2k^2), & \text{if } a = +1 \\ (\pi(2k+1))^2, & \text{if } a = -1, k = 0, 1, 2, \ldots. \end{cases}$$

To get a better overview of the situation we introduce coordinates x, y, z in the three-dimensional space of a real 2×2 matrix X of trace zero, by setting

$$X = x\begin{bmatrix} 1 & 0 \\ 0 & -1 \end{bmatrix} + y\begin{bmatrix} 0 & 1 \\ 1 & 0 \end{bmatrix} + z\begin{bmatrix} 0 & -1 \\ 1 & 0 \end{bmatrix}.$$

Then $\det X = -\frac{1}{2}\operatorname{tr} X^2 = -x^2 - y^2 + z^2$. The region $\det X > 0$ inside the cone $\det X = 0$ consists of matrices of elliptic type. Under exp, they get mapped onto the region $\det a = 1$, $-1 < \frac{1}{2}\operatorname{tr} a < 1$ in a periodic way; the 'periods' are separated by the two-sheeted hyperboloids $\det X = (\pi k)^2$, $k = 1, 2, \ldots$, which themselves get collapsed to the points ± 1. The region $\det X < 0$ outside the cone consists of matrices of the hyperbolic type; they get mapped onto the region $\det a = 1$, $\frac{1}{2}\operatorname{tr} a > 1$ in a bijective way. The cone $\det X = 0$ consists of nilpotent matrices ($X^2 = 0$; parabolic type); they get mapped bijectively onto the unipotent matrices ($(a-1)^2 = 0$) (Figure 1).

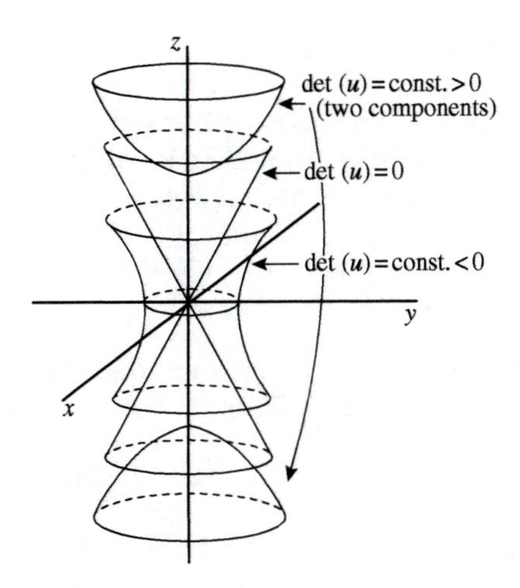

Fig. 1

These observations illustrate several results of this section. The equation

$$\exp(cXc^{-1}) = c(\exp X)c^{-1}$$

implies that exp maps the similarity class $\{cXc^{-1}|c \text{ an invertible } 2 \times 2 \text{ matrix}\}$ of X to the similarity class of $a = \exp X$. The similarity classes in $\{\operatorname{tr} X = 0\}$ are the surfaces $\det X = $ const., except for $X = 0$ on the cone $\det X = 0$. They are represented by the matrices

$$\begin{bmatrix} 0 & \theta \\ -\theta & 0 \end{bmatrix}, \quad \begin{bmatrix} \xi & 0 \\ 0 & -\xi \end{bmatrix}, \quad \begin{bmatrix} 0 & 1 \\ 0 & 0 \end{bmatrix}, \quad \begin{bmatrix} 0 & 0 \\ 0 & 0 \end{bmatrix} \tag{5}$$

with $\theta, \xi \neq 0$. The similarity classes in $\{\det a = 1\}$ are the surfaces $\frac{1}{2} \operatorname{tr} a =$ const., except for $a = \pm 1$ on $\frac{1}{2} \operatorname{tr} a = \pm 1$. They are represented by the matrices

$$\begin{bmatrix} \cos\theta & -\sin\theta \\ \sin\theta & \cos\theta \end{bmatrix}, \quad \pm \begin{bmatrix} \alpha & 0 \\ 0 & \alpha^{-1} \end{bmatrix}, \quad \pm \begin{bmatrix} 1 & 1 \\ 0 & 1 \end{bmatrix}, \quad \pm \begin{bmatrix} 1 & 0 \\ 0 & 1 \end{bmatrix} \tag{6}$$

with $\theta \neq 2\pi k$, $k = 0, \pm 1, \pm 2, \ldots$ and $\alpha > 0$. The critical points of exp, those where its differential is singular, comprise the two-sheeted hyperboloids $\det X = (\pi k)^2$, $k = 1, 2, \ldots$ (represented by $\theta = \pi k$ in (5)); they get collapsed to the points ± 1. These are exactly the points where exp fails to be locally invertible, in agreement with the Inverse Function Theorem. The condition $\frac{1}{2} \operatorname{tr} a > -1$ or $a = -1$ excludes those similarity classes of 2×2 matrices a with $\det a = 1$ from the image of exp that are represented by

$$\begin{bmatrix} \alpha & 0 \\ 0 & \alpha^{-1} \end{bmatrix}, \quad -1 \neq \alpha < 0, \quad -\begin{bmatrix} 1 & 1 \\ 0 & 1 \end{bmatrix}.$$

In particular, $\exp : \{\operatorname{tr} X = 0\} \to \{\det a = 1\}$ is far from being surjective.

Problems for §1.2

1. A matrix $X \in M$ is called *nilpotent* if $X^k = 0$ for some k (equivalently: all eigenvalues of X are equal to 0); $a \in M$ is called *unipotent* if $(1-a)^k = 0$ for some k (equivalently: all eigenvalues of a are equal to 1). Show:

   $X \to \exp X$ maps the nilpotent matrices bijectively onto the unipotent matrices with inverse $a \to \log a$.

   [Proposition 1 does not apply directly, nor does the Substitution Principle as stated; a minor adjustment will do.]

2. A matrix a is called *semisimple* if it is diagonalizable over $\mathbb{C}$. Show:

   (a) $X \to \exp X$ maps semisimple matrices to semisimple matrices.

   (b) If a is an invertible semisimple matrix, then there is a semisimple matrix X so that $a = \exp X$ and no two distinct eigenvalues of X differ by a multiple of $2\pi i$.

   (c) Assume X and X' are both semisimple and no two distinct eigenvalues of X differ by a multiple of $2\pi i$. Show that $\exp X = \exp X'$ if and only if X and X' are *simultaneously* diagonalizable with diagonal entries differing by multiples of $2\pi i$.

   Any matrix X can be *uniquely* written as $X = Y + Z$ where Y is semisimple, Z is nilpotent, and Y and Z *commute*. Furthermore, Y and Z are linear combinations of powers of X. $X = Y + Z$ is called the *Jordan decomposition* of X. [See Hoffman–Kunze (1961) Theorem 8, page 217, for example.]

3. Show:

   (a) Any invertible matrix a can be *uniquely* written as $a = bc$ where b is semisimple, c is unipotent, and b and c *commute*. Furthermore, b and c are linear combination of powers of a.

   (b) If $X = Y + Z$ is the decomposition of X as in (a), then $\exp X = \exp Y \exp Z$ is the decomposition of $\exp X$ as in (b).

   (c) Assume that no two distinct eigenvalues of X differ by a multiple of $2\pi i$. (See problem 2(b)). Show that $\exp X = \exp X'$ if and only $Z = Z'$ and Y and Y' are simultaneously diagonalizable with diagonal entries differing by integral multiples of $2\pi i$. Deduce Remark 3(c).

4. Let λ be a non-zero complex number. Show that the matrix

$$\begin{bmatrix} \lambda & 1 & 0\cdots & 0 \\ 0 & \lambda & 1\cdots & 0 \\ \vdots & \vdots & \vdots & \vdots \\ 0 & 0 & 0\cdots & \lambda \end{bmatrix}$$

is the exponential of a complex matrix. Deduce that every invertible complex matrix is an exponential. [Suggestion: For the second part, use the Jordan canonical form, which is the explicit version of the Jordan decomposition mentioned before problem 3.]

5. Show: not every matrix of determinant 1 is an exponential (real or complex) of a matrix of trace 0. Suggestion: Consider

$$\begin{bmatrix} -1 & 1 \\ 0 & -1 \end{bmatrix}.$$

6. Let $f(X) = \sum_k \alpha_k X^k$ be any matrix-power series. Use the Cayley–Hamilton Theorem to show that $f(X)$, can formally be rewritten as

$$f(X) = c_0(X) + c_1(X)X + \cdots + c_{n-1}(X)X^{n-1},$$

where the $c_j(X)$ are (multiple) power series in the matrix entries of X which are invariant under conjugation, i.e. $c_j(aXa^{-1}) = c_j(X)$. ['Formally' means 'do not worry about convergence when rearranging the series'].

7. Suppose $X \in M$ satisfies $\|X\| < 2\pi$. Show that for $v \in E$,

$$\exp(X)v = v \quad \text{if and only if} \quad Xv = 0.$$

[Suggestion: the power series for $z/(\exp z - 1)$ converges for $|z| < 2\pi$].

8. (a) Prove the *Jacobi Identity*

$$[[X,Y],Z] + [[Z,X],Y] + [[Y,Z],X] = 0.$$

Deduce that
(b) $(\operatorname{ad} Z)[X,Y] = [(\operatorname{ad} Z)X, Y] + [X, (\operatorname{ad} Z)Y]$,
(c) $\operatorname{ad}([X,Y]) = [\operatorname{ad} X, \operatorname{ad} Y]$.
(The bracket on the right side of (c) is that of linear transformations of the matrix space M.)

9. Show that for all $X \in M$,

$$\exp X = \lim_{k\to\infty} \left(1 + \frac{1}{k}X\right)^k.$$

[The formula has a 'physical' interpretation: subdivide the time interval $0 \le \tau \le 1$ into a large number of subintervals k; the fluid particle travelling on the trajectory $p(\tau) = \exp(\tau X)p_0$, with velocity $Xp(\tau)$ at $p(\tau)$, will move from p_0 to approximately $p_0 + (1/k)Xp_0 = (1 + (1/k)X)p_0$ in the first time interval, on to $(1 + (1/kX))^2 p_0$ in the second, etc., until at $\tau = 1$ it reaches approximately $(1 + 1/kX)^k p_0$, which must therefore approximate $\exp(X)p_0$].

10. Prove the *Trotter Product Formula*: for all $X, Y \in M$,

$$\exp(X+Y) = \lim_{k\to\infty}\left(\exp\left(\frac{1}{k}X\right)\exp\left(\frac{1}{k}Y\right)\right)^k.$$

[Suggestion: start with $\exp(\tau X)\exp(\tau Y) = \exp(\tau(X+Y)+o(\tau)]$.

11. Let $\mathbb{R} \to M$, $\tau \to a(\tau)$, be a continuous map satisfying $a(\sigma+\tau) = a(\sigma)a(\tau)$ and $a(0) = 1$. Show that $a(\tau) = \exp \tau X$ for some $X \in M$. [This shows that the differentiability hypothesis on $a(\tau)$ in part (d) of Proposition 1 of §1.1 can be replaced by continuity. Suggestion for the proof: consider $X(\tau) = \log(a(\tau))$ for small τ. Show that $X(\rho\tau) = \rho X(\tau)$, first for $\rho = p/q$ rational, then for all $\rho \in \mathbb{R}$].

12. Let $X \in M$ and let L be a subspace of M satisfying $[X, L] \subset L$. Show:

    (a) $\exp(\tau X)L\exp(-\tau X) \subset L$ for all $\tau \in \mathbb{R}$.

    (b) If $LL \subset L$ then $\exp(-X)\exp(X+L) \in 1+L$.

    [Suggestion for (b). Consider successive derivatives at $\tau = 0$ of $a(\tau) = \exp(-X)\exp(X+\tau Y)$.]

13. Show that for any differentiable matrix valued function $a = a(\tau)$ of a real variable τ with $\det a(\tau) \neq 0$,

$$\frac{d}{d\tau}\det a = (\det a)\operatorname{tr}\left(a^{-1}\frac{da}{d\tau}\right).$$

[Suggestion: assume first $a(0) = 1$ and prove this formula at $\tau = 0$.]

1.3 The Campbell–Baker–Hausdorff series

In a neighbourhood of the identity matrix in M where exp has an inverse (e.g. on $\|a-1\| \leq 1$, where the log series converges) one may think of exp as providing a coordinate system: the matrix X (restricted to a neighbourhood of 0) serves as coordinate point of the matrix $a = \exp X$ (restricted to a neighbourhood of 1). We shall refer to these coordinates as *exponential coordinates*, without (for now) giving any further meaning to the term 'coordinates' in general. We need a formula for matrix multiplication in exponential coordinates; that is, we need a formula for the solution of the equation

$$\exp X \exp Y = \exp Z$$

for Z in terms of X and Y, at least for X, Y, Z in a neighbourhood of 0 in M, where this equation does have a unique solution, namely

$$Z = \log(\exp X \exp Y). \tag{1}$$

Expanding the right side of (1) we get

$$
\begin{aligned}
Z &= \sum_{k=1}^{\infty} \frac{(-1)^{k-1}}{k} \left\{ \left(\sum_{i=0}^{\infty} \frac{X^i}{i!} \right) \left(\sum_{j=0}^{\infty} \frac{Y^j}{j!} \right) - 1 \right\}^k \\
&= \sum_{k=1}^{\infty} \frac{(-1)^{k-1}}{k} \left\{ \sum_{i,j\geq 0, i+j\geq 1} \frac{X^i Y^j}{i!j!} \right\}^k,
\end{aligned}
$$

which leads to

$$
Z = \sum \frac{(-1)^{k-1}}{k} \frac{X^{i_1}Y^{j_1}\dots X^{i_k}Y^{j_k}}{i_1!j_1!\cdots i_k!j_k!} \tag{2}
$$

where the sum is over all finite sequences $(i_1, j_1, \dots, i_k, j_k)$ of integers ≥ 0 satisfying $i_r + j_r \geq 1$. The order of summation is unimportant as long as the series converges in norm when summed in *some* order, which is the case when $\sum(1/k)(e^{\|X\|}e^{\|Y\|} - 1)^k$ converges, i.e. when $\|X\| + \|Y\| < \log 2$.

The monstrosity (2) is extremely awkward to work with, but at least one can write out the first few terms:

$$
\begin{aligned}
Z &= (X + Y + XY + \tfrac{1}{2}X^2 + \tfrac{1}{2}Y^2 + \cdots) - \tfrac{1}{2}(XY + YX + X^2 + Y^2 + \cdots) + \cdots \\
&= X + Y + \tfrac{1}{2}(XY - YX) + \cdots .
\end{aligned}
$$

Again we meet the bracket $[X, Y] = XY - YX$, in terms of which

$$
Z = X + Y + \tfrac{1}{2}[X, Y] + \cdots . \tag{3}
$$

The formula (2) gives the Taylor expansion of Z as function of X and Y; from (3) one sees that to first order, Z is just the sum of X and Y, the second order term involves their bracket. It is truly a remarkable fact that the whole series (2) can be rewritten solely in terms of repeated brackets of X and Y. The proof, in hindsight quite simple, occupied several mathematicians: Campbell (1897), Poincaré (1899), Baker (1905), Hausdorff (1906), and Dynkin (1947). The first four addressed the question whether Z can be represented as a bracket series at all (which is the crux of the matter) without producing the explicit formula. (As Bourbaki (1972) puts it: "chacun considère que les démonstrations de ses prédécesseurs ne sont pas convaincantes'.) Dynkin finally gives the explicit formula, now generally (and somewhat paradoxically) known as 'Campbell–Baker–Hausdorff Series'. The proof given here follows Duistermaat–Kolk (1988); the essential ingredient is F. Schur's formula for the differential of exp (Theorem 5 of §1.2).

Theorem 1 (Dynkin's Formula). *For matrices $X, Y, Z \in M$ sufficiently close to 0, the equation*

$$
\exp X \exp Y = \exp Z
$$

has a unique solution for $Z = C(X, Y)$ as a convergent series in repeated brackets of X and Y, namely

$$
C(X, Y) = \sum \frac{(-1)^{k-1}}{k} \frac{1}{(i_1 + j_1) + \cdots (i_k + j_k)} \frac{[X^{(i_1)}Y^{(j_1)}\dots X^{(i_k)}Y^{(j_k)}]}{i_1!j_1!\cdots i_k!j_k!}. \tag{4}
$$

Explanations and remarks. The sum is over all $2k$-tuples $(i_1, \ldots, i_k, j_1, \ldots, j_k)$ of integers ≥ 0 satisfying $i_r + j_r \geq 1$ with $k = 1, 2, \ldots$. The brackets are defined as follows. For any finite sequence $X_1, \ldots, X_k$ of matrices we set

$$[X_1, X_2, \ldots, X_k] = [X_1, [X_2, \ldots, [X_{k-1}, X_k], \ldots]],$$

the brackets being inserted in the order shown. The expression

$$[X^{(i_1)} Y^{(j_1)} \ldots X^{(i_k)} Y^{(j_k)}]$$

in (4) means the sequence starts with $i_1 X_1$s, then $j_1 Y_1$s etc. The convergence of (4) is not in question: if (4) is equal to (2), in the sense that the terms involving a given finite number of factors X, Y are equal, then (4) must converge for $\|X\| + \|Y\| < \log 2$, because (2) does, and must represent Z as long as $\|Z\| < 1$ (so that $\log(\exp Z)$ exists). Therefore the phrase 'sufficiently close to 0' may be interpreted more precisely as

$$\|X\| + \|Y\| < \log 2, \|Z\| < 1. \tag{5}$$

(Alternatively: compare (4) directly with the series expansion of $\sum_{k=1}^{\infty}(1/k)$ $(e^{\alpha}e^{\beta} - 1)$ using $\|[X, Y]\| < 2\|X\| \, \|Y\|$. In this way, one need not know (2) = (4) to prove the convergence of (4).)

Actually, the exact expression of Z in terms of X and Y is of little importance and will not be used later. What is of importance is that Z can be written as a bracket series in X and Y in some way. The exact formula is given mainly to dispel the air of mystery from the qualitative statement.

Of course, at this point it is not at all clear what is to be gained by writing the complicated series (2) in the even more complicated form (4). Certainly, as far as actual computations are concerned, the formula (4) is hardly practical; its significance is theoretical: the whole edifice of Lie theory ultimately rests on this formula (at least on its qualitative form).

Proof. Fix X, Y and let $Z = Z(\tau)$ be the solution of

$$\exp(Z) = \exp(\tau X)\exp(\tau Y). \tag{6}$$

Differentiate using Theorem 5 of §1.2:

$$\exp(Z)\frac{1 - \exp(-\operatorname{ad} Z)}{\operatorname{ad} Z}\frac{dZ}{d\tau} = X\exp(Z) + \exp(Z)Y$$

or

$$\frac{dZ}{d\tau} = \frac{\operatorname{ad} Z}{\exp(\operatorname{ad} Z) - 1}X + \frac{\operatorname{ad} Z}{\exp(\operatorname{ad} Z) - 1}\exp(\operatorname{ad} Z)Y. \tag{7}$$

Equation (6) gives

$$\exp(\operatorname{ad} Z) = \exp(\operatorname{ad} \tau X)\exp(\operatorname{ad} \tau Y), \tag{8}$$

(Proposition 4 of §1.2). Now use the power series expansions

$$A = \log(1 + (\exp A - 1)) = \sum_{k=0}^{\infty} \frac{(-1)^k}{k+1} (\exp A - 1)^{k+1}. \tag{9}$$

Apply (9) to the numerator $A = \operatorname{ad} Z$ of the fractions in (7), divide by the denominator, and apply (8) to write $\exp(\operatorname{ad} Z)Y = \exp(\tau X)Y$ in (7). Then compute as in the derivation of (2):

$$\begin{aligned}
\frac{dZ}{d\tau} &= \sum_{k=0}^{\infty} \frac{(-1)^k}{k+1} \{(\exp(\operatorname{ad} \tau X) \exp(\operatorname{ad} \tau Y) - 1)^k X \\
&\quad + (\exp(\operatorname{ad} \tau X) \exp(\operatorname{ad} \tau Y) - 1)^k \exp(\operatorname{ad} \tau X) Y\} \\
&= \sum_{k=0}^{\infty} \frac{(-1)^k}{k+1} \left\{ \sum \tau^{i_1+j_1+\cdots+i_k+j_k} \frac{[X^{(i_1)}Y^{(j_1)} \ldots X^{(i_k)}Y^{(j_k)}X]}{i_1 j_1! \cdots i_k! j_k!} \right. \\
&\quad \left. + \tau^{i_1+j_1+\cdots+i_k+j_k+i_{k+1}} \frac{[X^{(i_1)}Y^{(j_1)} \ldots X^{(i_k)}Y^{(j_k)}X^{(i_{k+1})}Y]}{i_1! j_1! \cdots i_k! j_k! i_{k+1}!} \right\}.
\end{aligned} \tag{10}$$

Integrate expression term-by-term (using $Z(0) = 0$):

$$Z(1) = \int_0^1 \frac{dZ}{d\tau} d\tau.$$

This gives the desired relation (4) after shifting the index of summation. One also needs to observe that the general term of (4) is zero unless j_k equals 0 or 1, which corresponds to the two terms in (10). QED

We close this section with a few formulas, which follow from Campbell–Baker–Hausdorff, but are actually better proved directly.

Proposition 2.

(a) $\exp(X) \exp(Y) = \exp(X + Y + \frac{1}{2}[X, Y] + \cdots)$

(b) $\exp(X) \exp(Y) \exp(-X) = \exp(Y + [X, Y] + \cdots)$

(c) $\exp(X) \exp(Y) \exp(-X) \exp(-Y) = \exp([X, Y] + \cdots)$

where the dots indicate terms involving products of three or more factors X, Y.

Proof.

(a) is a restatement of equation (3), although a simpler proof may be given if this is all one wants: one only needs to compare terms of order ≤ 2 in $\exp(X) \exp(Y) = \exp Z$, Z written as a power series in X, Y with unknown coefficients.

(b) This is seen either by writing

$$\exp(X) \exp(Y) \exp(-X) = \exp(\exp(\operatorname{ad} X)Y)$$

and writing out the first few terms of the inner exp, or by using (a) twice:

$$\begin{aligned}\exp(X)\exp(Y)\exp(-X) &= \exp(X+Y+\tfrac{1}{2}[X,Y]+\cdots)\exp(-X)\ \text{[once]}\\ &= \exp(Y+[X,Y]+\cdots)\text{[twice]}\end{aligned}$$

(c) Same method. QED

Problems for §1.3

1. Prove part (c) of Proposition 2.
2. Use Dynkin's formula (4) to show that

$$C(X,Y) = X+Y+\tfrac{1}{2}[X,Y]+\tfrac{1}{12}[X,[X,Y]]+\tfrac{1}{12}[Y,[Y,X]]+\cdots.$$

Check that this agrees with what one obtains by writing out the terms up to order three of the series (4).

3. Prove that the series C(X,Y) can also be written in the following form:

$$C(X,Y) = \sum \frac{(-1)^k}{k+1}\frac{1}{i_1+\cdots+i_k+1}\frac{[X^{(i_1)}Y^{(j_1)}\ldots X^{(i_k)}Y^{(j_k)}X]}{i_1!j_1!\cdots i_k!j_k!}.$$

[Suggestion: Start with $Z = Z(\tau)$ defined by $\exp(Z) = \exp(\tau X)\exp(Y)$ instead of (6); imitate the proof. Comment: this formula might seem slightly simpler than (4), but is equally unmanageable and less symmetric. If one reverses the roles of X and Y in this procedure one obtains a formula reflecting the relation $C(-Y,-X) = -C(X,Y)$, which is evident form the definition of $C(X,Y)$].

4. Write $\exp(Z) = \exp(\tau X)\exp(\tau Y)$ as in (6). Let

$$Z = \sum_k \tau^k C_k,$$

be the expansion of Z in powers of τ. Derive the recursion formula

$$(k+1)C_{k+1} = -[C_k, X] + \sum \gamma_j [C_{k_1}\cdots[C_{k_j}, X+Y]\cdots],$$

where the γ_j are defined as the coefficients of the series

$$\frac{x}{1-e^{-x}} = \sum_j \gamma_j x^j.$$

(Compare with the *Bernoulli numbers* β_j defined by

$$\frac{x}{e^x-1} = \sum_j \beta_j \frac{x^j}{j!},$$

i.e. $\gamma_j = (-1)^j\beta_j/j!$).

[Suggestion: Show first that

$$\frac{dZ}{d\tau} = -\operatorname{ad}(Z)X + \frac{\operatorname{ad} Z}{1 - \exp(-\operatorname{ad} Z)}(X + Y).$$

Then substitute power series.]

A *linear Lie algebra* is a space $\mathfrak{n} \subset M$ of linear transformations that is closed under the bracket operation:

$$X, Y \in \mathfrak{n} \quad \text{implies} \quad [X, Y] \in \mathfrak{n}.$$

5. Let $\mathfrak{n}$ be a linear Lie algebra consisting of *nilpotent* matrices. Let $N = \{\exp \mathfrak{n} = \exp X | X \in \mathfrak{n}\}$. Show that N is a *group* under matrix multiplication, i.e.

$$a \in N \quad \text{implies} \quad a^{-1} \in N,$$
$$a, b \in N \quad \text{implies} \quad ab \in N.$$

[Suggestion: for $a = \exp X$ and $b = \exp Y$, consider $Z(\tau)$ defined by $\exp(\tau X)\exp(\tau Y) = \exp Z(\tau)$ as a power series in τ with matrix coefficients.]

6. Let $\mathfrak{n}_1$ and $\mathfrak{n}_2$ be two linear Lie algebras consisting of nilpotent matrices as in problem 5, N_1 and N_2 be the corresponding groups. Let $\varphi : \mathfrak{n}_1 \to \mathfrak{n}_2$ be a linear map. Show that the rule $f(\exp X) = \exp \varphi(X)$ defines a *group homomorphism* $f : N_1 \to N_2$ (i.e. a well-defined map satisfying $f(ab) = f(a)f(b)$ for all $a, b \in N_1$) if and only if $\varphi([X, Y]) = [\varphi X, \varphi Y]$ for all $X, Y \in \mathfrak{n}_1$.

Problems 7 and 8 are meant to illustrate problems 5 and 6. Assume known the results of those problems.

7. (a) Describe all subspaces $\mathfrak{n}$ consisting of nilpotent upper triangular matrices real 3×3 matrices

$$\begin{bmatrix} 0 & \alpha & \gamma \\ 0 & 0 & \beta \\ 0 & 0 & 0 \end{bmatrix}$$

which satisfy $[\mathfrak{n}, \mathfrak{n}] \subset \mathfrak{n}$. Describe the corresponding groups N.

(b) Give an example of a subspace $\mathfrak{n}$ of M with $[\mathfrak{n}, \mathfrak{n}] \subset \mathfrak{n}$ for which $N = \exp \mathfrak{n}$ is *not* a group. [Suggestion: Consider Example 9 of §1.2.]

8. Let $\mathfrak{n}$ be the space of all nilpotent upper triangular real $n \times n$ matrices

$$\begin{bmatrix} 0 & * & * \cdots & * \\ 0 & 0 & * \cdots & * \\ \vdots & \vdots & \vdots & \vdots \\ 0 & 0 & 0 \cdots & 0 \end{bmatrix}. \tag{1}$$

Then $N = \exp \mathfrak{n}$ consists of all unipotent upper triangular real $n \times n$ matrices

$$\begin{bmatrix} 1 & * & * \cdots & * \\ 0 & 1 & * \cdots & * \\ \vdots & \vdots & \vdots & \vdots \\ 0 & 0 & 0 \cdots & 1 \end{bmatrix}.$$

Let $\varphi : \mathfrak{n} \to \mathbb{R}$ be a linear functional. Show that the rule

$$f(\exp X) = e^{\varphi(X)}$$

defines a group homomorphism $f : N \to \mathbb{R}^\times$ of N into the group $\mathbb{R}^\times$ of non-zero reals under multiplication if and only if $\varphi(\mathfrak{n}') = 0$, where $\mathfrak{n}'$ is the subspace of $\mathfrak{n}$ of matrices with entries zero directly above the main diagonal.

9. Compare the series (2) and (4):

$$Z = \sum \frac{(-1)^{k-1}}{k} \frac{X^{i_1}Y^{j_1} \cdots X^{i_k}Y^{j_k}}{i_1!j_1! \cdot i_k!j_k!}, \tag{2}$$

$$C(X,Y) = \sum \frac{(-1)^{k-1}}{k} \frac{1}{(i_1+j_1)+\cdots(i_k+j_k)} \frac{[X^{(i_1)}Y^{(j_1)} \cdots X^{(i_k)}Y^{(j_k)}]}{i_1!j_1! \cdots i_k!j_k!}. \tag{3}$$

This problem 'explains' the evident formal relation between the two through an outline of Dynkin's original proof.

Let A denote the collection of all formal *polynomials* (finite formal series) in the non-commuting formal variables X, Y. A is a real vector space with a basis consisting of the (infinitely many) distinct *monomials* in the list $X^{i_1}Y^{j_1} \cdots X^{i_k}Y^{j_k}, i_r, j_r = 0, 1, 2, \ldots$. Elements of A are multiplied in the obvious way, and we define a bracket in A by the formula

$$[a, b] = ab - ba.$$

This bracket is real-bilinear, skew-symmetric, and satisfies the Jacobi Identity

$$[a, [b, c]] + [b, [c, a]] + [c, [a, b]] = 0.$$

Let L be the subspace of A spanned by all repeated brackets of X's and Y's, including the elements X and Y themselves. It is clear that $[a, b]$ is in L whenever a and b are. Define a map $\gamma : A \to L$ by

$$\gamma(1) = 0, \qquad \gamma(X) = X, \qquad \gamma(Y) = Y,$$

and generally

$$\gamma(X^{i_1}Y^{j_1} \cdots X^{i_k}Y^{j_k}) = [X^{(i_1)}Y^{(j_1)} \cdots X^{(i_k)}Y^{(j_k)}],$$

the right side being interpreted as explained in connection with formula (4). For each $a \in A$ define a linear map $\delta(a) : A \to A$ by

$$\delta(1)c = c, \qquad \delta(X)c = [X, c], \qquad \delta(Y)c = [Y, c],$$

and generally

$$\delta(X^{i_1}Y^{j_1} \cdots X^{i_k}Y^{j_k})c = \delta(X)^{i_1}\delta(Y)^{j_1} \cdots \delta(X)^{i_k}\delta(Y)^{j_k}c.$$

These operations obey the rules

$$\delta(ab) = \delta(a)\delta(b),$$
$$\gamma(ab) = \delta(a)\gamma(b),$$

which follow directly from the definitions. Prove:

(a) If $a \in L$, then $\delta(a)b = [a, b]$.
(b) If $a, b \in L$, then $\gamma([a, b]) = [\gamma(a), b] + [a, \gamma(b)]$.
(c) If $a \in L$ is homogeneous of degree m, then $\gamma(a) = ma$.

Explanation. $a \in A$ is *homogeneous of degree* m if it is a linear combination of terms

$$X^{i_1}Y^{j_1} \cdots X^{i_k}Y^{j_k} \tag{4}$$

with $i_1 + j_1 + \cdots + i_k + j_k = m$. [Suggestion: It suffices to consider homogeneous elements. Use induction on the degree.]

Written as $a = (1/m)\gamma(a)$, the rule (c) says: a homogeneous polynomial of degree m which can be written as a bracket series in *some* way, remains unchanged if we replace each monomial therein by the corresponding bracket monomial and divide by m. If this recipe is applied to the homogeneous terms of the series (2), assuming known that this series does lie in L (as was indeed the case historically), there results the series (4).

2

Lie theory

2.1 Linear groups: definitions and examples

A *linear group* G is any family of invertible matrices with the property that

$$a \in G \quad \text{implies} \quad a^{-1} \in G,$$
$$a, b \in G \quad \text{implies} \quad ab \in G.$$

Example 1 (The rotation group $\mathrm{SO}(3)$**).** This is the *special orthogonal group* in three dimensions. It consists of all linear transformations of a three-dimensional real vector space E that leave invariant a positive-definite inner product and have determinant $+1$, i.e.

$$\mathrm{SO}(3) = \{a \in M_3(\mathbb{R}) | a^* a = 1, \det a = +1\}.$$

We shall discuss this group in some detail, subdividing the discussion into lemmas. First we determine which matrices generate one-parameter subgroups of $\mathrm{SO}(3)$:

Lemma 1A. *Let* $X \in M_3(\mathbb{R})$. $\exp(\tau X) \in \mathrm{O}(3)$ *for all* $\tau \in \mathbb{R}$ *if and only if* $X^* = -X$.

Proof. $\exp \tau X \in \mathrm{SO}(3)$ if and only if

$$(\exp \tau X)^* = \exp(-\tau X). \tag{1}$$

Since $(\exp X)^* = \exp(X^*)$, as one sees by using the additivity of the adjoint operation on the series defining exp, (1) means that

$$\exp(\tau X^*) = \exp(-\tau X). \tag{2}$$

If this holds for all real τ, then we must have $X^* = -X$, since exp is one-to-one in a neighborhood of zero. QED

In the present context the space of skew-symmetric 3×3 matrices is denoted $\mathsf{so}(3, \mathbb{R})$:

$$\mathsf{so}(3) = \{X \in M_3(\mathbb{R}) \mid X^* = -X\}. \tag{3}$$

$\mathsf{so}(3)$ is a subspace of $M_3(\mathbb{R})$ closed under the bracket operation as one checks immediately.

One will note that proof of the above lemma did not use the 'determinant $= +1$' condition: one arrives at the same space of skew-symmetric matrices starting from the full *orthogonal group* $\mathrm{O}(3) = \{a \in M_3(\mathbb{R}|a^*a = 1\}$. There is a simple explanation: a continuous curve in $\mathrm{O}(3)$, such as $a(\tau) = \exp(\tau X)$, has always $\det a(\tau) = \pm 1$ and cannot continuously pass from one value to the other, hence must lie entirely in $\mathrm{SO}(3)$ if it starts at $a(0) = 1$.

The sign of the determinant has a geometric interpretation: the ordered bases for a real vector space fall into two classes, characterized by the property that any two bases in the *same* class are related by a matrix with a *positive* determinant, while any two bases from *different* classes are related by a matrix with a *negative* determinant. We single out one of these classes and call the bases belonging to it *right-handed.*

Lemma 1B.

(a) For any $a \in \mathrm{SO}(3)$ there is a right-handed orthonormal basis $\mathrm{e}_1, \mathrm{e}_2, \mathrm{e}_3$ of E so that

$$\begin{aligned} a\mathrm{e}_1 &= \cos\alpha \mathrm{e}_1 + \sin\alpha \mathrm{e}_2, \\ a\mathrm{e}_2 &= -\sin\alpha \mathrm{e}_1 + \cos\alpha \mathrm{e}_2, \\ a\mathrm{e}_3 &= \mathrm{e}_3, \end{aligned} \tag{4}$$

for some $\alpha \in \mathbb{R}$.

(b) For any $X \in \mathsf{so}(3)$ there is a right-handed orthonormal basis $\mathrm{e}_1, \mathrm{e}_2, \mathrm{e}_3$ of E so that

$$X\mathrm{e}_1 = \alpha \mathrm{e}_2, \quad X\mathrm{e}_2 = -\alpha \mathrm{e}_1, \quad X\mathrm{e}_3 = 0, \tag{5}$$

for some $\alpha \in \mathbb{R}$.

(c) If $X \in \mathsf{so}(3)$ is given by (5) (for some right-handed orthonormal basis $\mathrm{e}_1, \mathrm{e}_2, \mathrm{e}_3$ of E), then $a = \exp X$ is given by (4) (for the same orthonormal basis).

(QED)

The linear transformation $a \in \mathrm{SO}(3)$ defined by (4) is called the *right-handed rotation around* e_3 with angle α. It is uniquely determined by α and e_3. The linear transformation $X \in \mathsf{so}(3)$ defined by (5) is called the *infinitesimal right-handed rotation around* e_3. The lemma implies that the exponential map $\mathsf{so}(3) \to \mathrm{SO}(3)$ is surjective, or, in geometric terms, that every element of $\mathrm{SO}(3)$ can be infinitesimally generated as a rotation. (Hence the name rotation group, and hence $\mathrm{SO}(3)$ rather than $\mathrm{O}(3)$.)

We now fix once and for all a right-handed orthonormal basis e_1, e_2, e_3 for E and define $E_1, E_2, E_3 \in \mathsf{so}(3)$ to be the infinitesimal rotations around the basis vectors, with matrices:

$$E_1 = \begin{bmatrix} 0 & 0 & 0 \\ 0 & 0 & -1 \\ 0 & 1 & 0 \end{bmatrix}, \quad E_2 = \begin{bmatrix} 0 & 0 & 1 \\ 0 & 0 & 0 \\ -1 & 0 & 0 \end{bmatrix}, \quad E_3 = \begin{bmatrix} 0 & -1 & 0 \\ 1 & 0 & 0 \\ 0 & 0 & 0 \end{bmatrix}. \tag{6}$$

These three matrices E_1, E_2, E_3 form a basis for the space $\mathsf{so}(3)$. Define a map $X \to \vec{X}$, sending the basis E_1, E_2, E_3 to the basis e_1, e_2, e_3.

One checks immediately that under this map $\mathsf{so}(3) \to E$ the inner product in E becomes

$$(\vec{X}, \vec{Y}) = \frac{1}{2} \operatorname{tr} Y^* X, \tag{7}$$

which is half of the usual inner product in the matrix space M. Furthermore, under the map $\mathsf{so}(3) \to E$, $X \to \vec{X}$ the action of the linear transformation $a \in \mathrm{SO}(3)$ on E corresponds to the adjoint action of $\mathrm{SO}(3)$ on elements of $\mathsf{so}(3)$:

Lemma 1C.

(a) For any $X, Y \in \mathsf{so}(3)$,

$$\overrightarrow{[X,Y]} = X\vec{Y}. \tag{8}$$

(b) For any $a \in \mathrm{SO}(3)$ and any $Y \in \mathsf{so}(3)$,

$$\overrightarrow{[\mathrm{Ad}(a)Y]} = a\vec{Y}. \tag{9}$$

Proof.

(a) is checked by calculation. (It suffices to take basis elements E_j for X and Y.)

(b) By part (c) of Lemma 1B we may write $a = \exp X$ with $X \in \mathsf{so}(3)$. Then $\mathrm{Ad}(a)Y = \exp(\operatorname{ad} X)Y$ and (9) follows by applying (8), written in the form $\overrightarrow{\operatorname{ad}(X)Y} = X\vec{Y}$, repeatedly to the terms of the series for $\exp(\operatorname{ad} X)Y$. QED

Remark 1D. Implicit in part (a) of Lemma 1C is that $\mathsf{so}(3)$ is closed under the bracket operation, as observed above. A simple calculation with the basis elements E_j shows that the bracket operation in $\mathsf{so}(3)$ corresponds to the cross product in $E \approx \mathbb{R}^3$ under the map $X \to \vec{X}$:

$$\overrightarrow{[X,Y]} = \vec{X} \times \vec{Y}. \tag{10}$$

We summarize:

Proposition 1E.

(a) $\exp\colon \mathsf{so}(3) \to \mathrm{SO}(3)$ *maps $X \in \mathsf{so}(3)$ to the right-handed rotation $a \in \mathrm{SO}(3)$ with angle $\alpha = \|\vec{X}\|$ around $\vec{X} \in E$.*

(b) $\exp\colon \mathsf{so}(3) \to \mathrm{SO}(3)$ *is surjective;* $\exp X = \exp Y$ *if and only if $\vec{X}$ and $\vec{Y}$ lie on a line through the origin in E, a distance $2\pi k$, $k = 0, 1, 2, \ldots$, apart.*

Proof.

(a) For $X \in \mathsf{so}(3)$, write $\vec{X} = \alpha c e_3$ with $c \in \mathrm{SO}(3)$ and $\alpha \geq 0$. Then necessarily $\alpha = \|\vec{X}\|$. From (9), $\vec{X} = \alpha c\vec{E} = \overline{\alpha\,\mathrm{Ad}(c)E_3}$, which gives $X = \alpha c E_3 c^{-1}$, and (a) follows.

(b) This follows from the explicit formula (4). QED

The proposition gives a very concrete picture of the group SO(3): its elements are in one-to-one correspondence with the closed ball $\|x\| \leq \pi$ in E when antipodal points on its surface $\|x\| = 1$ are identified. (This is a model for real projective 3-space.) Furthermore, exp maps $\mathsf{so}(3) \approx E$ onto SO(3) in a periodic way, sending a line through the origin to the one-parameter group of rotations around this line.

The adjoint operation of SO(3) on so(3) corresponds under exp to the conjugation operation of SO(3) on itself:

$$\exp(\mathrm{Ad}(c)X) = cXc^{-1}.$$

A subset of so(3) of the form $\{\mathrm{Ad}(c)X_o | c \in \mathrm{SO}(3)\}$ will be called an *adjoint orbit* of SO(3). It corresponds to the sphere $\{X | \|\vec{X}\| = \rho\}$ containing $\vec{X}_o$ and gets mapped onto the conjugacy class $\{c\exp(X_o)c^{-1} \mid c \in \mathrm{SO}(3)\}$. On an adjoint orbit, exp is one-to-one as long as the elements in the corresponding conjugacy class have distinct eigenvalues, which means $\|X\| \neq \pi k, k \in \mathbb{Z}$. There are two *singular conjugacy* classes in SO(3), represented by

$$\begin{bmatrix} 1 & 0 & 0 \\ 0 & 1 & 0 \\ 0 & 0 & 1 \end{bmatrix}, \quad \begin{bmatrix} 1 & 0 & 0 \\ 0 & -1 & 0 \\ 0 & 0 & -1 \end{bmatrix}.$$

They correspond to the spheres $\|X\| = 2k\pi$, $\|X\| = (2k+1)\pi$ in so(3), respectively.

We momentarily come back to O(3). This group comes in two pieces,

$$\mathrm{O}(3) = \mathrm{SO}(3) \bigcup (-\mathrm{SO}(3)),$$

which are its *connected components* in the sense that any two elements within the same piece can be joined by a continuous path, but no element from one to any element of the other. O(3) is *disconnected* and SO(3) is *connected component of the identity in* O(3).

The example $G = \mathrm{SO}(3)$ sets the pattern for the examples to follow, although the discussion will not be quite as detailed. For each group G we find the set g consisting of all matrices X for which the whole one-parameter group $\exp tX$ belongs to G. This set g will turn out to be closed under the operations of addition $X+Y$, scalar multiplication αX, and bracket $[X,Y]$, just as for SO(3). It is also invariant under the adjoint operation by G, i.e. $\mathrm{Ad}(c)\mathsf{g} = \mathsf{g}$ for all $c \in G$, as is clear from the formula

$$\exp(\mathrm{Ad}(c)X) = c\exp(X)c^{-1}.$$

The subset $\mathrm{Ad}(G)X_o = \{X \in \mathsf{g} \mid X = \mathrm{Ad}(c)X_o, c \in G\}$ of g is called the adjoint orbit of the element $X_o \in \mathsf{g}$. Under the exponential, it maps to the subset $\{ca_oc^{-1} \mid c \in G\}$, the conjugacy class of the element $a_o = \exp X_o$ in G. For $G = \mathrm{SO}(3)$ the adjoint orbits are the spheres we met above.

Example 2 (The special unitary group SU(2)). This is the group

$$\mathrm{SU}(2) = \{a \in M_2(\mathbb{C}) \mid aa^* = 1\}$$

where a^* is the Hermitian adjoint (conjugate transpose) of a. Explicitly, the elements of SU(2) are of the form

$$a = \begin{bmatrix} \alpha & -\bar{\beta} \\ \beta & \bar{\alpha} \end{bmatrix}, \qquad \alpha\bar{\alpha} + \beta\bar{\beta} = 1.$$

The group SU(2) is therefore just the 3–sphere S^3 in $M_2(\mathbb{C}) \approx \mathbb{R}^4$. We also set

$$\mathsf{su}(2) = \{X \in M_2(\mathbb{C}) \mid X^* = -X, \, \mathrm{tr}X = 0\}.$$

This is a *Lie algebra*, i.e. real vector space matrices closed under the bracket operation. (It is not a complex vector space.)

Lemma 2.A.
(a) Let $X \in M_2(\mathbb{C})$. $\exp(\tau X) \in \mathrm{SU}(2)$ for all $\tau \in \mathbb{R}$ if and only if $X \in \mathsf{su}(2)$.
(b) $\exp : \mathsf{su}(2) \to \mathrm{SU}(2)$ is surjective.

Proof.
(a) $(\exp X)^* = \exp(X^*)$ and $\det(\exp X) = e^{\mathrm{tr}X}$.
(b)By consideration of eigenvectors and eigenvalues one can check that any $a \in \mathrm{SU}(2)$ is conjugate to a matrix of the form

$$\begin{bmatrix} e^{i\theta} & 0 \\ 0 & e^{-i\theta} \end{bmatrix} = \exp \begin{bmatrix} i\theta & 0 \\ 0 & -i\theta \end{bmatrix}.$$

This implies (b) . QED

The three dimensional real vector space $E = \mathsf{su}(2)$ comes equipped with a positive definite inner product defined by $(X, X) = \frac{1}{2}\mathrm{tr}(X^*X)$. Explicitly, for $X \in \mathsf{su}(2)$

$$X = \begin{bmatrix} i\xi_3 & -\xi_1 + i\xi_2 \\ \xi_1 + i\xi_2 & -\, i\xi_3 \end{bmatrix}, \quad \frac{1}{2}\mathrm{tr}(X^*X) = \xi_1^2 + \xi_2^2 + \xi_3^2.$$

We now take SO(3) to be the rotation group of this space $\mathsf{su}(2)$. If $a \in \mathrm{SU}(2)$, then $\mathrm{Ad}(a)$ gives a linear transformation of $\mathsf{su}(2)$ which belongs to SO(3), still denoted $\mathrm{Ad}(a)$. The map $\mathrm{Ad}\colon \mathrm{SU}(2) \to \mathrm{SO}(3)$ is a homomorphism of groups. i.e. preserves products: $\mathrm{Ad}(ab) = \mathrm{Ad}(a)\mathrm{Ad}(b)$. Similarly, $X \in \mathsf{su}(2)$ gives $\mathrm{ad}(X) \in \mathsf{so}(3)$. The map $\mathrm{ad}\colon \mathsf{su}(2) \to \mathsf{so}(3)$ is a *homomorphism of Lie algebras*, i.e. preserves brackets: $\mathrm{ad}[X, Y] = [\mathrm{ad}X, \mathrm{ad}Y]$.

Lemma 2.B. (a) $\mathrm{ad}\colon \mathsf{su}(2) \to \mathsf{so}(3)$ is bijective.
(b) $\mathrm{Ad}\colon \mathrm{SU}(2) \to \mathrm{SO}(3)$ is surjective with kernel $\{\pm 1\}$.

Proof.
(a)Since $\mathrm{ad}\colon \mathsf{su}(2) \to \mathsf{so}(3)$ is a linear map between spaces of the same dimension (namely 3), it suffices to show that its kernel is zero. Suppose $X \in \mathsf{su}(2)$ satisfies $\mathrm{ad}(X)\mathsf{su}(2) = 0$. Any $Z \in M_2(\mathbb{C})$ is of the form $Z = U + iV$ where U, V are skew Hermitian, namely $U = \frac{1}{2}(Z - Z^*)$, $V = \frac{1}{2i}(Z - Z^*)$. Hence $M_2(\mathbb{C}) = \mathsf{su}(2) + i\mathsf{su}(2) + \mathbb{C}1$ and $\mathrm{ad}(X)\mathsf{su}(2) = 0$ implies $\mathrm{ad}(X)M_2(\mathbb{C}) = 0$, i.e. X commutes with all matrices. Hence X is a scalar matrix and $\mathrm{tr}X = 0$ implies $X = 0$.

(b) The surjectivity of $\mathrm{Ad}\colon \mathrm{SU}(2) \to \mathrm{SO}(3)$ follows from the surjectivity of the exponential maps and of $\mathrm{ad}\colon \mathsf{su}(2) \to \mathsf{so}(3)$: $\mathrm{Ad}(\exp \mathsf{su}(2)) = \exp(\mathrm{ad}\,\mathsf{su}(2) = \exp \mathsf{so}(3) = \mathrm{SO}(3)$. Suppose $a \in \mathrm{SU}(2)$ belongs to the kernel of $\mathrm{Ad}\colon \mathrm{SU}(2) \to \mathrm{SO}(3)$, i.e. $\mathrm{Ad}(a)X = X$ for all $X \in \mathsf{su}(2)$. As in the proof of (a) this implies that $a \in \mathrm{SU}(2)$ commutes with all $Z \in M_3(\mathbb{C})$. Hence a is a scalar matrix and $\det(a) = 1$ implies $a = \pm 1$. QED

Part (b) of the lemma says that the group homomorphism $\mathrm{SU}(2) \to \mathrm{SO}(3)$ is a *double covering* in the sense that each element a of $\mathrm{SO}(3)$ has exactly two preimages $\pm\tilde{a}$ in $\mathrm{SU}(2)$. The 'inverse map' $\mathrm{SO}(3) \rightsquigarrow \mathrm{SU}(2)$ is double-valued; it associates *two* unitary transformations $\psi \mapsto \pm\tilde{a}\psi$ of $\mathbb{C}^2$ to each $a \in \mathrm{SO}(3)$. This is the famous *spin representation* of $\mathrm{SO}(3)$. The elements ψ of $\mathbb{C}^2$ are referred to as *spinors* in this context, as in Weyl's *"Theory of Groups and Quantum Mechanics"* of 1931.

The lemma says that $\mathrm{Ad}(\mathrm{SU}(2)) \approx \mathrm{SO}(3)$: we may identify both $\mathsf{su}(2)$ and $\mathsf{so}(3)$ with $\mathbb{R}^3$ so that the adjoint actions of $\mathrm{SU}(2)$ and $\mathrm{SO}(3)$ become the usual rotation action of $\mathrm{SO}(3)$ on $\mathbb{R}^3$. In particular, the adjoint orbits of $\mathrm{SU}(2)$ are again spheres $\|\vec{X}\| = \rho$; but it takes the solid ball of radius 2π in $\mathbb{R}^3$ to cover $\mathrm{SU}(2)$, while the ball of radius π suffices for $\mathrm{SO}(3)$. $\mathrm{SU}(2)$ may also be realized as the group quaternions of norm 1. This may be seen as follows.

Recall that (real) quaternions are expressions of the form

$$a = \lambda 1 + \mu i + \nu j + \kappa k, \quad \lambda, \mu, \nu, \kappa \in \mathbb{R},$$

added and multiplied in the obvious way subject to

$$i^2 = j^2 = k^2 = ijk = -1.$$

The quaternions form a division algebra (non-commutative field) denoted $\mathbb{H}$. The *conjugate* $\bar{a}$ of $a \in \mathbb{H}$ is

$$\bar{a} = \lambda 1 - \mu i - \nu j - \kappa k,$$

and the square-norm is

$$|a|^2 = a\bar{a} = \lambda^2 + \mu^2 + \nu^2 + \kappa^2.$$

Identify complex numbers with quaternions of the form $\lambda + i\mu$. Any quaternion can be uniquely written as $\alpha + j\beta$ with $\alpha, \beta \in \mathbb{C}$. The map

$$\alpha + j\beta \to \begin{bmatrix} \alpha & -\bar{\beta} \\ \beta & \bar{\alpha} \end{bmatrix}$$

sets up a one-to-one correspondence between $\mathbb{H}$ and complex 2×2 matrices of the form indicated which turns multiplication of quaternions into matrix multiplication. (The matrix represents left multiplication by $\alpha + j\beta$ if $\mathbb{H}$ is considered a right vector space over $\mathbb{C}$ with basis $1, j$.) Under this correspondence the conjugate of a quaternion corresponds to the Hermitian adjoint of the matrix. It follows that the group of norm-one quaternions gets mapped isomorphically onto $\mathrm{SU}(2) = \{a \in M_2(\mathbb{C}) \mid aa^* = 1, \det a = 1\}$.

Example 3 (The real special linear group $\mathrm{SL}(2,\mathbb{R})$**).** This is the group of real 2×2 matrices of determinant 1:

$$\mathrm{SL}(2,\mathbb{R}) = \{a \in M_2(\mathbb{R}) | \det a = 1\}.$$

In analogy with the previous example we record:

Lemma 3A. *Let* $X \in M_2(\mathbb{R})$. $\exp(\tau X) \in \mathrm{SL}(2,\mathbb{R})$ *for all* $\tau \in \mathbb{R}$ *if and only if* $\operatorname{tr} X = 0$.

Proof. $\det(\exp X) = e^{\operatorname{tr} X}$. QED

We set

$$\mathsf{sl}(2,\mathbb{R}) = \{X \in M_2(\mathbb{R}) | \operatorname{tr} X = 0\},$$

again a space of matrices closed under the bracket operation.

The exponential map $\exp\colon \mathsf{sl}(2,\mathbb{R}) \to \mathrm{SL}(2,\mathbb{R})$ has been described in Example 9 of §1.2. Without repeating the discussion there, we note that Figure 1 of §1.2 gives a picture of the adjoint orbits of $\mathrm{SL}(2,\mathbb{R})$, and hence of those conjugacy classes in $\mathrm{SL}(2,\mathbb{R})$ that lie in the image of exp: the hyperboloids of one sheet correspond to conjugacy classes of hyperbolic elements; the sheets of the hyperboloids of two sheets correspond to conjugacy classes of hyperbolic elements; the two sheets of the double cone (without vertex) to conjugacy classes of unipotent elements; the origin to the identity element. On each adjoint orbit, except on those corresponding to the conjugacy classes of ± 1, exp is one-to-one.

The image of $\exp\colon \mathsf{sl}(2,\mathbb{R}) \to \mathrm{SL}(2,\mathbb{R})$ consists of the $a \in \mathrm{SL}(2,\mathbb{R})$ satisfying $\frac{1}{2} \operatorname{tr} a > -1$ together with $a = -1$. The remaining elements can evidently be obtained from these by multiplication by (-1), with some repetitions.

Figure 1 represents a path in $\mathrm{SL}(2,\mathbb{R})$ which runs through the conjugacy classes: every conjugacy class is visited once, except at the branch points ± 1, where the hyperbolic elements meet the elliplic elements; these branch points must be counted as triple points: the two additional points represent unipotent classes.

The discussion gives a picture of the decomposition of $\mathrm{SL}(2,\mathbb{R})$ into its conjugacy classes, but does not give a simple picture of $\mathrm{SL}(2,\mathbb{R})$ as a whole. To see $\mathrm{SL}(2,\mathbb{R})$ as a whole it is best to change point of view.

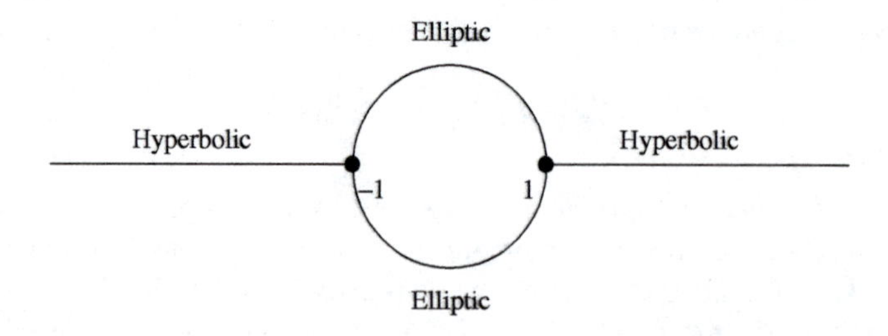

Fig. 1 The set of conjugacy classes in $\mathrm{SL}(2,\mathbb{R})$.

Lemma 3B. *Let K be the subgroup of* $\mathrm{SL}(2,\mathbb{R})$ *consisting of rotation matrices*

$$\begin{bmatrix}\cos\theta & -\sin\theta\\ \sin\theta & \cos\theta\end{bmatrix}$$

and let B be the subgroup consisting of triangular matrices with positive diagonal,

$$b=\begin{bmatrix}\alpha & \beta\\ 0 & \alpha^{-1}\end{bmatrix},\quad \alpha>0.$$

Then $\mathrm{SL}(2,\mathbb{R}) = KB$ *in the sense that every element* $a\in\mathrm{SL}(2,\mathbb{R})$ *can be uniquely witten as $a=kb$ with $k\in K$ and $b\in B$.*

Proof. $k\in K$ operates left translation $a\to ka$ through its rotation operation in each column. Hence we can find $k\in K$ so that the first column of ka is of the form $\left[\begin{smallmatrix}\alpha\\0\end{smallmatrix}\right]$ with $\alpha>0$, and k is then uniquely determined. Then $ka=b$ is of the desired form and $a=k^{-1}b$ the desired decomposition of a. QED

The map $K\times B\to\mathrm{SL}(2,\mathbb{R})$ may be used to identify $\mathrm{SL}(2,\mathbb{R})$ with $K\times B\approx S^1\times(\mathbb{R}^\times_{\text{pos}}\times\mathbb{R})\approx S^1\times\mathbb{R}^2$ as point-set (not as group: neither K nor B is normal in $\mathrm{SL}(2,\mathbb{R})$). If we grant that this map and its inverse are continuous (in a sense to be made precise in §2.3), we may conclude that $\mathrm{SL}(2,\mathbb{R})$ looks like $S^1\times\mathbb{R}^2$, topologically. In particular, $\mathrm{SL}(2,\mathbb{R})$ is connected in the sense that any two of its points can be joined by a continuous, even analytic, curve. The failure of its exponential map to be surjective cannot therefore be explained by disconnectedness, as in the case of $\mathrm{O}(3)$.

The decomposition $\mathrm{SL}(2,\mathbb{R})=KB$ has a geometric interpretation in terms of the embedding of $\mathrm{SL}(2,\mathbb{R})$ in the matrix space $M_2(\mathbb{R})$, which is curious if not of group-theoretical significance. Write elements of $M_2(\mathbb{R})$ explicitly as

$$x=\begin{bmatrix}\xi_1 & \xi_2\\ \xi_3 & \xi_4\end{bmatrix}.$$

Then $\mathrm{SL}(2,\mathbb{R})$ appears as (three-dimensional) quadratic surface

$$\xi_1\xi_4-\xi_2\xi_3=1. \tag{11}$$

The subgroup B is the intersection of this surface with the (three-dimensional) plane $\xi_3=0$, which may be written as $\mathrm{tr}(E_{12}x)=0$. The coset aB ($a\in\mathrm{SL}(2,\mathbb{R})$) is then evidently the intersection with the plane $\mathrm{tr}(E_{12}a^{-1}x)=0$. Since $\mathbb{R}E_{12}b=\mathbb{R}E_{12}$ for $b\in B$ this plane is also of the form $\mathrm{tr}(E_{12}k^{-1}x)=0$, when $a=kb$ i.e.

$$-\sin\theta\xi_1+\cos\theta\xi_3=0. \tag{12}$$

Thus the decomposition $\mathrm{SL}(2,\mathbb{R})=KB$ of $\mathrm{SL}(2,\mathbb{R})$ into cosets kB appears as a 'ruling' of the 3-dimensional quadratic surface (11) by its intersections with the 3-dimensional planes (12), which are parametrized by S^1. The 'lines' of the ruling are half-planes $\mathbb{R}^\times_{\text{pos}}\times\mathbb{R}$. There is a second such ruling obtained by replacing the upper triangular group B by the lower triangular group in $\mathrm{SL}(2,\mathbb{R})$. The embedding of $\mathrm{SL}(2,\mathbb{R})$ in the matrix space $M_2(\mathbb{R})$ is therefore rather analogous to the a hyperboloid of one sheet in $\mathbb{R}^3$ (Figure 2).

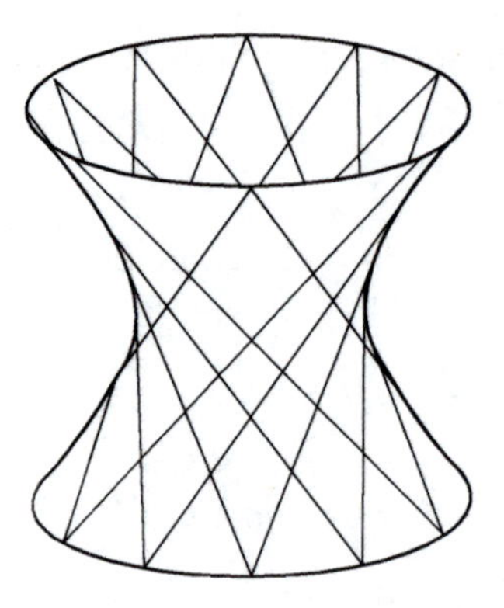

Fig. 2 SL(2, ℝ) as hyperboloid.

Example 4 (The complex special linear group $\mathrm{SL}(2,\mathbb{C})$**).** We shall be brief. Notation:

$$\mathrm{SL}(2,\mathbb{C}) = \{a \in M_2(\mathbb{C}) |\, \det a = 1\},$$
$$\mathfrak{sl}(2,\mathbb{C}) = \{X \in M_2(\mathbb{C}) |\, \operatorname{tr} X = 0\}.$$

The exponential map $\exp : \mathfrak{sl}(2,\mathbb{C}) \to \mathrm{SL}(2,\mathbb{C})$ is not surjective, but misses only one conjugacy class in $\mathrm{SL}(2,\mathbb{C})$, namely the one represented by

$$\begin{bmatrix} -1 & 1 \\ 0 & -1 \end{bmatrix}. \tag{13}$$

This may be seen as in Example 9 of §1.2 or as follows. If the eigenvalues α, α^{-1} of $a \in \mathrm{SL}(2,\mathbb{C})$ are distinct, then a is diagonalizable, hence an exponential of a matrix X of trace 0. If $\alpha = 1$, then $(a-1)^2 = 0$ and $X = (a-1)$ will do. If $\alpha = -1$ and a is not diagonalizable, then a is conjugate to the matrix above, which is not an exponential: if $\exp X$ has eigenvalues $-1, -1$, then the eigenvalues $\lambda, -\lambda$ of X are distinct ($\lambda = (k+\frac{1}{2})\pi i$ implies $\lambda \neq 0$), hence X is diagonalizable, and so is $\exp X$). But this is just not the case for the two elements above.

The conjuacy classes in $\mathrm{SL}(2,\mathbb{C})$ are represented by the matrices

$$\begin{bmatrix} \alpha & 0 \\ 0 & \alpha^{-1} \end{bmatrix}, \quad \alpha \neq 0, \quad \begin{bmatrix} 1 & 1 \\ 0 & 1 \end{bmatrix}, \quad \begin{bmatrix} -1 & 1 \\ 0 & -1 \end{bmatrix}.$$

A class of the first kind is the image under $\exp\colon \mathfrak{sl}(2,\mathbb{C}) \to \mathrm{SL}(2,\mathbb{C})$ of the complex quadratic surface $\det(X) = -\lambda^2$ when $e^\lambda = \alpha$ and the image is one-to-one for $\alpha \neq \pm 1$. The next class is the images of the complex cone $\det(X) = 0$, $X \neq 0$. The remaining class does not lie in the image of exp, as we just saw.

As in the previous example, $\mathrm{SL}(2,\mathbb{C})$ admits a decomposition

$$\mathrm{SL}(2,\mathbb{C}) = KB,$$

except that here $K = \mathrm{SU}(2)$ and B is the subgroup of triangular matrices with determinant 1 and positive real diagonal entries:

$$b = \begin{bmatrix} \alpha & \beta \\ 0 & \alpha^{-1} \end{bmatrix}, \quad \alpha \in \mathbb{R}^{\times}_{\text{pos}}, \ \beta \in \mathbb{C}.$$

The decompositon $a = kb$ is unique. Thus, topologically, $\mathrm{SL}(2,\mathbb{C})$ looks like $S^3 \times \mathbb{R} \times \mathbb{C}$. There is again an interpretation of the decomposition $\mathrm{SL}(2,\mathbb{C}) = KB$ as a 'ruling' of the complex quadratic surface $\det z = 1$ in $M_2(\mathbb{C})$, analogous to the real case.

Example 5 (The Euclidean motion group $\mathbb{R}^2 \rtimes \mathrm{SO}(2)$). Let E denote the Euclidean plane ($\mathbb{R}^2$ with its positive definite inner product). A *Euclidean transformation* of E is a transformation of the form $p \to ap + b$ with $a \in \mathrm{O}(2)$ (orthogonal group) and b $\in \mathbb{R}^2$. (It can be shown that these are the only transformations of E which preserve Euclidean distance.) Such a transformation may be implemented by matrix multiplication according to the formula

$$\begin{bmatrix} a & b \\ 0 & 1 \end{bmatrix} \begin{bmatrix} p \\ 1 \end{bmatrix} = \begin{bmatrix} ap + b \\ 1 \end{bmatrix}.$$

The *Euclidean motion group* of the plane consists of such transformations $p \to ap + b$ with $a \in \mathrm{SO}(2)$ and $b \in E$. It may be realized as the linear group of block matrices

$$\begin{bmatrix} a & b \\ 0 & 1 \end{bmatrix}, \quad a \in \mathrm{SO}(2), \ b \in \mathbb{R}^2.$$

The decomposition

$$\begin{bmatrix} a & b \\ 0 & 1 \end{bmatrix} = \begin{bmatrix} 1 & b \\ 0 & 1 \end{bmatrix} \begin{bmatrix} a & 0 \\ 0 & 1 \end{bmatrix}$$

of a Euclidean motion into a rotation about 0 and a translation is unique and will be written as $b \circ a$, $b \in \mathbb{R}^2$, $a \in \mathrm{SO}(2)$. The Euclidean motion group itself is written $\mathbb{R}^2 \rtimes \mathrm{SO}(2)$. This is a *semidirect product*, meaning the intersection of the two subgroups is $\{1\}$ and one of them (in this case the translation group $\mathbb{R}^2$) is a *normal* subgroup.

The exponential map appropriate to the Euclidean motion group takes the form

$$\exp : \begin{bmatrix} X & v \\ 0 & 0 \end{bmatrix} \to \begin{bmatrix} a & b \\ 0 & 1 \end{bmatrix}, \quad a = \exp(\tau X) \in \mathrm{SO}(2), \ b = \frac{\exp \tau X - 1}{X} v \in E. \tag{14}$$

For $\exp(\tau X)$ to lie in $\mathrm{SO}(2)$, X must be skew-symmetric, hence a multiple of the infinitesimal rotation

$$\begin{bmatrix} 0 & -1 \\ 1 & 0 \end{bmatrix}.$$

The exponential of such a matrix X is

$$\exp \begin{bmatrix} 0 & -\theta \\ \theta & 0 \end{bmatrix} = \begin{bmatrix} \cos\theta & -\sin\theta \\ \sin\theta & \cos\theta \end{bmatrix}.$$

We denote this rotation matrix $a(\theta)$; it belongs to the special orthogonal group SO(2). The image of exp in $\mathbb{R}^2 \rtimes \mathrm{O}(2)$ is therefore $\mathbb{R}^2 \rtimes \mathrm{SO}(2)$, called the *Euclidean motion group*, because its elements can be generated infinitesimally through exp.

The expression $(\exp \tau X - 1)/X$ in the formula for exp is the matrix

$$\begin{bmatrix} \dfrac{\sin\theta}{\theta} & \dfrac{\cos\theta - 1}{\theta} \\ \dfrac{-\cos\theta + 1}{\theta} & \dfrac{\sin\theta}{\theta} \end{bmatrix}.$$

The matrices on the left of (13), which get mapped onto $\mathbb{R}^2 \rtimes \mathrm{SO}(2)$ by exp, form a 3-dimensional vector space closed under the bracket operation, denoted $\mathbb{R}^2 \rtimes \mathsf{so}(3)$. $\mathbb{R}^2 \rtimes \mathrm{SO}(2)$ operates on this space by Ad:

$$\mathrm{Ad}\left(\begin{bmatrix} a & b \\ 0 & 1 \end{bmatrix}\right) \cdot \begin{bmatrix} X_o & v_o \\ 0 & 1 \end{bmatrix} = \begin{bmatrix} X_o & -X_o b + a v_o \\ 0 & 0 \end{bmatrix}.$$

Since the matrix X_o is determined by the parameter θ_o, elements of the space $\mathbb{R}^2 \rtimes \mathsf{so}(3)$ may be represented by pairs (θ_o, v_o). As $b \circ a(\theta)$ varies over $\mathbb{R}^2 \rtimes \mathrm{SO}(2)$, the element (14) varies over a subset of $\mathbb{R}^2 \times \mathsf{so}(2) \approx \mathbb{R}^3$ which is a plane $\theta = \theta_o$ if $\theta_o \neq 0$ or a circle $\|v\| = \|v_o\|$ in the plane $\theta = 0$ if $\theta_o = 0$. These subsets of $\mathbb{R}^2 \times \mathsf{so}(2)$ get mapped to conjugacy classes in $\mathbb{R}^2 \rtimes \mathrm{SO}(2)$ under exp. Two planes $\theta = \theta_o \neq 0$ get mapped to the same conjugacy class if their θ-levels differ by multiples of 2π, as one sees from the relation

$$\exp(\theta, v) = \exp(\theta', v') \quad \text{if and only if} \quad \theta' = \theta + 2\pi k, \ \ v'/\theta' = v/\theta.$$

Two circles $\|v\| = \|v_o\|$ in the plane $\theta = 0$ get mapped to the same conjugacy class if their radii differ by a multiple of 2π. Topologically, the group $\mathbb{R}^2 \rtimes \mathrm{SO}(2)$ looks, of course, like $\mathbb{R}^2 \times S^1$, just $\mathrm{SL}(2, \mathbb{R})$.

Example 6 (Some other groups). The *orthogonal group* $\mathrm{O}(n)$, the unitary group $\mathrm{U}(n)$, and the 'special' groups $\mathrm{SO}(n)$, $\mathrm{SU}(n)$, $\mathrm{SL}(n, \mathbb{R})$, $\mathrm{SL}(n, \mathbb{C})$ are self-explanatory.

The *triangular group* consists of matrices

$$\begin{bmatrix} a_1 & * & * & \cdots & * \\ 0 & a_2 & * & \cdots & * \\ \vdots & & & \ddots & \vdots \\ 0 & 0 & 0 & 0 & a_m \end{bmatrix} \tag{15}$$

with non-zero diagonal entries a_k. More generally, a block triangular subgroup consists of matrices of the same type but with each a_k itself an invertible square matrix of some fixed size. This is the group of linear transformations of E which leaves invariant a chain of subspaces $E_1 \subset E_2 \subset \cdots \subset E_m = E$. The *unipotent (block) triangular* group of such matrices has all diagonal entries equal to 1. The *(block) diagonal group* is self-explanatory. All of these groups have real, complex, and even quaternionic versions.

The *group of affine transformations* $p \to ap + b$ of E, discussed in problem 8 of §1.1, may also be realized as a linear group; it consists of block matrices of the form

$$\begin{bmatrix} a & b \\ 0 & 1 \end{bmatrix}$$

with $a \in \mathrm{GL}(E)$ and $b \in E$. When E is a Euclidean space (real with a positive definite inner product) this group contains the group of Euclidean transformations, which was discussed above for $\dim E = 2$.

The additive groups $\mathbb{R}^n$ and $\mathbb{C}^n$ may be considered as linear groups through their realization as the subgroups of affine groups consisting of translations: an element x of $\mathbb{R}^n$ or $\mathbb{C}^n$ is then identified with the matrix

$$\begin{bmatrix} 1 & x \\ 0 & 1 \end{bmatrix}.$$

The exponential which produces such matrices is

$$\exp \begin{bmatrix} 0 & x \\ 0 & 0 \end{bmatrix} = \begin{bmatrix} 1 & x \\ 0 & 1 \end{bmatrix},$$

so that the exponential map for $\mathbb{R}^n$ and $\mathbb{C}^n$ looks like the identity map $x \to x$.

For the record we mention that the multiplicative groups $\mathbb{R}^\times$ and $\mathbb{C}^\times$ are the linear groups $\mathrm{GL}(1, \mathbb{R})$ and $\mathrm{GL}(1, \mathbb{C})$. The *circle group* $\{\epsilon \in \mathbb{C} \mid |\epsilon| = 1\}$ of norm-one complexes is the linear group $\mathrm{U}(1)$, the group $\{\alpha \in \mathbb{H} \mid |\alpha| = 1\}$ of norm-one quaternions is $\mathrm{SU}(2)$. The circle group is also denoted $\mathbb{T}$; $\mathbb{T}^n$ denotes the group of diagonal $n \times n$ matrices with complex entries of modulus 1. Its elements will be written as n-tuples $(\epsilon_1, \ldots, \epsilon_n)$. $\mathbb{T}^n$ is called the *n-torus*, being a direct product of n copies of the circle $\mathbb{T} \approx S^1$. $\mathbb{T}^n$ may also be realized as $\mathbb{R}^n/\mathbb{Z}^n$: an element $(\theta_1, \ldots, \theta_n) \mod \mathbb{Z}^n$ of $\mathbb{R}^n/\mathbb{Z}^n$ may be identified with a diagonal matrix $(e^{2\pi i\theta_1}, \ldots, e^{2\pi i\theta_n})$. In this realization, the exponential map for $\mathbb{T}^n$ is the projection $\mathbb{R}^n \to \mathbb{R}^n/\mathbb{Z}^n$.

The *symmetric group* S_n of all permutations σ: $(1, 2, \ldots, n) \to (\sigma(1), \sigma(2), \ldots, \sigma(n))$ may be considered as a linear group, as follows. Relative to a basis $e_1, e_2, \ldots, e_n$ for E, a permutation σ corresponds to a *permutation matrix* s defined by

$$se_j = e_{\sigma(j)};$$

s sends the vector x with components

$$(\xi_1, \xi_2, \ldots, \xi_n)$$

to the vector sx with components

$$(\xi_{\sigma^{-1}(1)}, \xi_{\sigma^{-1}(2)}, \ldots, \xi_{\sigma^{-1}(n)})$$

mentioned only because of the inverse. No non-trivial exponential map can be defined for this group by the previous method, because the only matrix for which $\exp(\tau X)$ lies in this (finite) group is evidently $X = 0$. The same is true for certain infinite groups, such as the groups $\mathrm{GL}(n, \mathbb{Z})$ or $\mathrm{GL}(n, \mathbb{Q})$ of integral or rational matrices with inverses of the same kind.

Even though we shall be exclusively concerned with linear groups, it is sometimes appropriate to think in terms of *abstract groups*, when it is irrelevant that the elements of the groups in question are matrices. Such is the case, for example, when one defines subgroups or homomorphisms (although for linear groups homomorphisms will be required to be differentiable in a sense to be explained later). We assume known the rudiments of abstract group theory, such as can be found in the first few sections of any introduction to that subject (for example in Herstein (1964)). It should be remarked at this point that while subgroups and direct products of linear groups are again linear groups, such is not the case (in a general or natural way) for quotient groups. The *direct product* $G \times H$ of two linear groups G and H is in this context realized as the group of block-diagonal matrices

$$\begin{bmatrix} a & 0 \\ 0 & b \end{bmatrix}, \quad a \in G,\ b \in H.$$

Problems for §2.1

1. Prove Lemma 1B.

2. Check (7)

3. Check (8) and (10).

4. Define a kind of adjoint operation on $M_2(\mathbb{C})$ by $\begin{bmatrix} \alpha & \beta \\ \gamma & \delta \end{bmatrix}^+ = \begin{bmatrix} \bar{\alpha} & -\bar{\gamma} \\ -\bar{\beta} & \bar{\delta} \end{bmatrix}$ and let $\mathrm{SU}(1,1) = \{a \in M_2(\mathbb{C}) \mid aa^+ = 1, \det a = 1\}$.

   (a) Show that $(ab)^+ = b^+a^+$. Deduce that $\mathrm{SU}(1,1)$ is a group.

   (b) Formulate and prove an analog of Lemma 2A for $\mathrm{SU}(1,1)$.

5. (a) As in Example 2, identify complex numbers with quaternions of the form $\lambda + i\mu$. Show that, the map

$$\alpha + j\beta \to \begin{bmatrix} \alpha & -\bar{\beta} \\ \beta & \bar{\alpha} \end{bmatrix}$$

   sets up a one-to-one correspondence between $\mathbb{H}$ and complex 2×2 matrices of the form indicated which turns multiplication of quaternions into matrix multiplication. Verify that the conjugate of a quaternion corresponds to the Hermitian adjoint of the matrix. Deduce that the group of norm-one quaternions, $\mathrm{Sp}(1) = \{\alpha \in \mathbb{H} \mid |\alpha| = 1\}$, gets mapped isomorphically onto $\mathrm{SU}(2) = \{a \in M_2(\mathbb{C}) \mid aa^* = 1, \det a = 1\}$.

   (b) Show that any $\gamma \in \mathbb{H}$ satisfying $\bar{\gamma} = -\gamma$ can be written in the form $\gamma = \bar{\alpha} j \alpha$ for some $\alpha \in \mathbb{H}$.

6. Let $H(3,\mathbb{R})$ be the group of a real 3×3 matrices of the form

$$\begin{bmatrix} 1 & \alpha & \gamma \\ 0 & 1 & \beta \\ 0 & 0 & 1 \end{bmatrix}$$

(called the three-dimensional *Heisenberg group*).

(a) Describe the set $\mathsf{h}(3,\mathbb{R})$ of all matrices X for which $\exp(\tau X) \in H(3,\mathbb{R})$ for all $\tau \in \mathbb{R}$. Verify that $\mathsf{h}(3,\mathbb{R})$ is a three-dimensional vector space satisfying $[X,Y] \in \mathsf{h}(3,\mathbb{R})$ for all $X, Y \in \mathsf{h}(3,\mathbb{R})$.

(b) Prove that $\exp\colon \mathsf{h}(3,\mathbb{R}) \to H(3,\mathbb{R})$ is bijective.

(c) Describe the adjoint orbits in $\mathsf{h}(3,\mathbb{R})$.

7. Define the group of Euclidean motions in space, $\mathbb{R}^3 \rtimes \mathrm{SO}(3)$ in analogy with Example 5.

(a) Define an exponential map $\exp\colon \mathbb{R}^3 \times \mathsf{so}(3) \to \mathbb{R}^3 \rtimes \mathrm{SO}(3)$ and show that it is surjective.

(b) Describe the subsets of $\mathbb{R}^3 \times \mathsf{so}(3)$ which get mapped onto the conjugacy classes in $\mathbb{R}^3 \rtimes \mathrm{SO}(3)$ in analogy with Example 5.

8. Let $\mathrm{SO}(n) = \{a \in M_n(\mathbb{R}) \mid aa^* = 1 \text{ and } \det a = +1\}$. Define an exponential map $\exp\colon \mathsf{so}(n) \to \mathrm{SO}(n)$ and show that it is surjective. [Suggestion: For the second part, show first that every element of $\mathrm{SO}(n)$ is conjugate to a block-diagonal matrix with 2×2 blocks of the form

$$\begin{bmatrix} \cos\theta & -\sin\theta \\ \sin\theta & \cos\theta \end{bmatrix}$$

together with a single 1×1 block [1] when n is odd.]

9. (a) Show that $\mathrm{GL}(n,\mathbb{R}) = \mathrm{O}(n)B$, where B is the group upper triangular matrices with strictly positive diagonal entries. [Suggestion: write the Gram–Schmidt process

$$\begin{aligned} w_1 &= v_1 \\ w_2 &= v_2 - \frac{(v_2, w_1)}{(w_1, w_1)} w_1 \\ w_3 &= v_3 - \frac{(v_3, w_1)}{(w_1, w_1)} w_1 - \frac{(v_3, w_2)}{(w_2, w_2)} w_2, \\ &\cdots \end{aligned}$$

$$u_1 = \frac{w_1}{\|w_1\|}, \quad u_2 = \frac{w_2}{\|w_2\|}, \quad u_3 = \frac{w_3}{\|w_3\|}, \cdots$$

as a matrix equation

$$[v_1, v_2, v_3, \ldots] = [u_1, u_2, u_3, \ldots] b$$

with b upper triangular.]

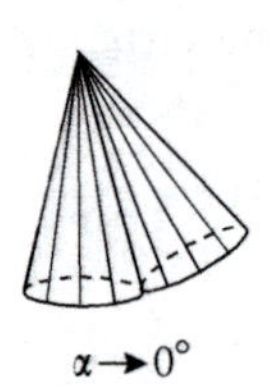

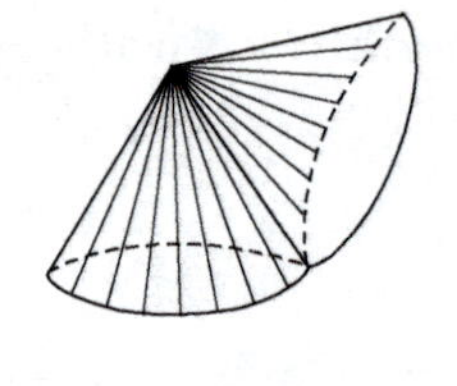

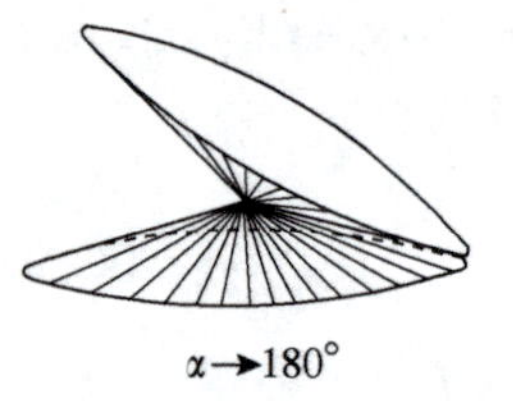

Fig. 3

(b) Show that $\mathrm{GL}(n,\mathbb{R})_+ = \{a \in \mathrm{GL}(n,\mathbb{R}) | \det(a) > 0\}$ and $\mathrm{SL}(n,\mathbb{R})$ are *connected* in the sense that any two of its elements can be joined by a continuous (even analytic) path.

10. Let $E_1 \subset E_2 \subset \cdots \subset E_m = E$ be a chain of subspaces of E. Let G be the linear group preserving this chain:

$$G = \{a \in \mathrm{GL}(E) \mid aE_k \subset E_k, \quad k = 1, 2, \ldots, m\}.$$

Show that, with respect to a suitable basis for E, G consists of all block-triangular matrices of the form (14) with invertible diagonal block a_k of size $\dim E_k - \dim E_{k-1}$.

11. (a) Derive a formula for the rotation $a(\tau) \in \mathrm{SO}(3)$ relating the position of the moving cone at time τ to its initial position. [Answer: $\exp(2\pi\tau X)\exp(\pi\tau\nu Y)$ where $\vec{X}$ and $\vec{Y}$ are the unit vectors along the axes of the cones at $\tau = 0$. The common vertex is taken as origin. $\nu =$? When is the motion periodic?]

(b) Assume the cones have the same (variable) opening angle α (Figure 3). Show:

- as $\alpha \to 0°$ the motion approaches a uniform rotation about the fixed axis at two revolutions per unit of time.
- as $\alpha \to 180°$ the motion approaches rest.

2.2 The Lie algebra of a linear group

Let G be a linear group. The *tangent space* $\mathbf{g}$ to G at 1 consists of all matrices X for which one can find a C^1 curve $a(\tau)$ that lies in G and satisfies $a(0) = 1$ and $a'(0) = X$.

Explanation. $a(\tau)$ is assumed to be defined for τ in some open interval around 0 in $\mathbb{R}$. The C^1 condition means that $a'(\tau)$ exists and is continuous as a matrix-valued function defined on this interval. The parameter τ is real, even though the matrix space M containing G may be complex.

A basic property of this tangent space g to a linear group G is:

Proposition 1.

(a) g *is a real vector space:*

$$X, Y \in \mathsf{g} \text{ implies } \alpha X + \beta Y \in \mathsf{g} \text{ for all } \alpha, \beta \in \mathbb{R}.$$

(b) g *is closed under the bracket operation:*

$$X, Y \in \mathsf{g} \text{ implies } [X, Y] \in \mathsf{g}.$$

Proof.

(a) Take $X, Y \in \mathsf{g}$ and $\alpha, \beta \in \mathbb{R}$. There are C^1 curves $a(\tau)$ and $b(\tau)$ that lie in G so that $a(0) = b(0) = 1$, $a'(0) = X$, and $b'(0) = Y$. Then the C^1 curve $c(\tau) = a(\alpha\tau)b(\beta\tau)$ also lies in G. It satisfies $c(0) = 1$ and

$$c'(0) = \Big(a'(\alpha\tau)\alpha b(\tau) + a(\alpha\tau)b'(\beta\tau)\beta\Big)\Big|_{\tau=0} = \alpha X + \beta Y.$$

This shows that $\alpha X + \beta Y \in \mathsf{g}$ as well.

(b) Again take $X, Y \in \mathsf{g}$ and choose $a(\tau)$ and $b(\tau)$ as above. For fixed $\sigma \in \mathbb{R}$ (sufficiently close to 0), $c_\sigma(\tau) = a(\sigma)b(\tau)a(\sigma)^{-1}$ is a C^1 curve in G. It satisfies $c_\sigma(0) = 1$ and $c'_\sigma(0) = a(\sigma)Ya(\sigma)^{-1}$. This shows that $a(\sigma)Ya(\sigma)^{-1} \in \mathsf{g}$ for σ sufficiently close to 0. As a curve in the vector space g, $\sigma \to a(\sigma)Ya(\sigma)^{-1}$ has its tangent vector in g. We compute (formula (A.8) of §1.1):

$$\frac{d}{d\sigma}\Big(a(\sigma)Ya(\sigma)^{-1}\Big)\Big|_{\sigma=0} = XY - YX = [X, Y].$$

This shows that $[X, Y] \in \mathsf{g}$, as required. QED

Any real vector space of matrices that is closed under the bracket operation is called a linear Lie algebra. The g belonging to a linear group G is called the Lie algebra of G.

Example 2 (The Lie algebras of SO(n) and SU(n)). *The special orthogonal group* SO(n) consists of all $n \times n$ matrices a satisfying

$$a^*a = 1, \quad \det a = 1. \tag{1}$$

If $a = a(\tau)$ is a curve in SO(n), then one finds by differentiating the first of these relations that

$$(a')^*a + a^*a' = 0.$$

If $a(0) = 1$ and $a'(0) = X$, this gives $X^* + X = 0$, i.e. X is skew-symmetric. Conversely, if X is skew-symmetric, then $a(\tau) = \exp \tau X$ is orthogonal and has $\det a = e^{\operatorname{tr} \tau X} = 1$, so $a(\tau) \in \mathrm{SO}(n)$ and therefore, $X = a'(0)$ is in the Lie algebra of SO(n). In short, the Lie algebra of SO(n) is the space of skew-symmetric matrices:

$$\mathsf{so}(n) = \{X \in M_n(\mathbb{R}) \mid X^* = -X\}. \tag{2}$$

Note that the condition $\det a = +1$ has not been used, so that this argument shows that the full orthogonal group $\mathrm{O}(n)$ has the same Lie algebra as $\mathrm{SO}(n)$. This comes from the fact that any continuous curve through 1 in $\mathrm{O}(n)$ lies already in $\mathrm{SO}(n)$, as one sees from the equation $\det a(\tau) = \pm 1$ by continuity considerations.

For the *special unitary group* $\mathrm{SU}(n)$, defined by

$$a^*a = 1, \quad \det a = 1 \tag{3}$$

the situation is slightly different. As above, differentiating the first of these equations at $\tau = 0$ one finds

$$X^* + X = 0, \tag{4}$$

i.e. X is skew-Hermitian. Differentiating the second one finds in addition

$$\operatorname{tr} X = 0. \tag{5}$$

(This may be seen by writing $a(\tau) = \exp X(\tau)$ and $\det a(\tau) = e^{\operatorname{tr} X(\tau)}$.)

Conversely, if X satisfies (4) and (5) then $a(\tau) = \exp \tau X$ satisfies (3) for real τ. Thus the Lie algebra of $\mathrm{SU}(n)$ consists of skew-Hermitian matrices of trace zero:

$$\mathsf{su}(n) = \{X \in M_n(\mathbb{C}) \mid X^* = -X, \operatorname{tr} X = 0\}. \tag{6}$$

For $\mathrm{U}(n)$ itself the second condition is omitted:

$$\mathsf{u}(n) = \{X \in M_n(\mathbb{C}) \mid X^* = -X\}.$$

We point out once more that the Lie algebra of a linear group is generally only a real vector space, even if the group itself lies in the complex matrix space $M_n(\mathbb{C})$. For example, the Lie algebras of $\mathrm{U}(n)$ or $\mathrm{SU}(n)$ are certainly not closed under multiplication by complex scalars.

A fundamental property of the Lie algebra of a linear group is that it gets mapped back into the group by the exponential map. We state this fact as:

Theorem 3. *Let G be a linear group, g its Lie algebra. Then* exp *maps* g *into* G.

Proof. The idea is this. The tangent space to G at 1 is g; the tangent space to G at an arbitrary point $a \in G$, defined similarly in terms of curves through a, is $a\mathsf{g}$. For $X \in \mathsf{g}$, the vector field $a \to aX$ on the matrix space M is therefore tangential to G at the points of G. One would then expect that the curve $a(\tau) = \exp \tau X$, which starts in G (because $a(0) = 1$) and is always tangential to G (because $a'(\tau) = a(\tau)X$), will stay entirely in G. The passage from the infinitesimal condition '$a(\tau)$ is tangential to G' to the global condition '$a(\tau)$ lies in G' is accomplished by constructing a (vector valued) function, denoted V in the proof below, with the property that $a(\tau) \in G$ for all τ in an interval about 0 if and only if $V(a(\tau)) \equiv 0$. The latter condition is equivalent to $(d/d\tau)V(a(\tau)) \equiv 0$, $V(1) = 0$, which can be verified using the fact that $a'(\tau) \in a(\tau)\mathsf{g}$. Now the details of the proof. Choose a basis $X_1, X_2, \ldots, X_m$ for g and set

$$g(\tau_1 X_1 + \tau_2 X_2 + \cdots + \tau_m X_m) = a_1(\tau_1) a_2(\tau_2) \cdots a_m(\tau_m).$$

where $a_k(\tau)$ is a C^1 curve with $a_k(0) = 1$ and $a_k'(0) = X_k$. Then $g : \mathsf{g} \to G \subset M$ and $dg_0 X = X$ for all $X \in \mathsf{g}$. Choose a subspace s of M complementary to g, so that $M = \mathsf{g} \oplus \mathsf{s}$, and a map[1] $h : \mathsf{s} \cdots \to M$ defined in a neighborhood of 0 in s and satisfying $h(0) = 1$ and $dh_0 Y = Y$ for all $Y \in \mathsf{s}$; for example $h(Y) = 1 + Y$. Define $f : \mathsf{g} \times \mathsf{s} \cdots \to M$ by $f(X, Y) = g(X)h(Y)$. Then f is defined and C^1 in a neighborhood of $(0, 0)$ in $\mathsf{g} \times \mathsf{s}$ and satisfies $df_{(0,0)}(X, Y) = X + Y$. In particular, this differential is invertible.

According to the Inverse Function Theorem, f itself has a local inverse $W : M \cdots \to \mathsf{g} \times \mathsf{s}$, $a \to (\mathrm{U}(a), V(a))$, defined for a in a neighborhood of 1 in M. Explicitly, this means that any a in a neighborhood of 1 in M is of the form $a = g(X)h(Y)$ for unique (X, Y) in a neighborhood of $(0, 0)$ in $\mathsf{g} \times \mathsf{s}$, namely $X = U(a)$, $Y = V(a)$. If $V(a) = 0$, then $a = g(X)h(0) = g(X)$ is in G.

$$V(a) = 0 \quad \text{implies} \quad a \in G. \tag{7}$$

Fix $(X, Y) \in \mathsf{g} \times \mathsf{s}$, sufficiently close to $(0, 0)$ so that W is defined near $a = f(X, Y)$ in M. Given $Z \in \mathsf{g}$, we have

$$V\big(g(X + \tau Z)h(Y)\big) = Y$$

for real τ close to 0. Differentiation with respect to τ at $\tau = 0$ gives

$$(dV)_a\{(dg_X Z)h(Y)\} = 0.$$

Substituting $h(Y) = g(X)^{-1}a$, we obtain

$$dV_a\{(dg_X Z)g(X)^{-1}a\} = 0. \tag{8}$$

The matrix $(dg_X Z)g(X)^{-1}$ is again in g, because it is the tangent vector at 1 of a C^1 curve in G:

$$\big((dg)_X Z\big)g(X)^{-1} = \frac{d}{d\tau} g(X + \tau Z)g(X)^{-1}\big|_{\tau=0}.$$

Thus we can define a map $A : \mathsf{g} \to \mathsf{g}$ by $AZ = (dg_X Z)g(X)^{-1}$. $A = A(X)$ is a linear transformation of g depending continuously on X. Equation (8) becomes

$$dV_a\{(AZ)a\} = 0, \tag{9}$$

where it is understood that

$$a = g(X)h(X), \quad A = A(X).$$

For $X = 0, A$ reduces to the identity transformation of g, because $g(0) = 1$ and $(dg)_0 Z \equiv Z$. In particular, $\det A \neq 0$ for $X = 0$. By continuity, $\det A \neq 0$ for X in a neighborhood of 0 in g. So if X is near 0 in g then A is invertible, and

[1] The dots indicate a partially defined map.

every element of $\mathbf{g}$ is of the form AZ for some $Z \in \mathbf{g}$. We can therefore replace AZ by Z in (9) and see that

$$dV_a(Za) = 0 \tag{10}$$

for all a in a neighborhood of 1 in M and all Z in $\mathbf{g}$.

We now return to the assertion that exp maps $\mathbf{g}$ into G. Take $X \in \mathbf{g}$ and set $a(\tau) = \exp \tau X$. According to (10),

$$dV_{a(\tau)}(Xa(\tau)) = 0,$$

which says that

$$\frac{d}{d\tau}V(a(\tau)) = 0; \tag{11}$$

and this holds for all τ in an interval about 0 in $\mathbb{R}$. Consequently, $V(a(\tau))$ is constant there. For $\tau = 0$ we have $V(a(0)) = 0$, and therefore $V(a(\tau)) \equiv 0$ for τ in an interval about 0 in $\mathbb{R}$. In view of (7) this implies that $a(\tau) = \exp \tau X$ lies in G for τ in that interval. Contemplating the identity $\exp X = (\exp X/k)^k$ one sees that $\exp X$ will be in G for *all* $X \in \mathbf{g}$. QED

The proof of the theorem gives some additional information worth recording:

Corollary 4. *Let $a(\tau)$ be a C^1 path in M with $a(0) = 1$ and $a'(\tau) \in \mathbf{g}a(\tau)$ for all τ in an interval about 0. Then there is a C^1 path $X(\tau)$ in $\mathbf{g}$ so that $a(\tau) = \exp X(\tau)$ for all τ in an interval about 0. (Necessarily $X(\tau) = \log a(\tau)$ for τ in an interval about 0.)*

Proof. In the above argument take for g the map $g(X) = \exp X$, which we now know to lie G for $X \in \mathbf{g}$. Since $a'(\tau) \in \mathbf{g}a(\tau)$ we find that $V(a(\tau)) \equiv 0$ as before and this implies that

$$a(\tau) = g(X(\tau))h(0) = \exp X(\tau)$$

with $X(\tau) = U(a(\tau))$. QED

Corollary 5. *In a sufficiently small neighborhood of 1 in M, every element $a \in G$ that can be connected to the identity by a C^1 path which lies in G and stays in this neighborhood is of the form $a = \exp X$ for some $X \in \mathbf{g}$.*

Proof. Let $a(\tau)$ be a path as mentioned. $a'(\tau)a(\tau)^{-1}$ lies in $\mathbf{g}$, being the tangent vector at $\sigma = 0$ of the C^1 curve $\sigma \to a(\tau + \sigma)a(\tau)^{-1}$. Therefore, $a'(\tau) \in \mathbf{g}a(\tau)$ and one can apply the previous lemma. QED

The property of exp: $\mathbf{g} \to G$ just proved is a bit subtle: for some groups G it may happen that in arbitrarily small neighborhoods of 1 in M, one can find elements a that lie in G, but that are not exponentials of elements of $\mathbf{g}$ (although, near 1, they are of course exponentials of some matrices in M). The condition that a can be connected to the identity by a path in G is essential, as one sees already with the multiplicative group of non-zero rationals. It is also essential that the path stays is a neighborhood of 1 in M: consider exp for $\mathrm{SL}(2, \mathbb{R})$, for example. There are also linear groups G for which exp: $\mathbf{g} \to G$ is

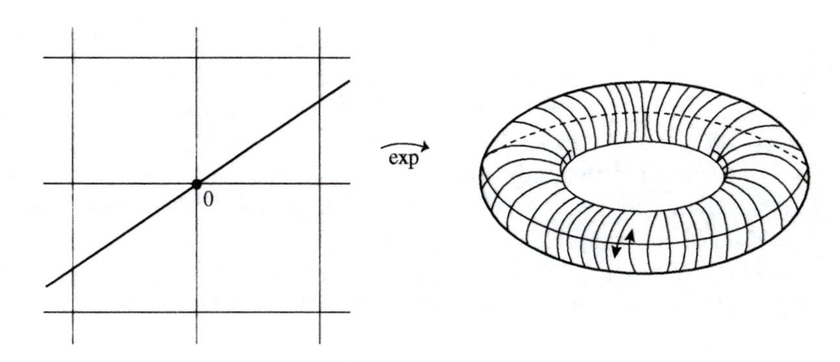

Fig. 1

surjective, while there are nevertheless elements of G, arbitrarily close to 1 in M, which cannot be joined to 1 by a continuous path in G that stays close to 1. See Example 8 below.

Remark 6. The statement of Corollary 5 remains correct when C^1 is replaced by C^0 (continuous), but this cannot be seen in the same way and is considerably harder to prove. The only proof known, due to Goto (1962), uses the Brouwer fixed point theorem.

As a consequence of the theorem, we have another characterization of the Lie algebra of a linear group:

Corollary 7. *The Lie algebra* $\mathbf{g}$ *of a linear group* G *consists of all matrices* X *for which* $\exp \tau X$ *lies in* G *for all* $\tau \in \mathbb{R}$.

Proof. One inclusion comes from the theorem, the other one from the observation that $a(\tau) = \exp \tau X$ provides a path for X, as required by the definition of $\mathbf{g}$. QED

Example 8 (Lines on the torus). The 2-torus $\mathbb{T}^2$ consists of all 2×2 diagonal matrices with complex entries of modulus 1, written $(\epsilon_1, \epsilon_2) = (e^{2\pi i\theta_1}, e^{2\pi i\theta_2})$ with θ_1, θ_2 real. The Lie algebra $\mathbf{t^2}$ of $\mathbb{T}^2$ consists of imaginary diagonal matrices $(2\pi i\theta_1, 2\pi i\theta_2)$, and the exponential map $\mathbf{t^2} \to \mathbb{T}^2$ is given by $\exp(2\pi i\theta_1, 2\pi i\theta_2) = (e^{2\pi i\theta_1}, e^{2\pi i\theta_2})$. It is a group homomorphism of the additive group $\mathbb{R}^2$ (realized as matrices $(2\pi i\theta_1, 2\pi i\theta_2)$ onto $\mathbb{T}^2$.

Consider the image in $\mathbb{T}^2$ of the line given parametrically by $\theta_1 = \alpha\tau$, $\theta_2 = \beta\tau$ for fixed α, β. Denote this group by $G_{\alpha\beta}$. The nature of the group depends on the ratio α/β. To simplify the statement we assume $\alpha, \beta > 0$.

Proposition 8′.

(a) If α/β *is rational, say* $\alpha/\beta = p/q(p, q$ *relatively prime), then* $G_{\alpha\beta}$ *is a closed curve on* $\mathbb{T}^2 = S^1 \times S^1$, *whose first component describes* p *full rotations as the second component describes* q *full rotations.*

(b) If α/β *is irrational, then* $G_{\alpha\beta}$ *winds around* $\mathbb{T}^2$ *infinitely often and is dense in* $\mathbb{T}^2$.

Proof.
(a) Assume $\alpha/\beta = p/q, p, q$ relatively prime, $0 < p \leq q$. The elements of $G_{\alpha\beta}$ are of the form $(e^{2\pi i\alpha\tau}, e^{2\pi i\beta\tau})$ with $\tau \in \mathbb{R}$. Starting at 1 for $\tau = 0$ the point $a(\sigma)$ travels around the torus $\mathbb{T}^2$, and returns to 1 when $\alpha\tau$ and $\beta\tau$ are both integers. The first time this happens is when $\tau = p/\alpha = q/\beta$. As τ varies from 0 to $p/\alpha = q/\beta$ the first component $e^{2\pi i\alpha\tau}$ of $a(\tau)$ travels p times around the circle and the second component $e^{2\pi i\beta\tau} q$ times.

(b) Assume α/β is irrational. It suffices to show that $G_{\alpha\beta}$ is dense in $\mathbb{T}^2$: given $(e^{2\pi i\xi}, e^{2\pi i\eta}) \in \mathbb{T}^2$ and $\epsilon > 0$ there is $(e^{2\pi i\varphi}, e^{2\pi i\psi}) \in G_{\alpha\beta}$ so that $|\xi - \varphi| < \epsilon$ and $|\eta - \psi| < \epsilon$.

By translation in $\mathbb{T}^2$ it suffices to consider $\xi = \eta = 0$. Thus, given $\epsilon > 0$, we have to find φ, ψ of the form $\varphi = \alpha\tau + m$, $\psi = \beta\tau + n$, with $\tau \in \mathbb{R}$ and $m, n \in \mathbb{Z}$, so that $|\varphi| < \epsilon$ and $|\psi| < \epsilon$. Try $\tau = k/\beta$ with $k \in \mathbb{Z}$ to be determined. Then $\beta\tau = k$ and $\alpha\tau = (\alpha/\beta)k$. $\psi = \beta\tau - k = 0$ will do for ψ. Since (α/β) is irrational one can choose $k \in \mathbb{Z}$ so that the fractional part of $(\alpha/\beta)k$ is $< \epsilon$: the fractional parts of the $(\alpha/\beta)k$ must have an accumulation point, so one can find two such fractional parts whose difference $(\alpha/\beta)(k_1 - k_2)$ is arbitrarily small. Then $\varphi =$(fractional part of $\alpha\tau$) $= (\alpha/\beta)k-$(integral part of $(\alpha/\beta)k$) will do for φ. QED

The group $G_{\alpha\beta}$ with α/β irrational is called the irrational line on the torus. It provides the example mentioned after Corollary 5: an element on this $G_{\alpha\beta}$ may be very close to 1 in the matrix space M, but may still have to travel several times around the torus before reaching 1 from within $G_{\alpha\beta}$. Nevertheless, the irrational line is not intrinsically pathological as a linear group: it is isomorphic with the real line $\mathbb{R}$ under addition by the bi-analytic map $\mathbb{R} \to G_{\alpha\beta}$ corresponding to the parametrization by τ (essentially the exponential map of $G_{\alpha\beta}$). This group just happens to be embedded in the matrix space M in a strange way.

We close this section with a few remarks about abstract Lie algebras. Just as there is a notion of 'abstract group', so there is a notion of '*abstract Lie algebra*', defined axiomatically as a vector space g (over some field, here always $\mathbb{R}$ or $\mathbb{C}$) together with a skew-symmetric, bilinear operation $\mathsf{g} \times \mathsf{g} \to \mathsf{g}$, $(X, Y) \to [X, Y]$, satisfying the Jacobi identity

$$[[X, Y], Z] + [[Z, X], Y] + [[Y, Z], X] = 0.$$

As with groups, it is sometimes appropriate to place oneself in this 'abstract' setting, even if one is exclusively concerned with linear Lie algebras. For example, a *homomorphism of algebras* $\varphi : \mathsf{g} \to \mathsf{h}$ of Lie algebras is simply a linear map that respects brackets:

$$\varphi[X, Y] = [\varphi X, \varphi Y].$$

An *isomorphism* is an invertible homomorphism, as usual. An *isomorphism* $\alpha : \mathsf{g} \to \mathsf{g}$ of a Lie algebra g with itself is called an automorphism of g. The collection of all automorphisms of a Lie algebra g is a subgroup of $\mathrm{GL}(\mathsf{g})$ and therefore, a linear group. It is denoted $\mathrm{Aut}(\mathsf{g})$.

When g is the Lie algebra of a linear group G, then every $a \in G$ defines an automorphism $\mathrm{Ad}\, a : \mathsf{g} \to \mathsf{g}$ by $\mathrm{Ad}(a)X = aXa^{-1}$. (To see that $\mathrm{Ad}(a)X$

belongs to g when X does one remarks that $\mathrm{Ad}(a)X$ is the tangent vector of the curve $a\exp(\tau X)a^{-1}$ at $\tau = 0$.) The map $G \to \mathrm{Aut}(\mathsf{g})$, $a \to \mathrm{Ad}\, a$, is a group homomorphism, called the *adjoint representation.*

A *Lie subalgebra of* g is a subspace closed under the bracket operation. A Lie subalgebra a is an *ideal* if $[\mathsf{a}, \mathsf{g}] \subset \mathsf{a}$. If g has no ideals other than $\{0\}$ and g, then g is called *simple Lie algebra*, except that a one-dimensional Lie algebra is not called simple, by convention. A Lie algebra g is *semisimple* if it is isomorphic with a direct product of simple Lie algebras. (The bracket in a *direct product* is naturally defined componentwise.)

In any Lie algebra one has an *adjoint representation* $\mathrm{ad} : \mathsf{g} \to \mathsf{gl}(\mathsf{g})$ which is the Lie algebra of the linear group $\mathrm{GL}(\mathsf{g})$ defined by

$$\mathrm{ad}(X)Y = [X, Y]$$

as before. The Jacobi identity says that ad is a homomorphism:

$$\mathrm{ad}([X, Y])Z = [\mathrm{ad}(X), \mathrm{ad}(Y)]Z.$$

$\mathrm{ad} : \mathsf{g} \to \mathsf{gl}(\mathsf{g})$ is not necessarily one-to-one. Nevertheless, *every Lie algebra is isomorphic with a linear Lie algebra.* This is Ado's Theorem, which can be found in Bourbaki (1960), for example. It may be taken as a justification for restricting attention to linear Lie algebras.

The Jacobi identity can also be read as saying

$$\mathrm{ad}(X)[Y, Z] = [\mathrm{ad}(X)Y, Z] + [Y, \mathrm{ad}(X)Z].$$

Any linear transformation δ of g with this property,

$$\delta[Y, Z] = [\delta Y, Z] + [Y, \delta Z],$$

is called a *derivation* of g. The collection of all derivations of a Lie algebra g, denoted $\mathrm{Der}(\mathsf{g})$, is a Lie subalgebra of $\mathsf{gl}(\mathsf{g})$, and therefore, a linear Lie algebra. In fact:

Proposition 9. $\mathrm{Der}(\mathsf{g})$ *is the Lie algebra of the linear group* $\mathrm{Aut}(\mathsf{g})$.

Proof. Using Corollary 7, we have to show that a linear transformation δ of g is a derivation, i.e. it satisfies

$$\delta[X, Y] = [\delta X, Y] + [X, \delta Y] \tag{12}$$

if and only if $\exp \tau\delta$ is an automorphism for all $\tau \in \mathbb{R}$, i.e. it satisfies

$$\exp(\tau\delta)[X, Y] = [\exp(\tau\delta)X, \exp(\tau\delta)Y]. \tag{13}$$

One way is clear: if (13) holds for all τ, then (12) follows by differentiation at $\tau = 0$. (Product rule for bilinear mappings; see (A.7), appendix to §1.1.)

Conversely, suppose (12) holds. To see that (13) then holds as well, write it in the form

$$\exp(-\tau\delta)[\exp(\tau\delta)X, \exp(\tau\delta)Y] = [X, Y]. \tag{14}$$

The derivative of the left side of (14) with respect to τ is

$$\exp(-\tau\delta)\big(-\delta[\exp(\tau\delta)X, \exp(\tau\delta)Y] + [\delta\exp(\tau\delta)X, \exp(\tau\delta)Y] \\ + [\exp(\tau\delta)X, \delta\exp(\tau\delta)Y]\big),$$

which is identically 0, in view of (12). Thus the left side of (14) is independent of τ, and therefore equal to the right side, as one sees by setting $\tau = 0$. QED

Problems for §2.2

1. Describe the Lie algebras of the following linear groups. (a) The group of invertible *block-triangular* matrices

$$\begin{bmatrix} a_1 & * & * & \cdots & * \\ 0 & a_2 & * & \cdots & * \\ \vdots & & \ddots & & \vdots \\ 0 & 0 & 0 & 0 & a_m \end{bmatrix}$$

   ($a_k \in M_{n_k}$, $\det a_k \neq 0$).

   (b) The group of *unipotent block-triangular* matrices (as above with $a_k = 1_k \in M_{n_k}$).

2. (a) Find a basis for $\mathsf{so}(n)$ and show that $\dim \mathsf{so}(n) = \frac{1}{2}n(n-1)$. (b) Find a basis for $\mathsf{su}(n)$ and show that $\dim \mathsf{su}(n) = n^2 - 1$.

3. Let $f \in M$ be a non-singular matrix. Let $G = \{a \in M \mid f^{-1}a^t fa = 1\}$.
   (a) Show that G is a group and find its Lie algebra.

   (b) Take for f the $2m \times 2m$ matrix

$$\begin{bmatrix} 0 & 1 \\ -1 & 0 \end{bmatrix}$$

   (blocks of size $m \times m$). Describe G and its Lie algebra explicitly (as block-matrices).

4. (a) Describe all Lie subalgebras of $\mathsf{so}(3)$. (b) Describe all *complex* Lie subalgebras of $\mathsf{sl}(2, \mathbb{C})$.

   [Suggestion for (b): classify the subalgebras up to conjugacy. Use the fact that every non-zero matrix of $\mathsf{sl}(2, C)$ is conjugate to

$$\begin{bmatrix} 1 & 0 \\ 0 & -1 \end{bmatrix} \quad \text{or} \quad \begin{bmatrix} 0 & 1 \\ 0 & 0 \end{bmatrix}.$$

   Find first all complex subspaces of $\mathsf{sl}(2, C)$ which are stable under bracketing (i.e. under ad) with one of these matrices. You may want to use the following basis for $\mathrm{SL}(2, C)$:

$$H = \begin{bmatrix} 1 & 0 \\ 0 & -1 \end{bmatrix}, \quad X_+ = \begin{bmatrix} 0 & 1 \\ 0 & 0 \end{bmatrix}, \quad X_- = \begin{bmatrix} 0 & 0 \\ 1 & 0 \end{bmatrix}.$$

These satisfy the famous relations

$$[H, X_+] = 2X_+, \quad [H, X_-] = -2X_-, \quad [X_+, X_-] = H,$$

which you may use.]

5. (a) Describe all Lie subalgebras of $\mathsf{sl}(2, \mathbb{R})$. (b) Determine which of these are isomorphic. (c) Determine which of these are conjugate by an element of $\mathrm{SL}(2, \mathbb{R})$.

6. Let A be a finite-dimensional associative algebra over $\mathbb{R}$ or $\mathbb{C}$, $A^\times$ the group of invertible elements of A. Consider an element $a \in A$ as linear transformation $A \to A, x \to ax$, so that $A^\times$ becomes a linear group. Show that its Lie algebra is A with bracket $[x, y] = xy - yx$.

7. Let A be a finite-dimensional, not necessarily associative, algebra over $\mathbb{R}$ or $\mathbb{C}$, i.e. a finite-dimensional vector space over $\mathbb{R}$ or $\mathbb{C}$ with a bilinear operation $A \times A \to A$, written $(x, y) \to x \cdot y$. Let $\mathrm{Aut}(A)$ be the automorphism group of A, i.e. the group of invertible linear transformations a of A satisfying $a(x \cdot y) = (ax) \cdot (ay)$. Show that the Lie algebra of $\mathrm{Aut}(A)$ consists of all derivations of A, i.e. all linear maps $D : A \to A$ satisfying $D(x \cdot y) = (Dx) \cdot y + x \cdot (Dy)$.

8. Fix $c \in \mathrm{GL}(E)$. Let $G = \{a \in \mathrm{GL}(E) \mid ac = ca\}$.

   (a) Check that G is a group and describe its Lie algebra $\mathbf{g}$.

   (b) Find an explicit matrix representation for the elements of G and of $\mathbf{g}$ in case c is diagonal (with respect to a suitable basis for E).

9. Let F be a subspace of E. Let $G = \{a \in \mathrm{GL}(E) \mid aF \subset F\}$.

   (a) Check that G is a group and describe its Lie algebra $\mathbf{g}$.

   (b) Find an explicit matrix representation for the elements of G and of $\mathbf{g}$ (with respect to a suitable basis for E).

2.3 Coordinates on a linear group

The exponential map $\exp \colon \mathbf{g} \to G$ of a linear group has many uses. First of all it can be used to introduce a *topology*[2] on G, which means specifying neighborhoods (or open sets). This is done much like for $\mathbb{R}^n$, as follows.

A *neighborhood of* $a \in G$ in G is any subset of G which contains $\{a \exp X \mid X \in \mathbf{g},\ \|X\| < \epsilon\}$ for some $\epsilon > 0$. (One gets the same notion of neighborhood using $\{(\exp X)a \mid X \in \mathbf{g},\ \|X\| < \epsilon\}$, although these sets themselves are generally not same.) A subset of G is *open in* G if it contains a neighborhood in G of each of its points, and *closed* in G if its complement is open in

[2]It is assumed this notion is familiar, at least from $\mathbb{R}^n$.

G. A *neighborhood of a subset* S of G is any set containing a neighborhood of each point of S. Convention: in the future we omit the qualification 'in G', with the understanding that *all topological notions pertaining to G (i.e. notions depending on notion of 'neighborhood' or 'open set') are understood in the sense just explained*, unless explicitly stated otherwise. This is of some importance, because the *group-topology* on G need not coincide with its *relative topology* in the matrix space M (meaning the open sets in G as defined above need not be exactly the intersections with G of open sets in M).

The exponential map can also be used to introduce coordinates on G, as follows. Choose $R > 0$ so that the log series converges on $\{\exp X \mid X \in M, \|X\| < R\}$, for example $R = \log 2$. The set

$$U = \{a \in G \mid a = \exp X, X \in \mathbf{g}, \|X\| < R\}$$

is then an open neighborhood of 1 in G on which $\log : U \to \mathbf{g}$ provides an inverse for $\exp\colon \mathbf{g} \to G$. Choose a basis $X_1, \ldots, X_m$ for $\mathbf{g}$. Then the equations

$$a = \exp X, \quad X = \xi_1 X_1 + \cdots + \xi_m X_m,$$

associate to each $a \in U$ an m-tuple $(\xi_1, \ldots, \xi_m) \in \mathbb{R}^m$ which serves as coordinates of $a \in U$. These coordinates $(\xi_1, \ldots, \xi_m)$ should be considered as a map $G \cdots \to \mathbb{R}^m$ with domain U; these are called *exponential* (or *canonical*) coordinates on G. One often thinks of $X \in \mathbf{g}$ itself as coordinate point of $a = \exp X \in U$, without introducing the basis $X_1, \ldots, X_m$.

This defines exponential coordinates in a neighborhood of 1 in G. Exponential coordinates in a neighborhood of an arbitrary point $a_o \in G$ are defined in the same way on the domain $a_o U$ by the equations

$$a = a_o \exp X, \quad X = \xi_1 X_1 + \cdots + \xi_m X_m.$$

It is of importance to know that the coordinate transformation $X \to \tilde{X}$ between two such coordinate systems is analytic. This transformation is defined by the equation

$$a_o \exp X = \tilde{a}_o \exp \tilde{X};$$

its domain consists of those $X \in \mathbf{g}, \|X\| < R$, for which this equation has a solution for $\tilde{X} \in \mathbf{g}$, $\|\tilde{X}\| < R$. This domain is evidently open in $\mathbf{g}$ and, if non-empty, the map $X \to \tilde{X}$ thereon is indeed analytic, being given by $\tilde{X} = \log(\tilde{a}_o^{-1} a_o \exp X)$.

In general, a *coordinate system* on a linear group G is a map

$$G \cdots \to \mathbb{R}^m, \quad a \to x(a) = (\xi_1(a), \ldots, \xi_m(a))$$

($m = \dim \mathbf{g}$) with the following properties:

(1) The domain of $a \cdots \to x(a)$ is an open subset U of G.

(2) $a \to x(a)$ maps U bijectively onto an open subset of $\mathbb{R}^m$.

(3) For every $a_o \in U$, the map $\mathrm{g} \cdots \to \mathbb{R}^m$, $X \to x(a_o \exp X)$ maps a neighborhood of 0 in g bi-analytically onto a neighborhood of $x(a_o)$ in $\mathbb{R}^m$.

A coordinate system in this sense may be written as a pair (U, x), but the coordinate domain U is usually suppressed. The same symbol $x = (\xi_1, \ldots, \xi_m)$ is used for the coordinate map as for a general element of $\mathbb{R}^m$, a convention familiar form the 'xy-coordinates' in the plane, for example. The n-tuple $x(a) = (\xi_1(a), \ldots, \xi_m(a))$ *is the coordinate point* corresponding to $a \in G$ in the given coordinate system, $\xi_1(a), \ldots, \xi_m(a)$ its *coordinates.* One may check that the coordinate transformation between any two coordinate systems is analytic, as has been done above for the exponential coordinates. The dimension m of g entering the definition of 'coordinates' is also called the *dimension* of the group G.

Coordinates on G allow one to define what it means for a function on G to be of class C^k: by definition, a function on a linear group G is said to be of class $C^k (0 \leq k \leq \infty,\ \omega)$ if it becomes such a function when elements of G are locally expressed in terms of coordinates $x = (\xi_1, \ldots, \xi_m)$, as just defined. ('Locally' means 'in some neighborhood of each point' in its domain, which is always understood to be an open subset of G.) This procedure evidently works for functions defined on an open subset of G, with values in $\mathbb{R}^p$, $\mathbb{C}^p$, or even in another linear group. In the latter case, the elements of both groups will have to be represented in coordinates, locally, and for this one has to require explicitly that f map sufficiently small neighborhoods of points in its domain into coordinate domains on the other group.

Note that with this terminology a coordinate system (U, x) on G is simply a bi-analytic map from an open subset U of G to an open subset of $\mathbb{R}^p$. There is a point to be made concerning these definitions, which is easily overlooked or taken for granted, but is not obvious:

Proposition 1. *A map f from an open subset of $\mathbb{R}^p$ into a linear group G is of class $C^k, 1 \leq k \leq \infty, \omega$, if and only if it is of class C^k as a map into the matrix space M (i.e. the matrix entries of $f(x)$ are C^k functions of x). The same is true for a map from another linear group into G.*

Proof. Consider $f : \mathbb{R}^p \cdots \to G$. According to the definition, f is of class C^k if for each x_o in the domain of f, the equation $f(x) = f(x_o) \exp X$ defines a C^k map $\mathbb{R}^p \cdots \to \mathrm{g}$, $x \to X$ for x near x_o and X near 0. With the 'near zero' restriction, the unique solution of the equation $f(x) = f(x_o) \exp X$ for X is $X = \log f(x_o)^{-1} f(x)$. If the matrix entries of $f(x)$ are C^k functions of x, so are those of X, and the converse is clear from the original equation $f(x) = f(x_o) \exp X$. The whole point is that it is not clear that $X = \log f(x_o)^{-1} f(x)$ lies in g. But when f is of class C^k as M-valued function, and $k \geq 1$, this follows from Corollary 5 of §2.2 upon consideration of the curve $a(\tau) = f(x_o)^{-1} f(x_o - \tau y)$ in G (for fixed x_o and y). The case of a map from another linear group into G reduces to $\mathbb{R}^p \cdots \to G$ via coordinates. QED

In view of this proposition the following should be obvious:

Corollary 2. *For any linear group G, the group operations $a \to a^{-1}$ and $(a, b) \to ab$ are analytic mappings.*

There is a version of the Inverse Function Theorem that is often useful for the construction of coordinates on linear groups:

Proposition 3. *Let $m = \dim G$, $\mathbb{R}^m \cdots \to G$, $x \to a(x)$, be an analytic map. If the differential of $x \to a(x)$ at x_o is injective (i.e. has rank m), then $x \to a(x)$ has a local analytic inverse $G \cdots \to \mathbb{R}^m$, which provides a coordinate system in a neighborhood of $a_o = a(x_o)$ on G.*

Explanation. $x \to a(x)$ may be considered as a map $\mathbb{R}^m \cdots \to M$ as far as analyticity and differential are concerned (Proposition 1).

Proof. It suffices to show that for such an x_o the equation

$$a(x) = a(x_o) \exp X$$

defines a bi-analytic map $x \to X$ from a neighborhood of x_o in $\mathbb{R}^m$ to a neighborhood of 0 in g. Write the equation as

$$\exp X = a(x_o)^{-1} a(x).$$

Since $X \to \exp X$ is bi-analytic from a neighborhood of 0 to a neighborhood of 1 in M, it is clear that the equation gives an analytic map $x \to X$, $\mathbb{R}^m \cdots \to M$, defined on a neighborhood of x_o in $\mathbb{R}^m$. By Corollary 5 of §2.2, this map takes on values in g for x in a neighborhood of x_o and therefore defines an analytic map $\mathbb{R}^m \cdots \to \mathsf{g}$ with a differential of rank m at x_o. The conclusion now follows from the Inverse Function Theorem. QED

Example 4 (Coordinates on SO(3)**).**
(a) *Exponential coordinates.* Recall that $\exp\colon \mathsf{so}(3) \to \mathrm{SO}(3)$ maps the solid ball $\{X \in \mathsf{so}(3) \mid \|X\| \le \pi\}$ onto SO(3) and is one-to-one on its interior. To show that its inverse defines a coordinate system on $U = \{\exp X \mid \|X\| < \pi\}$ it suffices to show that $\mathsf{so}(3) \to M_3(\mathbb{R})$, $X \to \exp -X_o \exp X$ has an injective differential on $V = \{X \in \mathsf{so}(3) \mid \|X\| < \pi\}$ (Proposition 3). That follows from the formula for $d\exp$ as in Proposition 7 of §1.2: the eigenvalues of $\operatorname{ad} X\,|_{\mathrm{SO}(3)}$ are 0 and $\pm i\|X\|$, hence cannot differ by a multiple of $2\pi i$ as long as $\|X\| < \pi$.

(b) *Euler angles.* There is a classical coordinate system on SO(3) that goes back to Euler (1775). (It was lost in Euler's voluminous writings until Jacobi (1827) called attention to it because of its use in mechanics.) It is based on the following lemma.

Lemma 5. *Every $a \in \mathrm{SO}(3)$ can be written in the form*

$$a = a_3(\theta) a_2(\phi) a_3(\psi)$$

where

$$a_3(\theta) = \exp(\theta E_3), \quad a_2(\phi) = \exp(\phi E_1)$$

and $0 \le \theta, \psi < 2\pi$, $0 \le \phi \le \pi$. Furthermore, (θ, ϕ, ψ) is unique as long as $\phi \ne 0, \pi$. (E_1, E_2, E_3 is the basis for $\mathsf{so}(3)$ introduced in Example 1 of §2.1.)

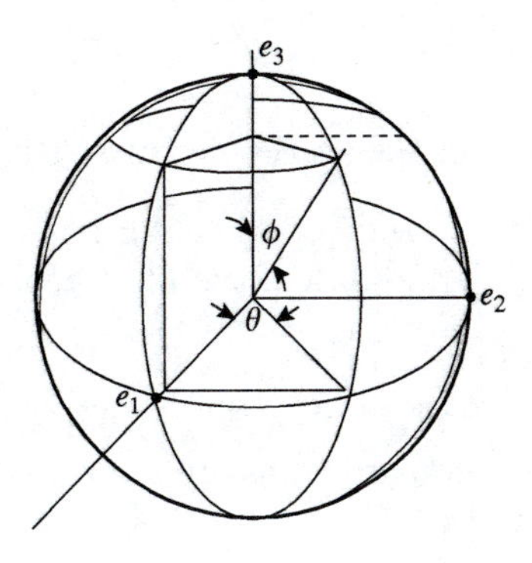

Fig. 1

Proof. Consider the rotation-action of SO(3) on the unit sphere $S^2 = \{x \in \mathbb{R}^3 \mid \|x\| = 1\}$. Write (θ, ϕ) for the longitude and co-latitude of a point $x \in S^2$, as shown in Figure 1: this amounts to writing

$$x = a_3(\theta)a_2(\phi)e_3.$$

Thus for any $a \in \mathrm{SO}(3)$ one has an equation

$$ae_3 = a_3(\theta)a_2(\phi)e_3.$$

for some (θ, ϕ) subject to the above inequalities and unique as long as $\phi \neq 0$. This equation implies that $a = a_3(\theta)a_2(\phi)b$ for some $b \in \mathrm{SO}(3)$ with $be_3 = e_3$. Such a $b \in \mathrm{SO}(3)$ is necessarily of the form $b = a_3(\psi)$ for a unique ψ, $0 \leq \psi \leq 2\pi$. QED

θ, ϕ, ψ are the *Euler angles* of the element $a = a_3(\theta)a_2(\phi)a_3(\psi)$. They form a coordinate system on SO(3) with domain consisting of those as whose Euler angles satisfy $0 < \theta, \psi < 2\pi$, $0 < \phi < \pi$ (strict inequalities). To prove this one has to show that the map $\mathbb{R}^3 \to M_3(\mathbb{R})$, $(\theta, \phi, \psi) \to a_3(\theta)a_2(\phi)a_3(\psi)$, has an injective differential for these (θ, ϕ, ψ). The partial derivatives of $a = a_3(\theta)a_2(\phi)a_3(\psi)$ are given by

$$\begin{aligned}
a^{-1}\frac{\partial a}{\partial \theta} &= \sin\phi\cos\psi E_1 - \sin\phi\sin\psi E_2 + \cos\phi E_3 \\
a^{-1}\frac{\partial a}{\partial \phi} &= \sin\psi E_1 + \cos\psi E_2 \\
a^{-1}\frac{\partial a}{\partial \psi} &= E_3
\end{aligned}$$

The matrix of coefficients of E_1, E_2, E_3 on the right has determinant $\sin\phi$, hence the three elements of $\mathsf{so}(3)$ given by these equations are linearly independent as long as $\sin\phi \neq 0$. This proves the desired injectivity of the differential on the domain in question.

Problems for §2.3

1. Check the calculation of the partial derivatives in Lemma 5.

2. Show that the group-topology of a linear group is *discrete* (meaning *every* subset of G is open), if and only if the Lie algebra of G is $\{0\}$.

3. (a) Show that the domain $a_o U = \{a_o \exp X \mid X \in \mathfrak{g}, \|X\| < R\}$ of the exponential coordinates around $a_o \in G$ is open in G.

   (b) Show that the coordinate transformation relating *any* two coordinate systems on G is analytic.

   (c) By definition, a function on a linear group G is said to be of class $C^k (0 \leq k \leq \infty, \omega)$ if it becomes such a function when elements of G are locally expressed in terms of coordinates. Show that this definition is independent of the coordinates chosen.

4. Let f be a real-valued function on a linear group. Show that if f extends to a C^k-function of the matrix entries in a neighborhood of each point in its domain, then f is a C^k-function on G. Give an example to show that the converse need not hold. (Any fixed k, $0 \leq k \leq \infty, \omega$.)

5. Show that every element of SU(2) can be written in the form

$$a = a_3(\theta) a_2(\phi) a_3(\psi)$$

   where now

$$a_3(\theta) = \begin{bmatrix} e^{i\theta/2} & 0 \\ 0 & e^{-i\theta/2} \end{bmatrix}, \quad a_2(\phi) = \begin{bmatrix} \cos\phi/2 & -\sin\phi/2 \\ \sin\phi/2 & \cos\phi/2 \end{bmatrix},$$

   and $0 \leq \theta, \psi < 2\pi, 0 \leq \phi < \pi$. Show that θ, ϕ, ψ form a coordinate system on the domain where the inequalities are all strict. [See Example 2 of §2.1, which also explains the analogy with Example 4(b).]

6. Show that every element of $\mathrm{SL}(2,\mathbb{R})$ can be written in the form

$$a = k(\theta) n(\sigma) a(\tau)$$

   where

$$k(\theta) = \begin{bmatrix} \cos\theta & -\sin\theta \\ \sin\theta & \cos\theta \end{bmatrix}, \quad n(\sigma) = \begin{bmatrix} 1 & \sigma \\ 0 & 1 \end{bmatrix}, \quad a(\tau) = \begin{bmatrix} e^{\tau} & 0 \\ 0 & e^{-\tau} \end{bmatrix},$$

   and $0 \leq \theta < 2\pi$, $\sigma, \tau \in \mathbb{R}$. Show that θ, σ, τ form a coordinate system on the domain where the first inequality is also strict. [Compare the decomposition $\mathrm{SL}(2,\mathbb{R}) = KB$ of Example 3 of §2.1.]

7. Show that every element of $\mathrm{SL}(2,\mathbb{R})$ can be written in the form

$$a = k(\theta) a(\tau) k(\phi)$$

with $k(\theta), a(\tau)$ as in the previous problem and $0 \leq \theta < \pi$, $0 \leq \phi < 2\pi$, $\tau \geq 0$. Show that θ, σ, τ form a coordinate system on the domain where all inequalities are strict. [Suggestion: to prove existence of the decomposition $a = k(\theta)a(\tau)k(\phi)$ consider the adjoint-action of $\mathrm{SL}(2,\mathbb{R})$ on one sheet of the hyperboloid (adjoint orbit) in $\mathsf{sl}(2,\mathbb{R})$ given by

$$\det X = 1;$$

see Example 3, §2.1. Argue as in Example 4(b), with the help of a sketch. To determine the range of the variables (θ, τ, ϕ) note that $k(\theta+\pi)a(\tau)k(\phi+\pi) = k(\theta)a(\tau)k(\phi)$.]

8. Let N be a linear group consisting of unipotent matrices, n its Lie algebra. Show that $\exp \mathsf{n}$ is open in N and that $a = \exp X \to X$ defines a coordinate system on all of $\exp \mathsf{n}$. [Suggestion: problem 1, §1.2.]

9. Let $f : G \to H$ be a differentiable homomorphism of linear groups. Show that for all $X \in \mathsf{g}$ (the Lie algebra of G),

$$f(\exp X) = \exp \varphi(X),$$

where

$$\varphi(X) = \frac{d}{d\tau} f(\exp \tau X)\big|_{\tau=0}.$$

[Suggestion: show that $f(\exp \tau X))$ satisfies the differential equation of exp.]

10. Let G be a linear group. Suppose there are arbitrarily small neighborhoods U of 1 in the matrix space M so that $G \cap U$ is C^1-connected, i.e. every element of $G \cap U$ can be joined to 1 by a C^1-curve in $G \cap U$. Show:

(a) There is a neighborhood U of 1 in M and a bi-analytic map $f : U \to V$ of U onto V an open ball in $\mathbb{R}^N$ ($N = \dim M$) which carries $G \cap U$ onto the open ball in the m-dimensional subspace $\mathbb{R}^m$ of $\mathbb{R}^N$. ($m = \dim G$; $\mathbb{R}^m$ is considered as the subspace of $\mathbb{R}^N$ where the last $N - m$ coordinates are 0.) [Suggestion: review the proof of Theorem 3, §2.2.]

(b) Every element $a \in G$ has a neighborhood in G of the form $G \cap U$ where U is a neighborhood of a in M. The open sets in G are exactly the intersections with G of open sets in M. [This means that the group-topology on G is the relative topology. Neighborhoods in G are – as always – understood in the sense defined in the text.]

(c) A function on G (defined on an open subset with values in $\mathbb{R}$ or in a linear group) is of class $C^k (0 \leq k \leq \infty, \omega)$ if and only if it extends to a function of class C^k in a neighborhood in the matrix space M of each point $a_o \in G$ in its domain.

11. Show that the following groups satisfy the hypothesis of problem 10.

(a) $\mathrm{SL}(n,\mathbb{R})$, $\mathrm{SL}(n,\mathbb{C})$,

(b) SU(n), SO(n),

(c) Block-triangular and block-diagonal groups (Example 6, §2.1).

Give an example of a linear group G which does not satisfy the hypothesis of problem 10 and specify explicitly a continuous function on G which does not extend to a continuous function on M, even locally as in problem 10 (c).

12. Let G be a linear group. Suppose G can be defined by a system of polynomial equations

$$f_1(a) = 0, \dots, f_k(a) = 0.$$

(The $f_j(a)$ are polynomials in the matrix entries of a with real or complex coefficients.) Show that G satisfies the hypothesis of problem 10. [This gives another solution of problem 11, since the groups there obviously satisfy the hypothesis. [Suggestion: for $X \in M$, define an operator X_R on polynomial functions f on M by the formula

$$X_R f(a) = (df_a)(aX) = \frac{d}{d\tau} f(a \exp \tau X)\big|_{\tau=0}.$$

Let f be a polynomial function on M. Show that for $\|X\|$ sufficiently small

$$X_R f = \frac{X_R}{\exp X_R - 1}(\exp X_R - 1) f.$$

Choose $\epsilon > 0$ so that this applies to all $f = f_j$ when $\|X\| < \epsilon$. Deduce that for $\|X\| < \epsilon, \exp X \in G \Leftrightarrow f_j(\exp \tau X) \equiv 0 \Leftrightarrow X \in \mathfrak{g}$.

Remark. Supplemented by a suitable convergence argument, the proof applies when, in some neighborhood U of 1 in M, $G \cap U$ is given by $f_j(a) = 0$ with f_j defined and analytic in U. But in fact it suffices that, for some neighborhood U of 1 in M, $G \cap U$ be closed in U. This we shall see by a completely different argument in §2.7.]

13. Let G be a linear group. Suppose G can be defined by an system of C^1 equations

$$f_1(a) = 0, \dots, f_k(a) = 0.$$

with linearly independent differentials at $a = 1$.

(a) Show that G satisfies the hypothesis of problem 10. [Suggestion: Inverse Function Theorem. It actually suffices that there be a neighborhood U of 1 in M so that the elements of $G \cap U$ form the solution set of such a system where the f_j are defined and C^1 on U and have linearly independent differentials at $a = 1$, but see the remark after problem 12.]

(b) Show that the Lie algebra $\mathfrak{g}$ of G consists of all $X \in M$ satisfying the linear equations

$$(df_1)_1(X) = 0, \quad \dots, \quad (df_k)_1(X) = 0.$$

14. Verify that the groups of problem 11 satisfy the hypothesis of problem 13. (This gives still another solution of problem 11.)

15. Show that every linear group has a neighborhood of the identity which contains no subgroup $\neq \{1\}$.

2.4 Connectedness

The topological notion of 'connectedness' will be of crucial importance in Lie theory. For this reason we shall discuss it in some detail.

Proposition 1. *The following conditions on a linear group G are equivalent:*

(a) Any two elements of G may be joined by a C^k-path in G.

(b) G is not the disjoint union of two non-empty open sets.

(c) G is generated by any neighborhood of 1.

(d) G is generated by $\exp \mathsf{g}$.

Explanation. In accordance with our standing convention, 'open' in (b) and 'neighborhood' in (c) refers to the group topology on G, defined using $\exp\colon \mathsf{g} \to G$. In (a) one may take any k, $0 \leq k \leq \infty, \omega$, and (a) says in more detail that for any two points a_o and a_1 in G, there is a C^k path $[0,1] \to G$, $\tau \to a(\tau)$ with $a(0) = a_o$ and $a_1 = a(1)$. Here the function $a(\tau)$ defined on the closed interval [0,1] is understood to be of class C^k if it extends to a C^k-function also in neighborhoods of the endpoints. Furthermore, 'C^k' is taken in the sense of G-valued functions, defined using exponential coordinates. For $k \geq 1$, however, 'C^k' could also be interpreted in terms of matrix entries, by Corollary 5 of §2.2. A subset of G *generates* G if every element of G is a (finite) product of elements of the subset and their inverses. In (d) this means that every element of G is of the form $\exp X_1 \exp X_2 \cdots \exp X_k$ for some $X_1, X_2, \ldots, X_k$ in the Lie algebra g of G.

Proof. (a) $\Rightarrow$ (b). Assume (a). If G could be split into two non-empty open subsets, then a C^0 path joining an element from one to an element of the other would give a partition of its interval of definition into two disjoint open subsets, which is impossible. Hence (b).

(b) $\Rightarrow$ (c). Assume (b). Let G_o be the subgroup of G generated by some neighborhood of 1. Then G_o is open in G, since it contains a neighborhood in G of each of its points. The same is true for any coset aG_o of G_o in G. This implies that $G = G_o$, because otherwise G would be the disjoint union of G_o and the union of the cosets other than G_o. Hence (c).

(c) $\Rightarrow$ (d) is clear.

(d) $\Rightarrow$ (a). Assume (d). Then every element of G is of the form $\exp X_1 \exp X_2 \cdots \exp X_k$ for some $X_1, X_2, \ldots, X_k$ in the Lie algebra g of G. In particular, given two elements a_o and a_1 of G, we can write $a_1 = a_o \exp X_1 \exp X_2 \cdots \exp X_k$. Then $a(\tau) = a_o \exp \tau X_1 \exp \tau X_2 \cdots \exp \tau X_k$ is a C^ω path in G with $a(0) = a_o$ and $a(1) = a_1$. QED

A linear group G is said to be *connected* if it satisfies the equivalent conditions of the proposition.

Remark 2. The condition (a) could be replaced by: (a′) Every element of G can be joined to 1 by a C^k path in G. For if $a(\tau)$ joins $a(0) = 1$ to $a(1) = a_o^{-1}a_1$, then $a(\tau)a_o$ joins a_o to a_1.

The condition (c) could be replaced by (c′) : G is generated by every open set.

The elements of a linear group G that can be joined to a given $a \in G$ by a continuous path form the *connected component* of G containing a. The connected component of the identity is simply called the *identity component* of G, denoted G_o. It is evident that G_o has the same Lie algebra as G: the Lie algebra sees only the connected component through exp. (But the whole group G affects its Lie algebra through Ad.)

Proposition 3. *Let G be a linear group,* $\mathbf{g}$ *its Lie algebra. The identity component G_o of G is the subgroup generated by* $\exp X, X \in \mathbf{g}$. *It is an open, normal subgroup of G, and the unique open connected subgroup of G. The connected component of any $a \in G$ is the coset aG_o.*

Proof. This should be clear from the proof of Proposition 1. (QED)

The quotient group G/G_o is called the *component group* of G. It is not a linear group, just an abstract group.

Example 4 (The identity component of $\mathrm{SO}(2,1)$**).** $\mathrm{SO}(2,1)$ consists of all linear transformations of $\mathbb{R}^3$ that preserve the indefinite bilinear form

$$\langle x, y\rangle = \xi_1\eta_1 + \xi_2\eta_2 - \xi_3\eta_3$$

and have determinant $+1$. The ξ_k and η_k are the components of x and y with respect to the standard basis e_1, e_2, e_3. This group is not connected. To describe its identity component, we start with a geometrically obvious lemma:

Lemma 4A. *The complement of the cone $\langle x, x\rangle = 0$ in $\mathbb{R}^3$ is the disjoint union of the following three connected sets:*

$$\begin{aligned} &(a)\ \langle x, x\rangle > 0, \\ &(b)\ \langle x, x\rangle < 0, \langle x, e_3\rangle > 0, \\ &(c)\ \langle x, x\rangle < 0, \langle x, e_3\rangle < 0. \end{aligned}$$

(see Figure 1):

Lemma 4B. *The following conditions on an element $a \in \mathrm{SO}(2,1)$ are equivalent:*

(a) a maps each of the three sets in Lemma 4A into itself.

(b) a maps the set in Lemma 4A(b) into itself.

(c) $\langle ae_3, e_3\rangle < 0$.

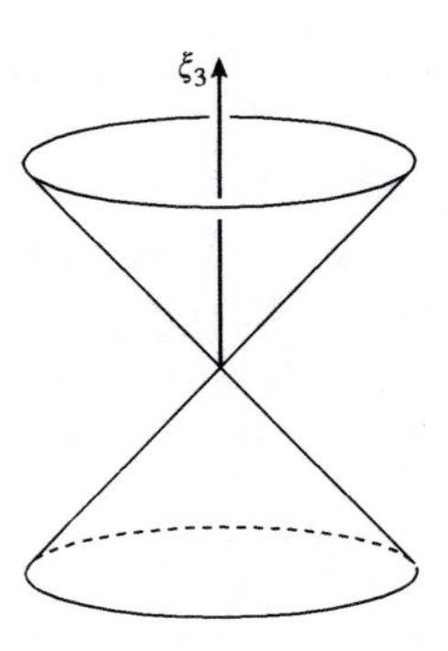

Fig. 1

Proof. (a) $\Rightarrow$ (b) is obvious.

(b) $\Rightarrow$ (a). If a maps the region inside the forward cone into itself, then it also maps the region inside the backward cone into itself ($x \to -x$).

(b) $\Rightarrow$ (c). This is seen by taking $x = -e_3$ in (b) of Lemma 4A.

(c) $\Rightarrow$ (b). Assume $\langle ae_3, e_3 \rangle < 0$. Since $\langle x, e_3 \rangle = 0$ implies $\langle x, x \rangle \geq 0$,

$$\langle x, x \rangle < 0 \quad \text{implies} \quad \langle x, e_3 \rangle \neq 0. \tag{1}$$

For the same reason (replace x by ax):

$$\langle x, x \rangle < 0 \quad \text{implies} \quad \langle x, a^{-1}e_3 \rangle \neq 0. \tag{2}$$

Consider the sign of the real number

$$\langle ax, e_3 \rangle = \langle x, a^{-1}e_3 \rangle \tag{3}$$

for x in the set

$$\langle x, x \rangle < 0, \quad \langle x, e_3 \rangle > 0. \tag{4}$$

The number (3) is never equal to 0 (because of the first condition in (4)) and depends continuously on x. Since the xs in (4) form a connected set, the number (3) always keeps the same sign for the xs in (4). For $x = -e_3$ the number (3) is $-\langle ae_3, e_3 \rangle > 0$. Thus the number (3) is > 0 for all x in (4), i.e. (4) implies $\langle ax, e_3 \rangle > 0$. Of course (4) implies $\langle ax, ax \rangle > 0$ as well. Hence (b). QED

We denote by $\mathrm{SO}_o(2,1)$ the subgroup of $\mathrm{SO}(2,1)$ described by the equivalent conditions of Lemma 4B.

A *Euclidean rotation* in $\mathbb{R}^3$ is a linear transformation a of the form

$$\begin{aligned} au &= \cos(\alpha)u - \sin(\alpha)v, \\ av &= \sin(\alpha)u + \cos(\alpha)v, \\ ax &= x \quad \text{if } \langle x, u \rangle = \langle x, v \rangle = 0, \end{aligned} \tag{5}$$

where u, v are two vectors satisfying

$$\langle u, u \rangle = 1, \quad \langle v, v \rangle = 1, \quad \langle u, v \rangle = 0$$

and $\alpha \in \mathbb{R}$.

A *hyperbolic rotation* in $\mathbb{R}^3$ is a linear transformation of a of the form

$$\begin{aligned} au &= \cosh(\alpha)u + \sinh(\alpha)v, \\ av &= \sinh(\alpha)u + \cosh(\alpha)v, \\ ax &= x \quad \text{if } \langle x, u\rangle = \langle x, v\rangle = 0, \end{aligned} \tag{6}$$

where u, v are two vectors satisfying

$$\langle u, u\rangle = 1, \quad \langle v, v\rangle = -1, \quad \langle u, v\rangle = 0$$

and $\alpha \in \mathbb{R}$.

Lemma 4C. *Euclidean and hyperbolic rotations belong to* $\mathrm{SO}_o(2,1)$. *Every element of* $\mathrm{SO}_o(2,1)$ *is of the form* $a = kh$ *where* k *is a Euclidean rotation fixing* e_3 *and* h *a hyperbolic rotation in a plane containing* e_3.

Proof. Let a be a Euclidean or hyperbolic rotation belonging to vectors $u, v \in \mathbb{R}^3$ and $\alpha \in \mathbb{R}$ as in the definition. Since any vector y is a linear combination of u, v and an x orthogonal to them, it is clear that $\langle ay, ay\rangle = \langle y, y\rangle$. Thus $a \in \mathrm{O}(2,1)$. Writing $a = a(\tau)\mid_{\tau=\alpha}$ with $a(\tau)$ defined by the formula (5) or (6) with τ in place of α one sees that the curve $a(\tau)$ lies in O(2,1). Hence $\det a(\tau) = \pm 1$ for all τ. By continuity, $\det a(\tau) = +1$ for all τ, and $a(\tau) \in \mathrm{SO}(2,1)$. Also, if x is a point in the set $\{\langle x, x\rangle \neq 0\}$, then $a(\tau)x$, is also therein and connects x to $a(\alpha)x$. Hence $a(\alpha)x$ lies in the same connected component of $\{\langle x, x\rangle \neq 0\}$ as x. Therefore $a(\alpha) \in \mathrm{SO}_o(2,1)$, as desired.

It remains to show that any $a \in \mathrm{SO}_o(2,1)$ is a product of a Euclidean rotation and a hyperbolic rotation as indicated. So let $a \in \mathrm{SO}_o(2,1)$. If a does not fix e_3, then e_3 and ae_3 are linearly independent ($ae_3 = -e_3$ is excluded by $\langle ae_3, e_3\rangle < 0$). These two vectors span a plane on which the form $\langle x, y\rangle$ is of the hyperbolic type $(+-)$: this plane contains the negative vector e_3 as well as the positive vector $ae_3 + \langle ae_3, e_3\rangle$. Since $\langle ae_3, e_3\rangle < 0$, ae_3 and e_3 lie on the same branch of the hyperbola $\langle x, x\rangle = -1$ in this plane. Hence one can find a hyperbolic rotation h in this plane with $hae_3 = e_3$. Then ha fixes e_3 and induces an orthogonal transformation of determinant 1 in the positive-definite subspace orthogonal to e_3, hence is a Euclidean rotation. QED

Proposition 4D. $\mathrm{SO}_o(2,1)$ *is the identity component of* $\mathrm{SO}(2,1)$ *as well as of* $\mathrm{O}(2,1)$.

Proof. From Lemma 4C one sees that $\mathrm{SO}_o(2,1)$ is connected: Euclidean and hyperbolic rotations (and products thereof) may be joined to 1 by making the α in (5) and (6) variable. And from the definition of $\mathrm{SO}_o(2,1)$ one sees that it contains a neighborhood of 1 in $\mathrm{O}(2,1)$ by continuity considerations. This implies the assertion. QED

Remark 5. The discussion above carries over immediately to the group $\mathrm{O}(p,1)$ of linear transformations of $\mathbb{R}^{p+1}$ which preserve the indefinite form

$$\langle x, y\rangle = \xi_1\eta_1 + \xi_2\eta + cd\xi_p\eta_p - \xi_{p+1}\eta_{p+1}.$$

The connected component $SO_o(p,1)$ is described by Lemma 4B with e_3 replaced by e_{p+1}.

Problems for §2.4

1. Prove Proposition 3 in detail.
2. Show that an open subgroup of a linear group in also closed.
3. Describe the connected components of $O(2,1)$ and the component group $O(2,1)/SO_o(2,1)$.
4. Show that the following groups are connected (see §2.1).
   (a) $SO(3)$ and $SU(2)$,
   (b) $SL(2,\mathbb{R})$ and $SL(2,\mathbb{C})$,
   (c) The group of Euclidean motions in the plane.
5. Show that $U(n)$ and $SU(n)$ are connected. (Assume known that every unitary matrix is similar to a diagonal matrix with diagonal entries $e^{i\theta}$.)
6. Show that $SO(n)$ is connected, while $O(n)$ has two connected components. [Suggestion: show first that every real orthogonal matrix is similar to a block-diagonal matrix with 2×2 blocks

$$\begin{bmatrix} \cos\theta & -\sin\theta \\ \sin\theta & \cos\theta \end{bmatrix}$$

   and 1×1 blocks ± 1.]
7. Let N be a *connected* group consisting of unipotent matrices, n its Lie algebra. Show that $\exp\colon \mathsf{n} \to N$ is bijective. [Problem 1 of §1.2 may be used if solved previously. Otherwise work that problem first and complete the proof.]

   Connected, dense, discrete sets. Let G be a linear group, A a subset of G.

   A is *connected* if it cannot be covered by two disjoint open subset of G without being contained in one of them.

   A is C^k*-connected* if any two points of A can be joined by a C^k path which lies in A.

   A is *dense* in G if every neighborhood of an element of G contains elements of A. Equivalently, the closure of A (i.e. the smallest closed subset of G containing A) is all of G. A is *discrete* in G if every neighborhood of A has a neighborhood in G that contains no other elements of A.

   A is *compact* if every cover of A by open subsets of G has a finite subcover.

(These concepts from general topology are assumed to be familiar, at least from $\mathbb{R}^n$, and their elementary properties are used without further comment. In particular: a continuous image of a compact (resp. connected) set is compact (resp. connected). This follows directly from the definitions).

8. If a linear group G has a connected dense subset A, then G itself is connected.

9. Show that a subgroup H of a connected linear group G is dense in G if and only if the closure of H contains a neighborhood of 1 in G. Give an example with H connected, $H \neq G$.

10. If a linear group G has an Abelian dense subgroup H, then G itself is Abelian. Give an example with $H \neq G$.

11. Show that a subgroup H of a linear group G is discrete in G if and only if there is a neighborhood of 1 in G that contains no other elements of H.

12. Show that a discrete normal subgroup of a connected linear group G is contained in the center of G. [Suggestion: for a continuous path $a(\tau)$ in G and an element c in a discrete normal subgroup, consider $a(\tau)ca(\tau)^{-1}$ as a function of τ.]

13. Show that a linear group G is discrete (in its group topology, as defined above) if and only if its Lie algebra $\mathfrak{g}$ is zero. [Careful: $\mathrm{GL}(n, \mathbb{Q})$ is discrete in its group topology, but is not discrete in $\mathrm{GL}(n, \mathbb{R})$; it is dense in $\mathrm{GL}(n, \mathbb{R})$.]

14. Prove Remark 5. [First prove Lemma 3A for the form $\langle x, y \rangle$ on $\mathbb{R}^{p+1}$. Check that the remaining argument carries over from $n = 3$ to $n = p + 1$.]

2.5 The Lie correspondence

We shall prove in this section that the passage from a linear group to its Lie algebra sets up a one-to-one correspondence between *connected* linear groups and linear Lie algebras, the inverse being the passage from a linear Lie algebra to the group generated by its exponentials. This is the essence of *Lie theory*, as Lie must have understood it, even though Lie's conception was on the one hand broader, in that he considered groups of transformations which were not necessarily linear, but on the other hand less complete, in that he took a local point of view. The global Lie correspondence is a refinement that is rather routine once the appropriate topological notions are available. The version of the Lie correspondence stated above follows from *Satz* 1 of Freudenthal (1941) and can be found in Bourbaki (1960) in the context of general Lie groups.

The essence of the Lie correspondence, in turn, is the Campbell–Baker–Hausdorff formula in its qualitative form, saying that in exponential coordinates the group multiplication is given by a bracket series and therefore completely determined by the Lie algebra, at least in a neighborhood of the identity. (Actually, Lie himself might object, if he could: he was not fond of any such algebraic formulation of his theory, which he conceived of as being essentially geometric and analytic. Even today the Lie correspondence is often established without Campbell–Baker–Hausdorff; but the principle that 'the Lie algebra determines the group' is certainly most simply and forcefully expressed by this formula.)

To succinctly state the Lie correspondence we use the following notation. The Lie algebra of a linear group G will be denoted $L(G)$ rather than $\mathbf{g}$ when it is necessary to bring out its dependence on G. Furthermore, we shall use the characterization of $L(G)$ in terms of exp:

$$L(G) = \{X \in M \mid \exp \tau X \in G \quad \text{for all} \quad \tau \in \mathbb{R}\}.$$

On the other hand, given a linear Lie algebra $\mathbf{g}$, we denote by $\Gamma(\mathbf{g})$ the linear group generated by $\exp \mathbf{g}$:

$$\Gamma(\mathbf{g}) = \{\exp X_1 \cdot \exp X_2 \cdots \exp X_k \mid X_1, X_2, \ldots, X_k \in \mathbf{g}\}.$$

$\Gamma(\mathbf{g})$ is simply called the linear group generated by $\mathbf{g}$.

Theorem 1 (The Lie Correspondence). *There is a one-to-one correspondence between connected linear groups G and linear Lie algebras $\mathbf{g}$ given by*

$$G \leftrightarrow \mathbf{g}$$

if

$$\mathbf{g} = L(G) \quad \textit{or equivalently} \quad G = \Gamma(\mathbf{g}).$$

Proof $\Gamma(L(G)) = G$. This says that G is generated by $\exp L(G)$, which is (d) of Proposition 1, §2.4.

$\mathbf{g} = L(\Gamma(\mathbf{g}))$. Let $\mathbf{g}$ be a linear Lie algebra. Then $\Gamma(\mathbf{g})$ is connected because any element $\exp X_1 \exp X_2 \cdots \exp X_k$ of $\Gamma(\mathbf{g})$ can be joined to 1 by the continuous path $\exp \tau X_1 \exp \tau X_2 \cdots \exp \tau X_k, \tau \in \mathbb{R}$.

$\mathbf{g} \subset L(\Gamma(\mathbf{g}))$ is clear since $\exp(\tau X) \in \Gamma(\mathbf{g})$ for all $X \in \mathbf{g}$. The point is to show that $L(\Gamma(\mathbf{g})) \subset \mathbf{g}$. Let

$$U = \{X \in \mathbf{g} \mid \|X\| < \epsilon\} \quad \text{and} \quad \bar{U} = \{X \in \mathbf{g} \mid \|X\| \leq \epsilon\}$$

for small $\epsilon > 0$. From Campbell–Baker–Hausdorff we know that the equation

$$\exp Z = \exp X \exp Y$$

defines a map $Z = C(X, Y)$ from $\bar{U} \times \bar{U}$ to a neighborhood of 0 *in* $\mathbf{g}$: $C(X, Y)$ is the Campbell–Baker–Hausdorff series. We set $V = C(U, U)$, $\bar{V} = C(\bar{U}, \bar{U})$. Thus $\exp(V) = \exp(U)\exp(U)$. Since $C(X, Y)$ reduces to $C(0, Y) = Y$ for $X = 0$, the map $U \to \mathbf{g}$, $Y \to C(X, Y)$ (X fixed) has a differential of rank $= \dim \mathbf{g}$ at $Y = 0$, as is obvious if $X = 0$ and remains true for X in a neighborhood of 0 by continuity. The Inverse Function Theorem implies that $C(X, U)$ is an open neighborhood of X in $\mathbf{g}$ provided X is sufficiently close to 0 in $\mathbf{g}$ (which we may assume to be the case for $X \in \bar{V}$) and provided the ϵ defining U is sufficiently small. This we assume to be so.

The set $\bar{V} = C(\bar{U}, \bar{U})$ is covered by the open sets $C(X, U)$, $X \in \bar{V}$ (because certainly $X \in C(X, U)$). Since $\bar{V}$ is a compact subset of $\mathbf{g}$ (being a continuous image of the compact set $\bar{U} \times \bar{U}$) already finitely many $C(X, U)$ cover $\bar{V}$, say $C(X_j, U)$, $j = 1, \ldots, N, X_j \in \bar{V}$. Write $\exp X_j = a'_j a''_j$ with a'_j, $a''_j \in \exp \bar{U}$ and apply exp to $\bar{V} \subset \bigcup_j C(X_j, U)$ to find that

$$\exp \bar{U} \exp \bar{U} \subset \bigcup_j a'_j a''_j \exp \bar{U},$$

even with $\bar{U}$ replaced by U on the right. Let $\{b_j \colon j = 1, 2, \ldots\}$ be the (countable) set of all products of finite sequences from $\{a'_j, a''_j \colon j = 1, \ldots, N\}$ and write $(\exp \bar{U})^k$ for the set of k-fold products of elements of $\exp \bar{U}$. From the above inclusion one gets inductively that

$$(\exp \bar{U})^k \subset \bigcup_{j=1}^{\infty} b_j(\exp \bar{U})$$

for all $k \geq 1$. Hence $\Gamma(\mathbf{g}) = \bigcup_{k=1}^{\infty} (\exp \bar{U})^k$ is expressible as a countable union

$$\Gamma(\mathbf{g}) = \bigcup_{j=1}^{\infty} b_j(\exp \bar{U}) \tag{1}$$

for appropriate $b_j \in \Gamma(\mathbf{g})$. By Baire's Covering Lemma (proved below; see also the comment (*b*) thereafter):

some $b_j \exp U$ contains a neighborhood of some point a_o in $\Gamma(\mathbf{g})$. (2)

Say

$$b_j \exp U \supset a_o \exp \tilde{U},$$

where $\tilde{U} = \{\tilde{X} \in L(\Gamma(\mathbf{g})) \mid \|\tilde{X}\| < \tilde{\epsilon}\}$ for some $\tilde{\epsilon} > 0$. Then

$$\exp \tilde{U} \subset c \exp U, \tag{3}$$

where $c = a_o^{-1} b_j$. This implies that for all $\tilde{X} \in \tilde{U}$

$$\exp \tilde{X} = c \exp X \tag{4}$$

with $X \in U$. Furthermore, for ϵ and $\tilde{\epsilon}$ sufficiently small, $X \in U$ and $\tilde{X} \in \tilde{U}$ will be arbitrarily close to 0, hence $c = \exp(-X) \exp(\tilde{X})$ will be close to 1 and the unique solution of (4) for $X \in U$ is

$$X = \log\left(c^{-1} \exp \tilde{X}\right). \tag{5}$$

Replacing $\tilde{X}$ by $\tau \tilde{X}$ with τ close to 0 in $\mathbb{R}$ we see that

$$\exp \tau \tilde{X} = c \exp X(\tau)$$

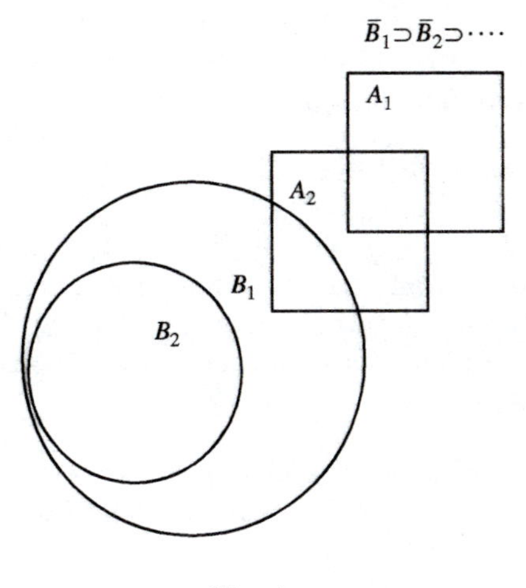

Fig. 1

with $X(\tau) \in U$ depending differentiably on τ. Setting $\tau = 0$ we find $c = \exp(-X(0))$ and therefore

$$\exp \tau \tilde{X} = \exp(-X(0)) \exp X(\tau).$$

Differentiating this equation at $\tau = 0$ we obtain

$$\tilde{X} = \frac{1 - \exp -(\mathrm{ad}(X(0))}{\mathrm{ad}(X(0))} X'(0),$$

which lies in $\mathbf{g}$ since $X(0), X'(0) \in \mathbf{g}$ and $\mathbf{g}$ is a Lie algebra. Thus the neighborhood $\tilde{U}$ of 0 in $L(\Gamma(\mathbf{g}))$ is contained in $\mathbf{g}$, hence $L(\Gamma(\mathbf{g})) \subset \mathbf{g}$ as required.

It remains to show that (1) implies (2). This will follow from the following general lemma, which will be useful more than once.

Lemma 2 (Baire's covering lemma). *Let $\{A_j\}$ be a* countable *family of subsets of G that cover G:*

$$G = \bigcup_{j=1}^{\infty} A_j. \tag{6}$$

Then the closure $\bar{A}_j$ of some A_j contains an open subset of G.

Comments.

(*a*) One could as well take the A_j to be closed in the first place.

(*b*) If A_j contains an open, dense subset of its closure $\bar{A}_j$, as does a ball, for example, then some A_j itself will have to contain an open subset of G.

(*c*) The lemma (and its proof) hold in more general spaces, in particular in manifolds, to be defined later.

Proof. (*By contradiction.*) Assume *no* $\bar{A}_j$ contains an open subset of G. The part of G outside of $\bar{A}_1$ is then certainly non-empty, hence (being open) contains a closed coordinate-ball[3] $\bar{B}_1$. For the same reason, the part of B_1 outside of $\bar{A}_2$ contains a closed coordinate-ball $\bar{B}_2$. Continuing in this way one gets a nested sequence of closed coordinate-balls in G:

$$\bar{B}_1 \supset \bar{B}_2 \supset \cdots .$$

The intersection of the $\bar{B}_j$ is non-empty, but lies outside of all $\bar{A}_j$, in contradiction to (6). (To see that $\cap_j \bar{B}_j$ is non-empty, one may consider any limit point of a sequence whose j-th term is taken from $\bar{B}_j$. Such a limit point always exists: since $\bar{B}_1$ may be identified with a ball in $\mathbb{R}^m$, we are essentially dealing with a bounded sequence in $\mathbb{R}^m$.) QED

The Lie correspondence provides a dictionary between connected linear groups and linear Lie algebras. We now list some of its entries, but its full power will only emerge gradually.

Proposition 3. *Let G be a connected linear group, g its Lie algebra. G is abelian if and only if g is abelian.*

Explanation. A group G is *Abelian* if $ab = ba$ for all $a, b \in G$; a Lie algebra g is Abelian if $[X, Y] = 0$ for all $X, Y \in \mathsf{g}$.

Proof. The proof is an illustration of the differentiation-exponentiation argument typical of the 'group $\leftrightarrow$ Lie algebra dictionary'; it runs as follows.

Assume G is abelian. Differentiating the equation

$$\exp \sigma X \exp \tau Y = \exp \sigma X \exp \tau Y$$

with respect to σ and τ at $\sigma = \tau = 0$ one gets $XY = YX$, i.e. $[X, Y] = 0$.

Conversely, if g is Abelian, then evidently

$$\exp X \exp Y = \exp Y \exp X$$

for all $X, Y \in \mathsf{g}$, hence $ab = ba$ for all $a, b \in G$, since G is generated by exponentials from g. QED

The relation between a linear group and its Lie algebra can be used to obtain a very explicit description of connected Abelian linear groups:

Proposition 4. *A connected abelian linear group is isomorphic with $\mathbb{T}^p \times \mathbb{R}^q$ for some p and q.*

Let A be a connected abelian linear group, a its Lie algebra. Since a is abelian, $\exp\colon \mathsf{a} \to A$ is a homomorphism from the additive group a to A. Thus $\exp \mathsf{a}$ is a subgroup of A. Since A is connected, $A = \exp \mathsf{a}$, by Proposition 1(d) of §2.4. Consider the kernel $\Gamma = \{X \in \mathsf{a} \mid \exp X = 1\}$ of $\exp\colon \mathsf{a} \to A$. It is a *discrete subgroup* of a, meaning every element of Γ has a neighborhood in a that

[3] A *coordinate-ball* consists of the elements with coordinates (relative to some coordinate system) in some ball in $\mathbb{R}^m$.

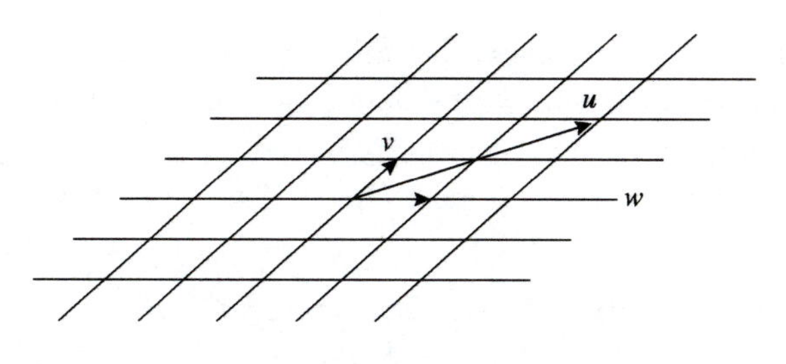

Fig. 2

contains no other element of Γ: this follows from the fact that $\exp\colon \mathsf{a} \to A$ is invertible in a neighborhood of 0.

We need a lemma, which holds in a more general context:

Lemma 5. *For any discrete subgroup Γ of an n-dimensional real vector space V there is a basis $u_1, u_2, \ldots, u_n$ of V so that Γ consists of exactly the elements of V of the form $n_1u_1 + \cdots + n_pu_p$ with $n_1, \ldots, n_p \in \mathbb{Z}$ (some $p \le n$).*

Assuming this, for the moment, we choose such a basis for $V = \mathsf{a}$ and Γ as above. The map $\mathbb{R}^n \to A$, $(\xi_1, \ldots, \xi_n \to \exp(\xi_1u_1 + \cdots + \xi_nu_n)$ induces the required isomorphism $\mathbb{T}^p \times \mathbb{R}^q \to A$ when we identify $\mathbb{T}^p$ with $\mathbb{R}^p/\mathbb{Z}^p$ $(n = p+q)$.

This completes the proof of proposition. QED

Proof of Lemma 5. The proof will provide an inductive construction of such a basis $\{u_k\}$. Let W be a subspace of V of so that $\Gamma \cap W = \mathbb{Z}u_1 + \cdots + \mathbb{Z}u_m$ for some basis $u_1, \ldots, u_m$ of W. (We could start with $W = \{0\}$.) Suppose there is an element u of Γ that does not lie in W. Consider the set of points in V of the form (Figure 2)

$$\xi_1u_1 + \cdots + \xi_mu_m + \xi u, \quad 0 \le \xi_k \le 1,\ 0 \le \xi \le 1. \tag{7}$$

Since this set is bounded (in a Euclidean norm on V), it can contain only finitely many points from the discrete set Γ. Hence there is a point v of the form (7) in Γ with minimal non-zero coefficient ξ, say $\xi = \mu > 0$. Any other element w of Γ of the form $\xi_1u_1 + \cdots + \xi_mu_m + \xi u$, will then have as coefficient ξ of u a multiple of the (minimal) coefficient μ. (Otherwise one could write $\xi = k\mu + \nu$ with $k \in \mathbb{Z}$ and $0 < \nu < \mu$ to obtain an element $w - kv$ of Γ of the form (8) whose coefficient $\xi = \nu$ is less than that of v. The coefficients ξ_k of this element may be made to satisfy $0 \le \xi_k \le 1$ after subtraction of suitable multiples of the u_k.) This means that

$$\Gamma \cap (W + \mathbb{R}u) = (\Gamma \cap W) + \mathbb{Z}v = \mathbb{Z}u_1 + \cdots + \mathbb{Z}u_m + \mathbb{Z}v.$$

$\{u_1, \ldots, u_m, v\}$ forms a basis for $W + \mathbb{R}v$. Replacing W by $W + \mathbb{R}v$ and repeating the argument one finds that $\Gamma = \mathbb{Z}u_1 + \cdots + \mathbb{Z}u_p$ for linearly independent vectors $u_1, \ldots, u_p$, which may be completed to a basis for V. QED

We continue with the 'group $\leftrightarrow$ Lie algebra dictionary'.

Proposition 6. *Let G be a connected linear group, g its Lie algebra. A connected subgroup of G is normal if and only if its Lie algebra is an ideal of g.*

Explanation. A subgroup H of G is *normal subgroup* if $aHa^{-1} = H$ for all $a \in G$; a Lie subalgebra h of g is an ideal if $[X, \mathsf{h}] \subset \mathsf{h}$ for all $X \in \mathsf{g}$ (briefly: $[\mathsf{g}, \mathsf{h}] \subset \mathsf{h}$).

Proof. Assume H is normal in G. Then $a \exp(\tau Y)a^{-1} \in H$ for all $Y \in \mathsf{h}$, $a \in G$. Differentiating at $\tau = 0$ we find that $\mathrm{Ad}(a)Y \in \mathsf{h}$ for all $a \in G$. Take $a = \exp \sigma X$ with $X \in \mathsf{g}$, $\sigma \in \mathbb{R}$ and differentiate with respect to σ at $\sigma = 0$ to see that $[X, Y] \in \mathsf{h}$.

Conversely, assume h is an ideal of g. Then for $X \in \mathsf{g}$ and $Y \in \mathsf{h}$, $\exp X \exp Y \exp -X = \mathrm{Ad}(\exp X) \exp Y = \exp(\exp(\mathrm{ad}\, X)Y) \in H$. Since G and H are generated by exponentials, one sees that $aba^{-1} \in H$ for all $a \in G$ and $b \in H$. QED

Proposition 7. *Let G be a connected linear group, g its Lie algebra. The commutator subgroup (G, G) of G is connected and its Lie algebra is the commutator subalgebra $[\mathsf{g}, \mathsf{g}]$ of g.*

Explanation. The *commutator subgroup* of G, denoted (G, G), is the subgroup generated by all $aba^{-1}b^{-1}$, $a, b \in G$. The *commutator subalgebra* of g is the subalgebra generated (namely spanned, in view of Jacobi) by all $[X, Y]$, $X, Y \in \mathsf{g}$.

Proof. Elements of (G, G) are products of commutators $aba^{-1}b^{-1}$ with $a, b \in G$. Since G is connected, a and b can be joined to 1 by continuous curves $a(\tau)$ and $b(\tau)$. Elements of (G, G) can be joined to 1 by the corresponding products of commutators $a(\tau)b(\tau)a(\tau)^{-1}b(\tau)^{-1}$. So (G, G) is connected.

$[\mathsf{g}, \mathsf{g}] \subset L((G, G))$. Let $X, Y \in \mathsf{g}$. Then $\exp \sigma X \exp \tau Y \exp -\sigma X \exp -\tau Y$ lies in (G, G), and differentiation at $\sigma = \tau = 0$ shows that $[X, Y] \in L((G, G))$.

$L((G, G)) \subset [\mathsf{g}, \mathsf{g}]$. We shall show the equivalent statement $(G, G) \subset \Gamma([\mathsf{g}, \mathsf{g}])$. (G, G) is generated by elements $\exp X \exp Y \exp -X \exp -Y$ with $X, Y \in \mathsf{g}$, so it suffices to show that these lie in $\Gamma([\mathsf{g}, \mathsf{g}])$. For this purpose, consider

$$a(\tau) = \exp -X \exp \tau Y \exp X \exp -\tau Y$$

with $\tau \in \mathbb{R}$. One can write

$$\begin{aligned} a(\tau) &= \exp -X \exp(\mathrm{Ad}(\exp \tau Y)X) \\ &= \exp -X \exp(\exp(\tau\, \mathrm{ad}\, Y)X) \\ &= \exp -X \exp X(\tau), \end{aligned}$$

where we have put $X(\tau) = (\exp \tau\, \mathrm{ad}\, Y)X$. Differentiating with respect to τ we get

$$a'(\tau) = a(\tau) \frac{1 - \exp(-\,\mathrm{ad}\, X(\tau))}{\mathrm{ad}\, X(\tau)} X'(\tau).$$

From the definition of $X(\tau)$ it is clear that $X'(\tau) \in [\mathsf{g}, \mathsf{g}]$, so

$$a'(\tau) \in a(\tau)[\mathsf{g}, \mathsf{g}].$$

By Corollary 4 of §2.2, applied to the group $\Gamma([\mathsf{g},\mathsf{g}])$, this implies that $a(\tau) \in \Gamma([\mathsf{g},\mathsf{g}])$ for all τ. In particular $a(1) = \exp -X \exp Y \exp X \exp -Y \in \Gamma([\mathsf{g},\mathsf{g}])$.
QED

The *centralizer* (in G) of a subset A of a linear group G consists of the elements of G that commute with those of A:

$$Z_G(A) = \{c \in G \mid ca = ac \quad \text{for all } a \in A\};$$

the *centralizer* (*in* g) of a subset a of its Lie algebra g consists similarly of the elements of G which commute with those of a:

$$Z_G(\mathsf{a}) = \{c \in G \mid \mathrm{Ad}(c)X = X \quad \text{for all } X \in \mathsf{a}\}.$$

In the Lie algebra of G one defines analogously:

$$\mathsf{z_g}(A) = \{Y \in \mathsf{g} \mid \mathrm{Ad}(a)Y = Y \quad \text{for all } a \in A\},$$
$$\mathsf{z_g}(\mathsf{a}) = \{Y \in \mathsf{g} \mid [X,Y] = 0 \quad \text{for all } X \in \mathsf{a}\},$$

called the *centralizer* (*in* g) of A and of a, respectively. (There is little lost if one considers only subgroups for A and subalgebras for a, since centralizers do not change by passing to the subgroup generated by A or to the subalgebra generated by a. But it is often convenient to use centralizers of single elements, for example.)

Proposition 8. *Let G be a linear group, g its Lie algebra, A a subset of G, and a a subset of g.*

(a) $Z_G(A)$ is a group with Lie algebra $\mathsf{z_g}(A)$.

(b) $Z_G(\mathsf{a})$ is a group with Lie algebra $\mathsf{z_g}(\mathsf{a})$.

(c) If A is a connected subgroup of G and a its Lie algebra, then $Z_G(A) = Z_G(\mathsf{a})$ and has Lie algebra $\mathsf{z_g}(A) = \mathsf{z_g}(\mathsf{a})$. (QED)

We omit the proof, which follows the by now familiar differentiation–exponentiation pattern. We record in particular the special case when $A = G$ and $\mathsf{a} = \mathsf{g}$: $Z_G(G)$, the subgroup whose elements commute with all of G, is called the *center* of G, denoted $Z(G)$; similarly, $\mathsf{z_g}(\mathsf{g})$, the subalgebra whose elements commute with all of g, is called the *center* of g, denoted $\mathsf{z}(\mathsf{g})$.

Upon taking $A = G$, $\mathsf{a} = \mathsf{g}$ in part (c) one finds:

Corollary 9. *Assume G is a connected linear group. The Lie algebra of the center $Z(G)$ of G is the center $\mathsf{z}(\mathsf{g})$ of its Lie algebra g.*

In analogy with centralizers one can define normalizers. We shall only consider two cases. The *normalizer* (*in* G) of a subgroup A of a linear group G is:

$$N_G(A) = \{c \in G \mid cAc^{-1} = A\}.$$

The *normalizer* (*in* g) of a subalgebra a of g is:

$$\mathsf{n_g(a)} = \{Y \in \mathsf{g} \mid [Y, \mathsf{a}] \subset \mathsf{a}\}.$$

Proposition 10. *Let G be a linear group, g its Lie algebra, A a connected subgroup of G, and a its Lie algebra. Then $N_G(A) = N_G(\mathsf{a})$ and has Lie algebra $n_{\mathsf{g}}(A) = n_{\mathsf{g}}(\mathsf{a})$.* (QED)

Again we omit the proof.

Problems for §2.5

1. (a) Prove Proposition 8, (b) prove Proposition 10.

2. Let $G \subset \mathrm{GL}(E)$ be a connected linear group, F a subspace of E. Show: F is G-stable [i.e. $aF \subset F$ for all $a \in G$] if and only if F is g-stable [i.e. $XF \subset F$ for all $X \in \mathsf{g}$].

3. *Lie algebra cocycles.* Let g be a real Lie algebra. A *cocycle* on g is a bilinear function ω: $\mathsf{g} \times \mathsf{g} \to \mathbb{R}$ that satisfies

$$\omega(X, Y) = -\omega(Y, X),$$
$$\omega([X, Y], Z) + \omega([Y, Z], X) + \omega([Z, X], Y) = 0.$$

A *coboundary* is a cocycle of the form $\omega(X, Y) = \varphi([X, Y])$ where $\varphi \in \mathsf{g}^*$ is a linear functional on g.

Let G be a linear group, g its Lie algebra, ω a cocycle on g. For any $a \in G$ define another cocycle ω^a on g by the formula

$$\omega^a(X, Y) = \omega(\mathrm{Ad}(a)X, \mathrm{Ad}(a)Y).$$

Show that for a *connected* linear group G, ω^a differs from ω by a coboundary for any $a \in G$. [Suggestion: consider a C^1 path $a = a(\tau)$ and differentiate $\omega^a(X, Y)$ with respect to τ; write $a'(\tau)$ as $a'(\tau) = a(\tau)Z(\tau)$ with $Z(\tau) \in \mathsf{g}$.]

4. If a linear group of dimension ≥ 2 has a dense one-parameter subgroup, then it is isomorphic with a torus $\mathbb{T}^n$.

Complex linear groups. A linear group $G \subset \mathrm{GL}(n, \mathbb{C})$ is complex if the Lie algebra g of G is a complex subspace of $\mathsf{gl}(n, \mathbb{C})$.

5. List *all* complex groups among the groups mentioned in §2.1. [Make sure your list is complete.]

6. The only connected complex abelian group which is a compact subset of the matrix space is the trivial group.

Comment. This is one of the few places where it is essential that one deals with matrix groups. For example, $\mathbb{C}/(\mathbb{Z}+i\mathbb{Z})$ is compact.

Semidirect products. Let G be a group, M and N subgroups of G with N normal. We say G is the semidirect product of M and N if every $a \in G$ can be uniquely written in the form $a = mn$ with $m \in M$ and $n \in N$. (Equivalently: $G = MN$, $M \cap N = \{1\}$.) We then write $G = MN$ (semidirect), or $G = M \ltimes N$. One could also write $G = NM$ (semidirect) or $G = N \rtimes M$). If not clear from the context, it must be specified which of the two groups M, N is normal. If they are both normal, the product is direct. When G is a linear group we also require that the map $G = MN \to M \times N$, $mn \to (m, n)$ be analytic. This requirement is superfluous if M and N have countably many connected components, see problem 9.

7. Suppose the linear group G is a semidirect product $G = MN$.

(a) Show that its Lie algebra is the direct sum $\mathsf{g} = \mathsf{m} \oplus \mathsf{n}$ of the Lie algebras m and n, m being a subalgebra, n an ideal of g. [One says that the Lie algebra g is the *semidirect product* of m and n.]

(b) Show that the exponential map $\exp\colon \mathsf{g} = \mathsf{m} + \mathsf{n} \to G = MN$ of G takes the form

$$\exp(X + Y) = \exp X \exp A(X)Y,$$

where $A(X)$ is a (generally non-linear) transformation of n depending on $X \in \mathsf{m}$. [It is understood that $X \in \mathsf{m}$, $Y \in \mathsf{n}$, both sufficiently close to 0.]

(c) Show that when N is abelian, $A(X) : \mathsf{n} \to \mathsf{n}$ is given by

$$A(X) = \frac{\exp(-\operatorname{ad} X) - 1}{\operatorname{ad} X}.$$

(d) Show that the group of affine transformations $x \to ax + b$, $a \in \mathrm{GL}(E)$, $b \in E$, is the semidirect product of the subgroup $\mathrm{GL}(E)$ of linear transformations $x \to ax$ and the subgroup E of translations $x \to x + b$. [See Example 6, §2.1.]

8. Let G be the group of block-triangular matrices of the form

$$\begin{bmatrix} a_1 & * & * & \cdots & * \\ 0 & a_2 & * & \cdots & * \\ & & \ddots & & \\ \vdots & & & & \vdots \\ 0 & 0 & 0 & 0 & a_m \end{bmatrix}.$$

Each a_k is an invertible block of some fixed size. Let M be the subgroup of block-diagonal matrices

$$\begin{bmatrix} a_1 & 0 & 0 & \cdots & 0 \\ 0 & a_2 & 0 & \cdots & 0 \\ & & \ddots & & \\ \vdots & & & & \vdots \\ 0 & 0 & 0 & 0 & a_m \end{bmatrix},$$

N the subgroup of unipotent block-triangular matrices

$$\begin{bmatrix} 1 & * & * & \cdots & * \\ 0 & 1 & * & \cdots & * \\ \vdots & & \ddots & & \vdots \\ \vdots & & & \ddots & \vdots \\ 0 & 0 & 0 & 0 & 1 \end{bmatrix}.$$

Show that G is the semidirect product $G = MN$.

9. Show that the analyticity requirement in the definition of 'semidirect product' is superfluous if M and N have countably many connected components. [Suggestion: use Lemma 2 to show that $\mathsf{g} = \mathsf{m} + \mathsf{n}$.]

10. Let G be a linear group, A, B two subgroups of G, and $\mathsf{g}, \mathsf{a}, \mathsf{b}$ their algebras. Show:

    (a) Assume $\mathsf{g} = \mathsf{a} + \mathsf{b}$, G is connected, and AB is closed in G. Then $G = AB$.

    (b) Give an example with $\mathsf{g} = \mathsf{a} + \mathsf{b}$, G connected, but $G \neq AB$ (even with A and B closed).

    (c) Assume $G = AB$ and A, B have countably many components. Then $\mathsf{g} = \mathsf{a} + \mathsf{b}$. [Suggestion for (b). Take $G = \mathrm{SL}(2, R)$, $A = \{$upper triangular$\}$, and find a suitable B. Suggestion for (c): write $\mathsf{g} = \mathsf{a}' \oplus \mathsf{b} \oplus \mathsf{c}$ with $\mathsf{a}' \subset \mathsf{a}$ and consider $\mathsf{a}' \times \mathsf{b} \times \mathsf{c} \to \exp \mathsf{a}' \exp \mathsf{b} \exp \mathsf{c} \subset G$. Use Baire's Lemma to show that $\exp \mathsf{a}'_\epsilon \exp \mathsf{b}_\epsilon$ contains an open set for any ϵ–balls in a', b. Argue that $\mathsf{c} = 0$.]

11. Let G be a connected linear group with the property that $G^* = G$. (a^* is the adjoint of a with respect to a positive definite form.) Fix a self-adjoint element $Z \in \mathsf{g}$: $Z^* = Z$. For $\lambda \in \mathbb{R}$, let $\mathsf{g}_\lambda = \{X \in \mathsf{g} \mid \mathrm{ad}(Z)X = \lambda X\}$. Show:

    (a) $\mathsf{g} = \sum_\lambda \mathsf{g}_\lambda$ (direct sum).

    (b) $[\mathsf{g}_\lambda, \mathsf{g}_\mu] \subset \mathsf{g}_{\lambda+\mu}$.

    (c) $(\mathsf{g}_\lambda)^* = \mathsf{g}_{-\lambda}$.

(d) Let $\mathsf{k} = \{X \in \mathsf{g} \mid X^* = -X\}$, $\mathsf{q} = \sum_{\lambda \geq 0} \mathsf{g}_\lambda$. Then k and q are subalgebras of g and $\mathsf{g} = \mathsf{k} + \mathsf{q}$.

(e) Let $K = \{k \in G \mid k^* k = 1\}$, $Q = N_G(\mathsf{q})$. Then $L(K) = \mathsf{k}$, $L(Q) = \mathsf{q}$, and $G = KQ$. [Suggestion. For (c) and (d) use $(\mathrm{ad}\, Z)^* = \mathrm{ad}(Z^*)$. For $G = KQ$ in (e) use the previous problem. To show that KQ is closed, use the compactness of K to show that the limit of a convergent sequence $k_j q_j$ belongs to KQ.]

12. In problem 11 take $G = \mathrm{SL}(2, \mathbb{R})$,

$$Z = \begin{bmatrix} 1 & 0 \\ 0 & -1 \end{bmatrix}.$$

Find K and Q.

13. In problem 11, take $G = \mathrm{GL}(n, \mathbb{R})_o$ and $Z = (1, 0, \ldots, 0)$ (diagonal matrix). Find K and Q.

14. In problem 11, take $G = \mathrm{GL}(n, \mathbb{R})_o$ and $Z = (\lambda_1, \lambda_2, \ldots, \lambda_n)$ with $\lambda_1 > \lambda_2 > \cdots > \lambda_n$ (diagonal matrix). Find K and Q. [Compare your answer with exercise 9, §2.1. Suggestion: to find the g_λ, review the proof of Lemma 8, §1.2.]

15. Let J be the $n \times n$ Jordan block

$$\begin{bmatrix} 0 & 1 & 0 & \cdots & 0 \\ 0 & 0 & 1 & \cdots & 0 \\ \vdots & & \ddots & & \vdots \\ \vdots & & & \ddots & \vdots \\ 0 & 0 & 0 & 0 & 0 \end{bmatrix}.$$

(a) Show that the centralizer of J in $\mathsf{gl}(n, \mathbb{R})$ or $\mathsf{gl}(n, \mathbb{C})$ consists of all matrices of the form

$$\begin{bmatrix} a_1 & a_2 & \cdots & a_{n-1} & a_n \\ 0 & a_1 & a_2 & \cdots & a_{n-1} \\ \vdots & & \ddots & & \vdots \\ \vdots & & & \ddots & \vdots \\ 0 & 0 & 0 & 0 & a_1 \end{bmatrix}.$$

(b) Describe the one-parameter group generated by J and its centralizer in $\mathrm{GL}(n, \mathbb{R})$ or $\mathrm{GL}(n, \mathbb{C})$.

2.6 Homomorphisms and coverings of linear groups

The Lie correspondence 'groups $\leftrightarrow$ Lie algebras' works not only on the level of sets, but also on the level of maps. The main point is:

Theorem 1. *Let $f : G \to H$ be a differentiable homomorphism of linear groups, $\varphi : \mathsf{g} \to \mathsf{h}$ its differential at 1. Then φ is a homomorphism of Lie algebras and*

$$f(\exp X) = \exp \varphi(X)$$

for all $X \in \mathsf{g}$.

Proof. The proof uses the familiar 'differentiation-exponentiation' method. We first prove that $f(\exp X) = \exp \varphi(X)$. We have:

$$\varphi(X) = \frac{d}{d\tau} f(\exp \tau X)\bigg|_{\tau=0}$$

and generally

$$\begin{aligned} \frac{d}{d\sigma} f(\exp \sigma X) &= \frac{d}{d\tau} f(\exp(\sigma + \tau)X)\bigg|_{\tau=0} \\ &= \frac{d}{d\tau} f(\exp \sigma X \exp \tau X\bigg|_{\tau=0} \\ &= f(\exp \sigma X) \frac{d}{d\tau} f(\exp \tau X)\bigg|_{\tau=0} \\ &= f(\exp \sigma X)\varphi(X). \end{aligned}$$

This shows that $a(\sigma) = f(\exp \sigma X)$ satisfies the differential equation $a'(\sigma) = a(\sigma)\varphi(X)$ and of course also the initial condition $a(0) = 1$. Therefore $a(\sigma) = \exp \sigma\varphi(X)$ and in particular $f(\exp X) = \exp \varphi(X)$ as required.

To show that $\varphi : \mathsf{g} \to h$ is a Lie algebra homomorphism we must show that

$$\varphi[X, Y] = [\varphi X, \varphi Y]$$

for all $X, Y \in \mathsf{g}$. To prove this, start with the equation

$$f(\exp \tau X \exp \sigma Y \exp -\tau X) = \exp(\tau\varphi X) \exp(\sigma\varphi Y) \exp(-\tau\varphi X),$$

valid for all $\sigma, \tau \in \mathbb{R}$, by what has just been proved. Differentiate with respect to σ at $\sigma = 0$:

$$\varphi(\exp(\tau X) Y \exp(-\tau X)) = \exp(\tau\varphi X)\varphi(Y) \exp(-\tau\varphi X).$$

Now differentiate with respect to τ at $\tau = 0$:

$$\varphi[X, Y] = [\varphi X, \varphi Y].$$

QED

Terminology and notation. A *homomorphism* of linear groups $f : G \to H$ is required to be differentiable, by definition. Its differential at 1, $\varphi : \mathsf{g} \to \mathsf{h}$ is called its *Lie map*, denoted $L(f)$ or Lf.

Remark 2. One sees from the theorem that a differentiable homomorphism between matrix groups is in fact analytic. It can be shown that the same is true even for a continuous homomorphism. While the proof is not difficult, we shall not give it here, as its only purpose would be a justification for restricting oneself to differentiable homomorphisms in the first place. The proof may be found in §11 of Freudenthal and de Vries (1969), for example.

Example 3 (Aut, Der, Ad, **and** ad)**.** Let G be a linear group. For $a \in G$, define the *conjugation by* a to be the transformation $c(a) : G \to G$, $b \to aba^{-1}$. $c(a)$ is an *automorphism* of G, and is also called the *inner automorphism* determined by a. Since

$$\frac{d}{d\tau}\left(a\exp(\tau X)a^{-1}\right)\bigg|_{\tau=0} = aXa^{-1},$$

the Lie map of $c(a) : G \to G$, $b \to aba^{-1}$ is $\operatorname{Ad} a : \mathsf{g} \to \mathsf{g}, X \to aXa^{-1}$. The theorem says that $\operatorname{Ad} a$ is a Lie algebra homomorphism and that

$$a\exp(X)a^{-1} = \exp(aXa^{-1}),$$

which is clear.

Note that $\operatorname{Ad} a$ is invertible (with inverse $\operatorname{Ad}(a^{-1})$) and

$$\operatorname{Ad}(ab) = \operatorname{Ad}(a)\operatorname{Ad}(b),$$

so that $a \to \operatorname{Ad} a$ is a homomorphism from G to the group $\operatorname{Aut}(\mathsf{g})$ of automorphisms of the Lie algebra g:

$$\operatorname{Ad} : G \to \operatorname{Aut}(\mathsf{g}).$$

The Lie algebra of $\operatorname{Aut}(\mathsf{g})$ is $\operatorname{Der}(\mathsf{g}) = \{\text{derivations of } \mathsf{g}\}$ (Proposition 9, §2.2). The Lie map of $\operatorname{Ad} : G \to \operatorname{Aut}(\mathsf{g})$ is therefore a homomorphism $\mathsf{g} \to \operatorname{Der}(\mathsf{g})$. We have

$$\begin{aligned}
\frac{d}{d\tau}\operatorname{Ad}(\exp\tau X)Y\bigg|_{\tau=0} &= \frac{d}{d\tau}(\exp\tau X)Y(\exp -\tau X)\bigg|_{\tau=0} \\
&= XY - YX \\
&= [X, Y] \\
&= \operatorname{ad}(X)Y.
\end{aligned}$$

Thus the Lie map of $\operatorname{Ad} : G \to \operatorname{Aut}(\mathsf{g})$ is $\operatorname{ad} : \mathsf{g} \to \operatorname{Der}(\mathsf{g})$. The theorem says that $\operatorname{ad} : \mathsf{g} \to \operatorname{Der}(\mathsf{g})$ is a homomorphism, i.e.

$$\operatorname{ad}[X, Y] = [\operatorname{ad} X, \operatorname{ad} Y],$$

which is a form of the Jacobi identity, and that

$$\operatorname{Ad}(\exp X) = \exp\operatorname{ad}(X),$$

which we know.

Example 4. Recall that $\mathbb{T}^n = \{(\epsilon_1, \ldots, \epsilon_n) \mid \epsilon_k = e^{2\pi i\theta_k}, \theta_k \in \mathbb{R}\}$ (Example 6 of §2.1). Let

$$f : \mathbb{T}^n \to \mathbb{C}^\times$$

be a homomorphism of matrix groups. The Lie map of f is an $\mathbb{R}$-linear functional

$$\varphi : \mathsf{t}^n = \{(2\pi i\theta_1, \ldots, 2\pi i\theta_n) \mid \theta_k \in \mathbb{R}\} \to \mathbb{C},$$

Hence of the form

$$\varphi(X) = 2\pi i(l_1\theta_1 + \cdots + l_n\theta_n)$$

for certain $l_k \in \mathbb{R}$. The equation

$$f(\exp X) = e^{\varphi(X)}$$

implies that $\varphi(X) \in 2\pi i\mathbb{Z}$ when $\exp X = 1$, i.e. $(l_1\theta_1 + \cdots + l_n\theta_n) \in \mathbb{Z}$ when $\theta_k \in \mathbb{Z}$, i.e. $l_k \in \mathbb{Z}$.

Conversely, if $l_k \in \mathbb{Z}$ for $k = 1, \ldots, n$, then the above equation defines a homomorphism f. These homomorphisms f are therefore in one-to-one correspondence with $\mathbb{Z}^n$.

Note that each such homomorphism $\mathbb{T} \to \mathbb{C}^\times$ extends uniquely to a *holomorphic* homomorphism $(\mathbb{C}^\times)^n \to \mathbb{C}^\times$, given by the same formula. Furthermore, every such holomorphic homomorphism is of this form, as one sees in the same way.

We return to the general situation of Theorem 1.

Corollary 5. *Let $f : G \to H$ be a homomorphism of linear groups, $L(f) : \mathsf{g} \to \mathsf{h}$ its Lie map.*

(a) $L(\ker f) = \ker L(f)$.

(b) $L(\operatorname{im} f) = \operatorname{im} L(f)$, *provided G has countably many connected components.*

Proof.

(a) $X \in L(\ker f) \Leftrightarrow f(\exp \tau X) = 1$ for all $\tau \in \mathbb{R} \Leftrightarrow \exp \tau Lf(X) = 1$ for all $\tau \in \mathbb{R} \Leftrightarrow Lf(X) = 0, \Leftrightarrow X \in \ker L(f)$.

(b) $(\supset) f(\exp \tau X) = \exp \tau Lf(X)$ implies $L(\operatorname{im} f) \supset \operatorname{im} L(f)$.

$(\subset)$ It suffices to show that $\Gamma(\operatorname{im} Lf)$ contains an open subset of $\operatorname{im} f$. This follows from Baire's Covering Lemma (Lemma 2, §2.5): write $G_o = \cup_{j=1}^{\infty} A_j$ as in eqn (1) of §2.5. Each A_j is a coordinate-ball in G_o. Since G has countably many components, we can choose a countable family $\{a_k\}$ of elements of G, one from each component. Replacing the family $\{A_j\}$ by $\{a_k A_j\}$ we may assume $G = \cup_{j=1}^{\infty} A_j$. Then $f(G) = \cup_{j=1}^{\infty} f(\bar{A}_j)$. Each $f(\bar{A}_j)$ is closed ($\bar{A}_j$ is compact), hence some $f(\bar{A}_j)$ contains an open subset of $f(G)$. QED

Corollary 6. *Let $f : G \to H$ be a homomorphism of linear groups.*

(a) f is locally injective if and only if Lf is injective.

(b) Assume H is connected and G has countably many connected components. Then f is surjective if and only if Lf is surjective.

Explanation. $f : G \to H$ is *locally injective* if every $a \in G$ has a neighborhood on which f is one-to-one. This is the case if and only if f is one-to-one on some neighborhood of $a = 1$, which happens if and only if $L(\ker f) = \{0\}$. For example, the homomorphism $\theta \to e^{2\pi i\theta}$, from $\mathbb{C}, +$ to $\mathbb{C}^\times, \times$ (or from $\mathbb{R}$ to $\mathbb{T}$) is locally injective, but certainly not globally injective.

$f : G \to H$ is *locally surjective* if any $a \in G$ has a neighborhood which gets mapped onto a neighborhood of $f(a)$. The statement (a) holds with 'locally injective' replaced by 'locally surjective', but the global statement (b) is usually more interesting. In any case, a locally surjective homomorphism into a connected group is surjective, because a connected group is generated by any open subset. This explains the asymmetry in the statements (a) and (b).

Proof.
(a) Since exp is one-to-one on a neighborhood of 0, the equation

$$f(\exp X) = \exp Lf(X)$$

shows that f is one-to-one in a neighborhood of 1 in $G \Leftrightarrow Lf$ is one-to-one in a neighborhood of 0 in g, which happens $\Leftrightarrow Lf$ is injective.

(b) $\operatorname{im} f = H \Leftrightarrow L(\operatorname{im} f) = \mathsf{h}$ [H is connected] $\Leftrightarrow \operatorname{im}(Lf) = \mathsf{h}$ [preceding corollary]. QED

Example 7 (Adjoint representation). Let G be a connected linear group. For fixed $a \in G$ we have the inner automorphism $c(a) : G \to G, b \to aba^{-1}$. Its Lie map is $\mathrm{Ad}(a) : \mathsf{g} \to \mathsf{g}$. The kernel of $c(a)$ is $Z_G(a) = \{b \in G \mid ab = ba\}$, the centralizer of a in G. By part (a) of Corollary 5, the Lie algebra of $Z_G(a)$ is $\mathsf{z}_{\mathsf{g}}(a) = \{X \in \mathsf{g} \mid \mathrm{Ad}(a)X = X\}$.

Now consider the homomorphism $\mathrm{Ad} : G \to \mathrm{Aut}(\mathsf{g})$. Its Lie map is $\mathrm{ad} : \mathsf{g} \to \mathrm{Der}(\mathsf{g})$. By part (b) of Corollary 5, the Lie algebra of $\mathrm{Ad}(G)$ is $\mathrm{ad}(\mathsf{g})$, the Lie algebra of *inner derivations* of g. The kernel of $\mathrm{Ad} : G \to \mathrm{Aut}(\mathsf{g})$ is $Z(G) = \{c \in G | ca = ac \text{ for all } a \in G\}$, the center of G. By part (a) of the same corollary, the Lie algebra of $Z(G)$ is the kernel of $\mathrm{ad} : \mathsf{g} \to \mathrm{Der}(\mathsf{g})$, which is $\mathsf{z}(\mathsf{g}) = \{X \in \mathsf{g} | [X, Y] = 0 \text{ for all } Y \in \mathsf{g}\}$, the *center* of g.

A homomorphism $f : G \to H$ is *locally bijective* if every $a \in G$ has a neighborhood that gets mapped bijectively onto a neighborhood of $f(a)$.

Lemma 8. *The following conditions on a homomorphism of linear groups, $f : G \to H$, are equivalent.*

(a) f is locally bijective.

(b) $Lf : L(G) \to L(H)$ is bijective.

For such a homomorphism, $\ker f$ *is a discrete subgroup of G.*

Proof. (a) $\Rightarrow$ (b). If f is locally bijective, then Lf must be bijective, because $Lf(X) = 0$ implies $f(\exp X) = 1$.

(b) $\Rightarrow$ (a). If Lf is bijective, then f is locally bijective at 1 (Inverse Function Theorem), hence locally bijective at any point (because $f(a_o a) = f(a_o)f(a)$).

For such an f there is a neighborhood of 1 in G where $f(a) = 1$ implies $a = 1$, which says that $\ker f$ is discrete. QED

A locally bijective homomorphism $p : \tilde{G} \to G$ between connected linear groups called a *covering*. A covering is necessarily surjective with discrete kernel, and these properties characterize coverings. $\tilde{G}$ together with p is the called a *covering group* of G; the covering homomorphism p need not be mentioned explicitly.

The kernel of a covering $\tilde{G} \to G$ is necessarily contained in the center of $\tilde{G}$, because generally *a discrete normal subgroup Z of a connected group H is contained in the center of H*: by assumption, any $a \in H$ may be connected to 1 by a continuous path $a(\tau)$; for $z \in Z$, the continuous function $\tau \to a(\tau)za(\tau)^{-1}$ on $[0,1]$ takes on values in the discrete set Z, hence must be constant. Thus $a(\tau)za(\tau)^{-1} \equiv z$ and in particular $aza^{-1} = z$. As a consequence, the groups G covered by a given group $\tilde{G}$ are isomorphic with $\tilde{G}/Z$ (as abstract groups), where Z is a discrete central subgroup of $\tilde{G}$.

We now return to the situation of Theorem 1. The equation

$$f(\exp X) = \exp \varphi(X)$$

may be interpreted as saying that in exponential coordinates a group homomorphism $f : G \to H$ becomes a Lie algebra homomorphism $\varphi : \mathsf{g} \to \mathsf{h}$. Suppose, conversely, we start with a homomorphism $\varphi : \mathsf{g} \to \mathsf{h}$ between linear Lie algebras. Then the equation $f(\exp X) = \exp \varphi(X)$ defines a map from a neighborhood of 1 in the connected linear group $G = \Gamma(\mathsf{g})$ to $H = \Gamma(\mathsf{h})$. Furthermore, it is clear from the Campbell–Baker–Hausdorff Formula that

$$f(ab) = f(a)f(b).$$

for a, b in some neighborhood of 1 in G. Such a map f is called a *local homomorphism* from G to H. There remains the question if such a local homomorphism $G \cdots \to H$ extends to a global homomorphism $G \to H$, defined on all of G. In general the answer is *no*. For example, $a \to \log a$ defines a local homomorphism of $\mathbb{C}^\times, \times$ to $\mathbb{C}, +$ (or from $\mathbb{T}$ to $\mathbb{R}$) that does not extend to a global homomorphism.

It is here that coverings come in. We introduce the following terminology. Let G and H be connected linear groups, $\varphi : \mathsf{g} \to \mathsf{h}$ a homomorphism between their Lie algebras. Let $p : \tilde{G} \to G$ be a covering of G. We say $\varphi : \mathsf{g} \to \mathsf{h}$ *lifts* to $f : \tilde{G} \to H$ (or simply *lifts* to $\tilde{G}$) if there is a homomorphism $f : \tilde{G} \to H$ so that $L(f) = \varphi \circ L(p) : \tilde{\mathsf{g}} \to \mathsf{h}$. This means the Lie map of f is the given φ when one identifies $\tilde{\mathsf{g}}$ with g by the isomorphism $L(p) : \tilde{\mathsf{g}} \to \mathsf{g}$. If such an f exists it is unique since a homomorphism between connected linear groups is determined by its Lie map. If the local homomorphism $G \cdots \to H$ corresponding to $\varphi : \mathsf{g} \to \mathsf{h}$ does extend to a global homomorphism $f : G \to H$, then we may take $\tilde{G} = G$ and $\varphi : \mathsf{g} \to \mathsf{h}$ lifts to $f : G \to H$. This is not always the case, as we observed, but φ always lifts to *some* covering group $\tilde{G}$ of G. This is an important fact, which we list as a theorem, even though it is very easy to prove:

Theorem 9. *Let G and H be linear groups, G connected, $\varphi : \mathsf{g} \to \mathsf{h}$ a homomorphism between their Lie algebras. Then there is a linear covering group $\tilde{G}$ of G so that φ lifts to $f : \tilde{G} \to H$.*

Proof. Given G, H, and $\varphi : \mathsf{g} \to \mathsf{h}$, let $\tilde{\mathsf{g}} = \{(X, Y) \in \mathsf{g} \times \mathsf{h} \mid Y = \varphi(X)\}$, the graph of φ. $\tilde{\mathsf{g}}$ is a linear Lie algebra. Let $\tilde{G}$ be the corresponding connected linear group. It is a subgroup of $G \times H$; let $p : \tilde{G} \to G$ be the restriction to $\tilde{G}$ of the projection $G \times H \to G$, and $f : \tilde{G} \to H$ the restriction of the projection $G \times H \to H$. The Lie map $L(p)$ of p is the map $\tilde{\mathsf{g}} \to \mathsf{g}, (X, \varphi(X)) \to X$, which is an isomorphism. Thus $p : \tilde{G} \to G$ is a covering. The Lie map of f is the map $\tilde{\mathsf{g}} \to \mathsf{h}, (X, \varphi(X)) \to \varphi(X)$, from which one sees that $f : \tilde{G} \to H$ lifts $\varphi : \mathsf{g} \to \mathsf{h}$. QED

One will ask whether every connected group admits a *universal covering group*, meaning a single covering group $\tilde{G}$ of G so that every Lie algebra homomorphism $\varphi : \mathsf{g} \to \mathsf{h}$ lifts to this group. If such a universal covering group exists, it is unique up to an isomorphism compatible with the covering map: if $\tilde{G}$ and $\tilde{G}'$ are two universal covering groups, then the Lie algebra homomorphisms $\mathsf{g} \to \tilde{\mathsf{g}}'$ and $\mathsf{g} \to \tilde{\mathsf{g}}$ lift to homomorphisms $\tilde{G} \to \tilde{G}'$ and $\tilde{G}' \to \tilde{G}$, which are inverses of each other because the Lie map of their composite is the identity $\mathsf{g} \to \mathsf{g}$. One may therefore speak of *the* universal covering group. Universal covering groups always exist when one admits general Lie groups (to be defined in §4.3) for the covering groups, but not when the covering groups are themselves required to be linear groups. For example, it can be shown that $\mathrm{SL}(2, \mathbb{R})$ does *not* admit a *linear* group as universal covering group (Example 7, §4.3).

Within the category of general Lie groups, universal covering groups are characterized by the topological property of being *simply connected*, meaning any continuous, closed path can be continuously deformed into a point. The precise definition is not needed here, but we take for granted the following fact from the theory of covering spaces (see Massey (1967), for example): *A simply connected topological space admits no non-trivial coverings*, i.e. the only coverings of such a space are homeomorphisms. (The space is here assumed locally arcwise connected, i.e. every neighborhood of a point contains an arcwise connected neighborhood, as is evidently the case for linear groups). In particular, the only coverings of simply connected linear groups are isomorphisms. (Coverings of linear groups, as defined above are coverings in the sense of algebraic topology.) Theorem 9 therefore implies:

Theorem 10. *Let G and H be linear groups, G simply connected, $\varphi : \mathsf{g} \to \mathsf{h}$ a homomorphism between their Lie algebras. Then φ lifts to a homomorphism $f : G \to H$.*

Example 11 (SU(2) **is simply connected).** SU(2) consists of the matrices

$$\begin{bmatrix} \alpha & -\bar{\beta} \\ \beta & \bar{\alpha} \end{bmatrix}$$

with $\alpha\bar{\alpha} + \beta\bar{\beta} = 1$ and is therefore homeomorphic with the sphere S^3 in $\mathbb{C}^2 = \mathbb{R}^4$, hence simply connected.

Example 12 (The covering $\mathrm{SU}(2) \to \mathrm{SO}(3)$**).** Recall the homomorphism $p : \mathrm{SU}(2) \to \mathrm{SO}(3)$ of Example 2, §2.1, which is just the adjoint representation when $\mathsf{su}(2)$ is identified with $\mathbb{R}^3$ by a suitable basis. We have seen that its image is SO(3) and its kernel $\{\pm 1\}$, hence a covering of SO(3). We shall prove this once more using Lie theory to illustrate the concepts introduced above.

Proposition 12A. *The homomorphism* $p : \mathrm{SU}(2) \to \mathrm{SO}(3)$ *is a covering with kernel* $\{\pm 1\}$.

Proof. SU(2) is connected, as we know. Thus the image of SU(2) by p is connected as well, hence contained in the connected component SO(3) of O(3). Let $\pi : \mathsf{su}(2) \to \mathsf{so}(3)$ be the Lie map of p. Since $p = \mathrm{Ad}$, $\pi = \mathrm{ad}$ and therefore the kernel of π consists of the $X \in \mathsf{su}(2)$ satisfying $XY - YX = 0$ for all skew Hermitian $Y \in M_2(\mathbb{C})$ of trace $= 0$, hence for all skew Hermitian $Y \in M_2(\mathbb{C})$, hence for all $Y \in M_2(\mathbb{C})$ (as one sees by writing $Y = iY' + Y''$ with Y' and Y'' skew Hermitian). Thus X is a scalar and therefore equals 0, because $\mathrm{tr}\, X = 0$. Since the kernel of π is $\{0\}$, π is an isomorphism by dimension-count. It follows that p is a covering. Its kernel consists of all $a \in \mathrm{SU}(2)$ satisfying $aYa^{-1} = Y$ for all skew Hermitian Y, hence are scalar, hence equal to ± 1. QED

Example 13. SO(3) is not simply connected, but admits a simply connected double covering. There are two ways of seeing this: (a) $\mathrm{SU}(2) \to \mathrm{SO}(3)$ is the required double covering.

(b) SO(3) is homeomorphic with the closed unit ball in $\mathbb{R}^3$, antipodal points on its surface being identified (Example 1, §2.1). This is a model of real projective 3-space, well-known to have a simply connected double covering.

It follows from general facts of homotopy theory that there must be a continuous loop C_o on SO(3) that cannot be continuously deformed into a point, while $2C_o$ (C_o traversed twice) *can* so be deformed; and any loop may be deformed either into C_o or into the constant loop. The one-parameter group of rotations about a fixed axis, traversed once, provides a model for such a path C_o. That $2C_o$ may be deformed into the constant loop may be seen thus, according to Weyl (1939):

Take two solid straight circular cones of aperture α with common vertex and touching each other along a generator, the one fixed in space, the other rolling on the first. The roller describes a closed motion which is $2C_o$ for $\alpha = 90°$ and approaches rest as $\alpha \to 180°$. By continuous variation of the parameter α one thus deforms $2C_o$ into the point 1.

(The initial motion $2C_o$ in this picture is not exactly a double rotation about a fixed axis but homotopic thereto ; see problem 11, §2.1.)

Example 14. $\mathrm{SL}(2,\mathbb{C})$: the covering $\mathrm{SL}(2,\mathbb{C}) \to \mathrm{SO}_o(3,1)$. $\mathrm{SL}(2,\mathbb{C}) = \{a \in M_2(\mathbb{C}) |\, \det a = 1\}$ operates on the real vector space of Hermitian 2×2 matrices $\{X \in M_2(\mathbb{C}) | X^* = X\}$ by $X \to aXa^*$. Writing

$$X = \begin{bmatrix} \xi_4 + \xi_3 & \xi_1 - i\xi_2 \\ \xi_1 + i\xi_2 & \xi_4 - \xi_3 \end{bmatrix},$$

and $\vec{X} = (\xi_1, \xi_2, \xi_3, \xi_4)$, one obtains an operation of $a \in \mathrm{SL}(2,\mathbb{C})$ on $\mathbb{R}^4$ denoted $p(a)$:

$$p(a)\vec{X} = \overrightarrow{aXa^*}.$$

Since

$$-\det X = \xi_1^2 + \xi_2^2 + \xi_3^2 - \xi_4^2 = \langle \vec{X}, \vec{X} \rangle,$$

and $\det aXa^* = \det X$ for $a \in \mathrm{SL}(2, \mathbb{C})$, the linear transformation $p(a)$ of $\mathbb{R}^4$ belongs to $\mathrm{O}(3,1)$. The map $p : \mathrm{SL}(2, \mathbb{C}) \to \mathrm{O}(3,1)$ is evidently a group homomorphism. $\mathrm{SL}(2, \mathbb{C})$ is connected, as we know from Example 4, §2.1. Thus the image of $\mathrm{SL}(2, \mathbb{C})$ by p is connected as well, hence contained in the connected component $\mathrm{SO}_o(3,1)$ of $\mathrm{O}(3,1)$. In fact:

Proposition 14A. *The homomorphism* $p : \mathrm{SL}(2, \mathbb{C}) \to \mathrm{SO}_o(3,1)$ *is a covering with kernel* $\{\pm 1\}$.

Proof. Let $\pi : \mathsf{sl}(2, \mathbb{C}) \to \mathrm{SO}(3,1)$ be the Lie map of p. Its kernel consists of the $X \in \mathsf{sl}(2, \mathbb{C})$ satisfying $XY + YX^* = 0$ for all Hermitian $Y \in M_2(\mathbb{C})$, hence for all $Y \in M_2(\mathbb{C})$. $Y = 1$ gives $X^* = -X$, and then $XY = YX$ for all $Y \in M_2(\mathbb{C})$. Thus X is a scalar and therefore equals 0, because $\operatorname{tr} X = 0$. Since the kernel of π is $\{0\}$, π is an isomorphism by dimension-count. It follows that p is a covering. Its kernel consists of all $a \in \mathrm{SL}(2, \mathbb{C})$ satisfying $aYa^* = Y$ for all Hermitian Y, hence are scalar, hence equal to ± 1. QED

We close with an example showing that the countability assumption in Corollary 5(b) cannot be omitted. The example is necessarily pathological.

Example 15. Choose a basis for $\mathbb{R}$ over $\mathbb{Q}$ ('Hamel basis') to identify $\mathbb{R}$ with the set $\mathbb{Q}^{(A)}$ of finite sequences (α_a) indexed by elements of some (uncountable) index set A. For any map $B \to A$ we get an induced $\mathbb{Q}$-linear map $\mathbb{Q}^{(B)} \to \mathbb{Q}^{(A)}$. Let B be a subset of A of the same cardinality. Choose a surjection $B \to A$ to obtain a $\mathbb{Q}$-linear surjection $f : \mathbb{Q}^{(B)} \to \mathbb{Q}^{(A)} = \mathbb{R}$. Consider the additive group $\mathbb{R}$ as a linear group (see Example 6, §2.1) and consider $\mathbb{Q}^{(B)}$ as a subgroup of $\mathbb{R}$ by the inclusion of $\mathbb{Q}^{(B)}$ into $\mathbb{R} = \mathbb{Q}^{(A)}$ induced by the inclusion of B into A. Assume $\mathbb{Q}^{(B)}$ does not contain any interval about 0 in $\mathbb{R}$ (as may be arranged by an appropriate choice of the subset B of A). Then $L(\mathbb{Q}^{(B)}) = \{0\}$ (so $L(\mathbb{Q}^{(B)})$ carries the discrete topology as matrix group). Hence certainly $L(f) = 0$. Thus $\operatorname{im} L(f) = \{0\}$; but $L(\operatorname{im} f) = \mathbb{R}$.

Problems for §2.6

1. Let $(\mathbb{C}^\times)^n = \{t = (\tau_1, \ldots, \tau_n) \mid 0 \neq \tau_k \in \mathbb{C}\}$. Show that every *holomorphic* homomorphism $f : (\mathbb{C}^\times)^n \to \mathbb{C}^\times$ is of the form

$$f(t) = (\operatorname{sgn} \tau_1)^{\delta_1} \cdots (\operatorname{sgn} \tau_n)^{\delta_n} |\tau_1|^{l_1} \cdots |\tau_n|^{l_n}$$

for certain $l_k \in \mathbb{Z}$ and certain $\delta_k \in \mathbb{Z}_2$.

2. Show that $\mathrm{Aut}(\mathbb{T}^n) \approx \mathrm{GL}(n, \mathbb{Z}) = \{a \in M_n(\mathbb{Z}) \mid a^{-1} \in M_n(\mathbb{Z})\} = \{a \in M_n(\mathbb{Z}) \mid \det a = \pm 1\}$.

[Suggestion: consider Lie maps.]

3. Let G be a connected linear group. Show that any finite number of homomorphisms $\varphi_k : \mathsf{g} \to \mathsf{h}_k$ of the Lie algebra g of G to Lie algebras h_k of linear groups H_k lift simultaneously to homomorphisms $f_k : \tilde{G} \to H_k$ of a single linear covering group $\tilde{G}$ of G.

4. Give direct proofs of the surjectivity of the maps $\mathrm{SU}(2) \to \mathrm{SO}(3)$ and $\mathrm{SL}(2,\mathbb{C}) \to \mathrm{SO}_o(3,1)$, without appealing to the connectedness of $\mathrm{SO}(3)$ and $\mathrm{SO}_o(3,1)$. [Suggestion: use Lemma 1.B of §2.1 and Lemma 4C, §2.4.]

5. Let F be the real vector space of all polynomial functions $f(x) = f(\xi_1, \ldots, \xi_n)$ of degree $\leq d$ on $\mathbb{R}^n$ (some d some n). Define a homomorphism $T : \mathrm{GL}(n,\mathbb{R}) \to \mathrm{GL}(F)$ of $\mathrm{GL}(n,\mathbb{R})$ into the group $\mathrm{GL}(F)$ of invertible linear transformations of F by the formula

$$T(a)f(x) = f(a^{-1}x).$$

Notation: for $a \in \mathrm{GL}(n,\mathbb{R})$ and $x \in \mathbb{R}^n$,

$$ax = \sum_{ij} a_{ij}\xi_j e_i \quad \text{if } x = \sum_j \xi_j e_j.$$

(a) Show that the Lie map $\tau : \mathsf{gl}(n,\mathbb{R}) \to \mathsf{gl}(F)$ of T is given by

$$\tau(X)f(x) = \sum_{ij} -X_{ij}\xi_j \frac{\partial f}{\partial \xi_i}$$

(b) Show that a polynomial function $f(x) = f(\xi_1, \xi_2, \xi_3)$ on $\mathbb{R}^3$ is invariant under the rotation group $\mathrm{SO}(3)$, i.e.

$$f(ax) = f(x) \quad \text{for all } a \in \mathrm{SO}(3) \quad \text{and all} \quad x \in \mathbb{R}^3$$

if and only if

$$\xi_i \frac{\partial f}{\partial \xi_j} - \xi_j \frac{\partial f}{\partial \xi_i} = 0$$

for all $i \neq j (i,j = 1,2,3)$. Can you generalize this?

6. Construct a double covering $\mathrm{SL}(2,\mathbb{C}) \to \mathrm{SO}(3,\mathbb{C})$.

7. (a) Construct a double covering $\mathrm{SL}(2,\mathbb{R}) \to \mathrm{SO}_o(2,1)$.

   (b) Let $\mathrm{SU}(1,1)$ be the group of all complex 2×2 matrices of the form $\begin{bmatrix} \alpha & \beta \\ \bar{\beta} & \bar{\alpha} \end{bmatrix}$, $\alpha\bar{\alpha} - \beta\bar{\beta} = 1$. Construct a double covering $\mathrm{SU}(1,1) \to \mathrm{SO}_o(2,1)$.

   (c) Find an isomorphism $\mathrm{SL}(2,\mathbb{R}) \approx \mathrm{SU}(1,1)$ compatible with (a) and (b).

   [Suggestion: for (c), consider conjugation by suitable 2×2 matrix.]

8. (a) *The double covering* $\mathrm{SL}(2,\mathbb{C}) \times \mathrm{SL}(2,\mathbb{C}) \to \mathrm{SO}(4,\mathbb{C})$. Construct this double covering.

   (b) *The double covering* $\mathrm{SL}(2,\mathbb{C}) \to \mathrm{SO}_o(1,3)$. Construct this double covering.

[Suggestion: For (a) consider $\mathrm{SL}(2,\mathbb{C})\times\mathrm{SL}(2,\mathbb{C})$ acting on $M_2(\mathbb{C})\approx\mathbb{C}^2$ by $X\mapsto aXb^{-1}$. For (b) consider $\mathrm{SL}(2,\mathbb{C})$ acting $\{X\in M_2(\mathbb{C})\mid X^*=X\}\approx\mathbb{R}^3$ by $X\mapsto aXa^*$. These actions preserve the quadratic form $\det(X)$. Show that the Lie maps of the group homomorphisms defined by these actions are 1–1. Note that $\mathrm{SL}(2,\mathbb{C})=KB$ is connected, see Example 4 of §2.1.]

9. *The double covering* $\mathrm{SL}(4,\mathbb{C})\to\mathrm{SO}(6,\mathbb{C})$. This problem assumes familiarity with the wedge product of alternating forms. [See Hoffmann-Kunze (1961), section 5.7, for example.]

   $\mathrm{SL}(4,\mathbb{C})$ acts in a natural way on the six-dimensional space F of all skew-symmetric bilinear forms on $\mathbb{C}^4$. Denote the linear transformation of F corresponding to $a\in\mathrm{SL}(4,\mathbb{C})$ by $f(a)$. F carries a non-degenerate symmetric bilinear form (φ,ψ) defined by

   $$\varphi\wedge\psi=(\varphi,\psi)e,$$

   where e is a fixed non-zero four-form. Show that $f(a)\in\mathrm{SO}(F)\approx\mathrm{SO}(6,\mathbb{C})$ and that the map $f:\mathrm{SL}(4,\mathbb{C})\to\mathrm{SO}(6,\mathbb{C})$ is a double covering with kernel $\{\pm1\}$. [Suggestion: start by showing that the Lie map of f is $1-1$; then count dimensions. Use connectedness.]

10. *The double covering* $\mathrm{Sp}(2,\mathbb{C})\to\mathrm{SO}(5,\mathbb{C})$. Denote by σ the non-degenerate skew-form on $\mathbb{C}^4$ defining $\mathrm{Sp}(2,\mathbb{C})$. Show that the map $\mathrm{SL}(4,\mathbb{C})\to\mathrm{SO}(6,\mathbb{C})$ defined in problem 9 induces a double covering $\mathrm{Sp}(2,\mathbb{C})\to\mathrm{SO}(5,\mathbb{C})$, $\mathrm{SO}(5,\mathbb{C})$ realized as the subgroup of $\mathrm{SO}(6,\mathbb{C})$ of transformations that preserve the five-dimensional hyperplane $(\varphi,\sigma)=0$ in $F=\mathbb{C}^6$ and have $\det=1$ therein. [Prove also that the group described is $\mathrm{SO}(5,\mathbb{C})$.]

2.7 Closed subgroups

As we know, the group topology of a linear group does not necessarily coincide with its relative topology in matrix space: one should keep in mind $\mathrm{GL}(n,\mathbb{Q})$ or the irrational line on the torus. More generally, the group topology on a subgroup H of a linear group G need not coincide with its relative topology in G, i.e. the open sets of H need not be exactly the intersections of open sets of G with H. This phenomenon cannot occur if H is closed in G, as may seem plausible from the examples mentioned. The essential point is the following theorem.

Theorem 1 (Closed subgroup theorem). *Let G be a linear group, H a closed subgroup of G. Let s be a vector space complement for the Lie algebra h of H in the Lie algebra g of $G:\mathsf{g}=\mathsf{s}\oplus\mathsf{h}$. There is an open neighborhood U of 0 in s so that the map*

$$U\times H\to G,\quad (X,b)\to\exp(X)b$$

is an analytic bijection onto an open neighborhood of H in G.

Proof. We first show that the differential of $\mathsf{s} \times H \to G$, $(X, b) \to \exp(X)b$, is invertible at every point (X, b) with X in a neighborhood U of 0 in s. It suffices to show this for points $(X, 1)$ with X in a neighborhood of 0 in s (by composing with right translation by b^{-1}), and in fact only for the point $(0, 1)$ (by continuity in X). But at $(0, 1)$ the differential is simply the map $\mathsf{s} \times \mathsf{h} \to \mathsf{g}$, $(X, Y) \to X + Y$, which is invertible by the choice of s. It follows from the Inverse Function Theorem that the map $U \times H \to G$ is *locally* bi-analytic onto a neighborhood of H in G.

To prove the theorem it remains to show that this map $U \times H \to G$, $(X, b) \to \exp(X)b$ is one-to-one when X is restricted to a suitable (possibly smaller) neighborhood $\{X \in \mathsf{s} \mid \|X\| < \epsilon\}$ of 0 in s. Suppose this were *not* the case, then for *any* $\epsilon > 0$ we would have an equation

$$\exp(X_1)b_1 = \exp(X_2)b_2$$

with $X_1 \neq X_2 \in \mathsf{s}$, $\|X_1\|, \|X_2\| < \epsilon$, $b_1, b_2 \in H$. Then

$$\exp(-X_1)\exp(X_2) = b_1 b_2^{-1} \in H.$$

Since $\mathsf{s} \times \mathsf{h} \to G$, $(X, Y) \to \exp(X)\exp(Y)$, has an invertible differential at $(0, 0)$, we can write uniquely (for small ϵ):

$$\exp(-X_1)\exp(X_2) = \exp(X)\exp(Y)$$

with (X, Y) in a neighborhood of $(0, 0)$ in $\mathsf{s} \times \mathsf{h}$. Thus $\exp(X) \in H$ as well, and taking ϵ small enough we may pick such an X in *any* neighborhood of 0 in s. Also, $X \neq 0$, since otherwise $\exp(X_2) = \exp(X_1)\exp(Y)$ would give $X_2 = X_1$, because of the uniqueness of the $\exp(X)\exp(Y)$-form. So we can choose a sequence $\{X_k\}$ in s with $X_k \neq 0$, $\|X_k\| \to 0$, and $\exp(X_k) \in H$. Passing to a subsequence, we may further assume that $X_k/\|X_k\|$ converges to some X, which is necessarily nonzero and in s. We shall show that also $X \in \mathsf{h}$, contradicting $\mathsf{s} \cap \mathsf{h} = 0$.

To see this, take $\tau \in \mathbb{R}$, and pick integers n_k so that $n_k\|X_k\| \to \tau$. (Take $n_k\|X_k\| \leq \tau < (n_k + 1)\|X_k\|$.) Then

$$(\exp X_k)^{n_k} = \exp(n_k X_k) = \exp\left(n_k\|X_k\|\frac{X_k}{\|X_k\|}\right) \to \exp(\tau X).$$

Since H is closed, $\exp \tau X \in H$, and this for all $\tau \in \mathbb{R}$. Thus $X \in \mathsf{h}$, which is the desired contradiction. QED

Corollary 2. *Let G be a linear group, H a closed subgroup of G. Any point $a_0 \in H$ has a neighborhood U in G that lies in the domain of a coordinate system $(\xi_1, \ldots, \xi_{m+1}, \ldots, \xi_n)$ on G so that the elements of H in U are precisely those for which $\xi_{m+1} = 0, \ldots, \xi_n = 0$. Here $n = \dim G$ and $m = \dim H$.*

Terminology. We say H is given by the equations $\xi_{m+1} = 0, \ldots, \xi_n = 0$ locally around a_0.

Proof. In a neighborhood of $a_0 \in H$ in G, write $a = \exp(X)\exp(Y)a_0$ with $X \in \mathsf{s}$ and $Y \in \mathsf{h}$ close to 0. Then X and Y (or their components with respect to bases) give a coordinate system in a neighborhood of a_0 on G in which H is given by the equation $X = 0$, as required. QED

Corollary 3. *Let G be linear group. H a closed subgroup of G. The open subsets of H are the intersections of open subsets of G with H, i.e. the group-topology of H is its relative topology in G.*

Proof. Evident from Corollary 2 and the definition of 'open'. QED

The theorem and its corollaries apply in particular to closed subgroups of $\mathrm{GL}(E)$. As a consequence, the group-topology on a closed subgroup of $\mathrm{GL}(E)$ is its relative topology in the matrix space M.

Recall that a subset C of $\mathbb{R}^N$ is *compact* if every open cover of C has a finite subcover. One can use the same definition to define *compact linear groups* (with 'open' understood in the sense of the group-topology), but such a group is necessarily a compact subset of the matrix space: the image of a compact space (in this sense) under a continuous map is again compact, as is clear from the definition. This fact may be applied to the inclusion of a compact group in the matrix space. The basic compactness criterion is the Heine–Borel Theorem: *A subset of $\mathbb{R}^N$ is compact if and only if it is closed and bounded.* This criterion applies in particular to linear groups as subset of the matrix space. For example, the $\mathrm{SO}(n)$ and $\mathrm{SU}(n)$ are evidently closed and bounded subset of the matrix space.

Problems for §2.7

1. Show that a locally closed subgroup of a linear group is closed. [A subset A of G is *locally closed* if every point of A has a neighborhood U in G so that $A \cap U$ is closed in U. Suggestion: assume $a_k \in H$, $a_k \to a$ in G. If H is locally closed in G so is aH.]

2. Prove the converse of Corollary 3: Let G be linear group, H a subgroup of G. Suppose the open subsets of H are the intersections of open subsets of G with H. Show that H is closed in G. [Suggestion: preceding problem and Inverse Function Theorem.]

3. (a) Let G be a linear group, suppose H is a subgroup of G that can be written in the form $AB = \{ab \mid a \in A, b \in B\}$ where A is a compact subgroup and B a closed subset of G. Show that H is closed in G. [Suggestion: Assume $a_k b_k \to c$. Choose a convergent subsequence of $\{a_k\}$.]

   (b) Show that the conclusion of (a) need not hold if A is only assumed to be closed in G. [Suggestion: consider upper and lower triangular matrices in $\mathrm{SL}(2, \mathbb{R})$.]

4. Let G be a linear group g its Lie algebra. Let h be a Lie sub-algebra of g with the property that the only elements $X \in \mathsf{g}$ satisfying $[X, \mathsf{h}] \subset \mathsf{h}$ are those of h. Show that $\Gamma(\mathsf{h})$ is closed in G.

5. Show that a one-parameter group $\{\exp(\tau X) \mid \tau \in \mathbb{R}\}$ of linear transformations fails to be closed in $\mathrm{GL}(E)$ if and only if X is similar over $\mathbb{C}$ to an imaginary diagonal matrix with two entries of irrational ratio. [Suggestion for 'only if': Let $X = Y + Z$ be the Jordan decomposition of X, see problem 3 for §1.2. Show first that $\exp(\tau X) \to a_0$ implies $\tau \to \tau_0$ unless $Z = 0$ and the eigenvalues of Y are all imaginary. In the latter case, argue that closedness is equivalent to $\exp(\tau X) = 1$ for some $\tau \neq 0$.]

6. Consider $\mathbb{R}^n$ as the Lie algebra of $\mathbb{T}^n = \mathbb{R}^n/\mathbb{Z}^n$. Let a be a subspace (i.e. subalgebra) of $\mathbb{R}^n$, $A = \Gamma(\mathsf{a}) = \mathsf{a}/(\mathsf{a} \cap \mathbb{Z}^n)$ the corresponding connected subgroup of $\mathbb{R}^n/\mathbb{Z}^n$. Show that the following conditions on a are equivalent:

   (i) A is closed in $\mathbb{T}^n$

   (ii) a is spanned by $\mathsf{a} \cap \mathbb{Z}^n$

   (iii) a is spanned by $\mathsf{a} \cap \mathbb{Q}^n$

   (iv) The system $(\mathsf{a}, X) = 0$ has $n - \dim \mathsf{a}$ $\mathbb{Q}$-linearly independent solutions $X \in \mathbb{Q}^n$.

   Deduce that the one-parameter subgroup $\{\exp(\tau X) \mid \tau \in \mathbb{R}\}$ is dense in $\mathbb{Q}^n$ if and only if the components of $X \in \mathbb{R}^n$ are linearly independent over $\mathbb{Q}$. [This is a theorem of Kronecker (1884).]

3

The classical groups

3.1 The classical groups: definitions, connectedness[1]

The classical groups could be introduced as further examples of linear groups, but that would not do them justice: they form the deepest and most beautiful part of the subject. (The connoisseur will add the exceptional groups, which do not concern us here.)

We shall consider bilinear and sesquilinear forms on vector spaces over $\mathbb{R}$, $\mathbb{C}$, and $\mathbb{H}$ (quaternions). The basic facts from linear algebra concerning such forms are proved in the appendix to this section; here we concentrate on their automorphism groups.

We start with the real case: E is an n-dimensional real vector space, φ a bilinear form on E. An *automorphism* of φ is an invertible linear transformation $a \in M = L(E)$ which preserves φ in the sense that

$$\varphi(ax, ay) = \varphi(x, y). \tag{1}$$

These automorphisms of φ form a group, the *automorphism group of* φ, denoted $\mathrm{Aut}(\varphi)$. Assume now that φ is non-degenerate. Then every $a \in M$ has an *adjoint* with respect to φ, denoted a^φ, defined by the condition that

$$\varphi(ax, y) = \varphi(x, a^\varphi y) \tag{2}$$

for all $x, y \in E$. From (1) and (2) one sees that in this case

$$\mathrm{Aut}(\varphi) = \{a \in \mathrm{GL}(E) \mid a^\varphi a = 1\}. \tag{3}$$

Fix a basis $\{e_k\}$ for E, and write

$$\varphi(x, y) = \sum \varphi_{ij} \xi_i \eta_j$$

[1]The results of this chapter are not required until §6.7, but familiarity with §3.1, at least, is desirable in any case.

in terms of the components ξ_i of x and η_j of y. Using matrices with respect to this basis

$$\varphi(x, y) = x^* f y, \quad a^\varphi = f^{-1} a^* f,$$

where f is the matrix (φ_{ij}). Thus

$$\mathrm{Aut}(\varphi) = \{a \in M \mid f^{-1} a^* f a = 1\} \tag{4}$$

when elements of M are taken in their matrix representation.

The Lie algebra of $\mathrm{Aut}(\varphi)$ is

$$\mathsf{aut}(\varphi) = \{X \in M \mid X^\varphi = -X\},$$

as one verifies as in Examples 1 and 2, §2.1.

Any bilinear form is uniquely a sum of a symmetric form and a skew-symmetric form; a linear transformation preserving the form must preserve both parts.

When φ is non-degenerate and symmetric, $\mathrm{Aut}(\varphi)$ is called the *(indefinite) orthogonal group of* φ, denoted $\mathrm{O}(\varphi)$, or, if the form φ is understood, $\mathrm{O}(E)$. (The optional 'indefinite' refers to the possibility of φ being indefinite.) Such a form can always be written as

$$\varphi(x, y) = \pm\xi_1\eta_1 \pm \xi_2\eta_2 \pm \cdots \pm \xi_n\eta_n. \tag{5}$$

in terms of the components ξ_i and η_j of x and y with respect to a suitable basis. A symmetric form $\varphi(x, y)$ is uniquely determined by the corresponding *quadratic form* $\varphi(x, x)$, which for the form (5) is

$$\varphi(x, x) = \pm\xi_1^2 \pm \xi_2^2 \pm \cdots \pm \xi_n^2. \tag{6}$$

The number p of $+$ signs (q of $-$ signs) is independent of the basis and evidently determines the form up to a change of basis. We call (p, q) the *signature* of φ. One may then take $E = \mathbb{R}^n$ with (5) as form and write $\mathrm{O}(p, q)$ for $\mathrm{O}(\varphi)$. When φ is positive definite, i.e. when all signs in (5) are $+$, one writes $\mathrm{O}(n)$ instead of $\mathrm{O}(n, 0)$.

When φ is non-degenerate and skew-symmetric, $\mathrm{Aut}(\varphi)$ is called the *symplectic group* of φ, denoted $\mathrm{Sp}(\varphi)$, or, if the form φ is understood, $\mathrm{Sp}(E)$. Such a form can always be written as

$$\varphi(x, y) = (\xi_1\eta_{m+1} - \xi_{m+1}\eta_1) + \cdots + (\xi_m\eta_{2m} - \xi_{2m}\eta_m) \tag{7}$$

in terms of the components ξ_i and η_j of x and y with respect to a suitable basis. The dimension $n = 2m$ of E is then necessarily even. If we take $E = \mathbb{R}^{2m}$ with this form, then $\mathrm{Sp}(\varphi)$ is denoted $\mathrm{Sp}(m, \mathbb{R})$. (Some would write $\mathrm{Sp}(2m, \mathbb{R})$ instead.)

We now turn to the complex case: E is an n-dimensional vector space over $\mathbb{C}$, φ a complex bilinear form on E. $\mathrm{Aut}(\varphi)$ is defined as above and, in the non-degenerate case, given by (3) or (4). When φ is non-degenerate and symmetric the signs in the normal form (5) may all be chosen to be $+$, so that

$$\varphi(x, y) = \xi_1\eta_1 + \xi_2\eta_2 + \cdots + \xi_n\eta_n. \tag{8}$$

The automorphism group of φ is called the *(complex) orthogonal group*, again denoted $\mathrm{O}(\varphi)$, or $\mathrm{O}(E)$; when E is taken to be $\mathbb{C}^n$ with the form (8) one writes $\mathrm{O}(n,\mathbb{C})$. When φ is non-degenerate and skew-symmetric it has the same normal form (7) as in the real case, and $\mathrm{Aut}(\varphi)$, the *(complex) symplectic group*, is denoted $\mathrm{Sp}(\varphi)$, $\mathrm{Sp}(E)$, or $\mathrm{Sp}(m,\mathbb{C})$.

In the complex case, one may also consider a *sesquilinear* form on E, meaning a bi-additive function φ on $E \times E$ that satisfies

$$\varphi(\alpha x, \beta y) = \bar{\alpha}\beta\varphi(x,y).$$

(Note that $\varphi(x,y)$ is conjugate–linear in the *first* variable x; this unusual convention is forced upon us if we wish to keep it in case $\mathbb{C}$ is replaced by the non–commutative field $\mathbb{H}$.) $\mathrm{Aut}(\varphi)$ is defined in the same way as above for a bilinear form. In the non-degenerate case, one has again adjoints a^φ with respect to φ. If one writes

$$\varphi(x,y) = \sum \varphi_{ij}\bar{\xi}_i\eta_j$$

in terms of a basis, then in matrix form

$$\varphi(x,y) = x^* f y, \quad a^\varphi = f^{-1}a^* f \tag{9}$$

and

$$\begin{aligned}\mathrm{Aut}(\varphi) &= \{a \in \mathrm{GL}(E) \mid a^\varphi a = 1\}\\ &= \{a \in M \mid f^{-1}a^* f a = 1\}.\end{aligned} \tag{10}$$

A sesquilinear form φ is Hermitian if $\varphi(x,y) = \overline{\varphi(y,x)}$. A non-degenerate Hermitian form has the normal form

$$\varphi(x,y) = \pm\bar{\xi}_1\eta_1 \pm \bar{\xi}_2\eta_2 \pm \cdots \pm \bar{\xi}_n\eta_n \tag{11}$$

with respect to a suitable basis. It is uniquely determined by its quadratic form $\varphi(x,x)$, given by

$$\varphi(x,x) = \pm|\xi_1|^2 \pm |\xi_2|^2 \pm \cdots \pm |\xi_n|^2.$$

The *signature* (p,q) is again independent of the basis. The group $\mathrm{Aut}(\varphi)$ is called the *(indefinite) unitary group of φ*, denoted $\mathrm{U}(\varphi)$, $\mathrm{U}(E)$, or $\mathrm{U}(p,q)$. When the form is positive definite, the notation is $\mathrm{U}(n)$.

In analogy with the real case, one might be led to consider skew-Hermitian forms as well; but this is superfluous as multiplication by i turns a skew-Hermitian form into a Hermitian one, and *vice versa*.

Exceptionally, we shall now take for E a vector space over the skew-field of *quaternions*, denoted $\mathbb{H}$. The definition of a vector space over $\mathbb{H}$ is identical to that over a commutative field, except that it is preferable to write the scalars on the right, so that the condition for a transformation $a : E \to E$ of E to be linear reads $a(x\alpha + y\beta) = (ax)\alpha + (ay)\beta$. The reason is that if one represents elements of E as column n-vectors by means of an $\mathbb{H}$-basis for E, then a linear transformation of E may be represented by left multiplication by a matrix with entries from $\mathbb{H}$, as usual. [Vector spaces over $\mathbb{H}$ have bases, as over commutative fields, and are here required to be finite dimensional. The properties of vector spaces over the non-commutative (skew) field $\mathbb{H}$ used here may be proved in the same way as in the case of vector spaces over $\mathbb{R}$ or $\mathbb{C}$, as is done, for example, in the first chapter of the book by Artin (1957) for vector spaces over general skew-fields.]

The *conjugate* $\bar{\alpha}$ of $\alpha = \alpha_1 1 + \alpha_i i + \alpha_j j + \alpha_k k (\alpha_1, \alpha_i, \alpha_j, \alpha_k \in \mathbb{R})$ is by definition

$$\bar{\alpha} = \alpha_1 1 - \alpha_i i - \alpha_j j - \alpha_k k.$$

One has $\overline{\alpha\beta} = \bar{\beta}\bar{\alpha}, \alpha\bar{\alpha}$ is real, and the *norm* of α is $|\alpha| = (\alpha\bar{\alpha})^{1/2}$.

A *sesquilinear form* on E is a bi-additive function φ on $E \times E$ that satisfies

$$\varphi(x\alpha, y\beta) = \bar{\alpha}\varphi(x, y)\beta. \tag{12}$$

It may be written as

$$\varphi(x, y) = \sum \bar{\xi}_i \varphi_{ij} \eta_j \tag{13}$$

in terms of the components ξ_i of x and η_j of y with respect to an $\mathbb{H}$-basis for E. (The notion of 'bilinear form' is of no use for non-commutative fields.) The automorphism group $\mathrm{Aut}(\varphi)$ of such a sesquilinear form is defined as before. If the form is non-degenerate, one again has adjoints, and $\mathrm{Aut}(\varphi)$ is given by (10).

A sesquilinear form φ is *Hermitian* if $\varphi(x, y) = \overline{\varphi(y, x)}$, as in the complex case. A non-degenerate Hermitian form has a normal form

$$\varphi(x, y) = \pm\bar{\xi}_1\eta_1 \pm \cdots \pm \bar{\xi}_n\eta_n. \tag{14}$$

When $E = \mathbb{H}^n$ with this form, then $\mathrm{Aut}(E)$ is denoted $\mathrm{Sp}(p, q)$. This group may also be realized as the subgroup $\mathrm{Sp}(n, \mathbb{C})$ preserving a complex-Hermitian form of signature $(2p, 2q)$, which explains the notation.

A sesquilinear form φ is *skew-Hermitian* if $\varphi(x, y) = -\overline{\varphi(y, x)}$, and a non-degenerate skew-Hermitian form has the normal form

$$\varphi(x, y) = \bar{\xi}_1 j \eta_1 + \cdots + \bar{\xi}_n j \eta_n \tag{15}$$

When $E = \mathbb{H}^n$ with this form, then $\mathrm{Aut}(E)$ is denoted $\mathrm{O}^*(2n)$. This group may also be realized as the subgroup of $\mathrm{O}(2n, \mathbb{C})$ which preserves a nondegenerate complex skew-Hermitian form, which explains the notation once more.

The *classical groups* are exactly the general linear groups over $\mathbb{R}$, $\mathbb{C}$, or $\mathbb{H}$, together with the automorphism groups of non-degenerate forms (symmetric or skew-symmetric, Hermitian or skew-Hermitian) discussed above. Usually, these groups are subjected to the additional restriction that the transformations in question have determinant equal to 1. (For the determinant of a linear transformation of a vector space over $\mathbb{H}$ we may here take its determinant as a $\mathbb{C}$-linear transformation.) We list these groups with the 'determinant = 1' restriction.

(a) *Special linear groups*: linear transformations over $\mathbb{R}$, $\mathbb{C}$, or $\mathbb{H}$ with determinant equal to 1. These are denoted:

$$\mathrm{SL}(n, \mathbb{R}), \quad \mathrm{SL}(n, \mathbb{C}), \quad \mathrm{SL}(n, \mathbb{H}).$$

Table 3.1 Automorphism groups of forms

Group	Field	Form
$\mathrm{SO}(p,q)$	$\mathbb{R}$	Symmetric
$\mathrm{SO}(n,\mathbb{C})$	$\mathbb{C}$	Symmetric
$\mathrm{Sp}(m,\mathbb{R})$	$\mathbb{R}$	Skew-symmetric
$\mathrm{Sp}(m,\mathbb{C})$	$\mathbb{C}$	Skew-symmetric
$\mathrm{SU}(p,q)$	$\mathbb{C}$	Hermitian
$\mathrm{Sp}(p,q)$	$\mathbb{H}$	Hermitian
$\mathrm{SO}^*(2n)$	$\mathbb{H}$	Skew-Hermitian

(b) *Automorphism groups of forms*: linear transformations over $\mathbb{R}$, $\mathbb{C}$, or $\mathbb{H}$ of determinant equal to 1 that preserve a non-degenerate form, which may be symmetric, skew-symmetric, Hermitian or skew-Hermitian, and of determinant equal to 1.

It should be pointed out, however, that on a real vector space Hermitian and skew-Hermitian are same as symmetric and skew-symmetric, respectively, on a complex vector space skew-Hermitian forms get turned into Hermitian forms by multiplication by i, and *vice versa.* Further, on a quaternionic vector space there are no non-zero bilinear forms at all. This leaves the possibilities shown in Table 3.1.

The forms may be taken as:

$$\begin{aligned}
&\text{Symmetric:} && \pm\xi_1\eta_1 \pm \xi_2\eta_2 \pm \cdots \pm \xi_n\eta_n,\\
&\text{Skew-symmetric:} && (\xi_1\eta_{m+1} - \xi_{m+1}\eta_1) + \cdots + (\xi_m\eta_{2m} - \xi_{2m}\eta_m),\\
&\text{Hermitian:} && \pm\bar{\xi}_1\eta_1 \pm \cdots \pm \bar{\xi}_n\eta_n,\\
&\text{Skew-Hermitian:} && \bar{\xi}_1 j\eta_1 + \cdots + \bar{\xi}_n j\eta_n.
\end{aligned}$$

The groups $\mathrm{SL}(n,\mathbb{H})$, $\mathrm{Sp}(p,q)$, and $\mathrm{SO}^*(2n)$, consisting of matrices over $\mathbb{H}$, may also be realized by matrices over $\mathbb{C}$, as follows.

Consider $\mathbb{H}^n$ (column vectors) as right vector space over $\mathbb{H}$. An $\mathbb{H}$-linear transformation of $\mathbb{H}^n$ is given by left multiplication by a matrix from $M_n(\mathbb{H})$. Think of $\mathbb{C}$ as consisting of quaternions of the form $\lambda + i\mu$. Then we may write $\mathbb{H}^n = \mathbb{C}^n + j\mathbb{C}^n$ and $M_n(\mathbb{H}) = M_n(\mathbb{C}) + jM_n(\mathbb{C})$. If $x + jy \in \mathbb{H}^n$ is written as $\left[\begin{smallmatrix}x\\y\end{smallmatrix}\right]$, then left multiplication by $a + jb$, $a, b \in M_n(\mathbb{C})$ is represented by the block matrix

$$\begin{bmatrix} a & -\bar{b} \\ b & \bar{a} \end{bmatrix}.$$

$M_n(\mathbb{H})$ is in this way identified with the subalgebra of $M_{2n}(\mathbb{C})$ consisting of matrices of this form. It is not a complex subspace of $M_{2n}(\mathbb{C})$, but carries its own multiplication by i given by $i(a + jb) = (ia) - j(ib)$ in terms of quaternion multiplication.

In its complex guise, $\mathrm{SL}(n,\mathbb{H})$ is denoted $\mathrm{SU}^*(2n)$; the notation for $\mathrm{Sp}(p,q)$ and $\mathrm{SO}^*(2n)$ is already taken from their realization by complex matrices.

These groups are described in more detail in the problems at the end of this section.

The groups listed in (a) and (b) above are more precisely called the *real classical groups*, even though they may consist of linear transformations of vector spaces over $\mathbb{C}$ or $\mathbb{H}$. (The terminology indicates that the Lie algebras of these groups are real vector spaces.) The *complex classical groups* are those of the three families $\mathrm{SL}(n,\mathbb{C})$, $\mathrm{SO}(n,\mathbb{C})$, $\mathrm{Sp}(m,\mathbb{C})$. (The terminology is explained analogously.) These complex classical groups have *compact real forms*, namely the subgroups which preserve the positive definite form (x,y) obtained by taking all signs to be $+$ in Hermitian form listed above. (We use the basis which reduces also $\varphi(x,y)$ to one of the normal forms listed.) These are the *compact classical groups* $\mathrm{SU}(n)$, $\mathrm{SO}(n)$, $\mathrm{Sp}(m)$. 'Compact' refers to the fact that these groups are compact (closed and bounded) subsets of the matrix space. (Their group topology is the same as their relative topology in the matrix space; this follows from Corollary 3, §2.7.) The 'real form' part is explained in problem 12. The unitary subgroup $\mathrm{Sp}(m)$ of $\mathrm{Sp}(m,\mathbb{C})$ is the complex realization of the unitary subgroup $\mathrm{Sp}(m,0)$ of $\mathrm{SL}(m,\mathbb{H})$. We list these groups in Table 3.2 together with their dimensions.

The following theorem concerning conjugacy classes in the compact classical groups is of central importance. Even though its proof is elementary, and may well be familiar, at least for $\mathrm{U}(n)$ and $\mathrm{O}(n)$, we shall give the argument in complete detail.

Theorem 1.
(a) Every element of $\mathrm{SU}(n)$ *is conjugate to a diagonal matrix*

$$(\epsilon_1, \epsilon_1, \ldots, \epsilon_n),$$

with $\epsilon_k = e^{2\pi i\theta_k}$, $\prod \epsilon_k = 1$.

(b) Every element of $\mathrm{SO}(n)$ *is conjugate to a block-diagonal matrix,*

$$\begin{aligned} &(t_1, t_2, \ldots, t_m) \quad \text{if } n = 2m \text{ is even,} \\ &(t_1, t_2, \ldots, t_m, 1) \quad \text{if } n = 2m+1 \text{ is odd.} \end{aligned}$$

Each t_k *is a* 2×2 *block*

$$\begin{bmatrix} \cos\theta & -\sin\theta \\ \sin\theta & \cos\theta \end{bmatrix}.$$

The 1 *for* $n = 2m+1$ *is a* 1×1 *block.*

Table 3.2 The complex and compact classical groups

Type	G	K	$\dim_{\mathbb{C}} G = \dim_{\mathbb{R}} K$
A_{n-1}	$\mathrm{SL}(n,\mathbb{C})$	$\mathrm{SU}(n)$	n^2-1
B_n	$\mathrm{SO}(2n+1,\mathbb{C})$	$\mathrm{SO}(2n+1)$	$2n^2+n$
C_n	$\mathrm{Sp}(n,\mathbb{C})$	$\mathrm{Sp}(n)$	$2n^2+n$
D_n	$\mathrm{SO}(2n,\mathbb{C})$	$\mathrm{SO}(2n)$	$2n^2-n$

(c) Each element of $\mathrm{Sp}(n)$ *is conjugate to a diagonal matrix of the form*

$$(\epsilon_1, \epsilon_2, \ldots, \epsilon_n) \quad \textit{(quaternionic realization)};$$

equivalently,

$$(\epsilon_1, \bar{\epsilon}_1, \epsilon_2, \bar{\epsilon}_2, \ldots, \epsilon_n, \bar{\epsilon}_n) \quad \textit{(complex realization)},$$

with $\epsilon_k = e^{2\pi i \theta_k}$.

Proof.
(a) $\mathrm{SU}(n)$. Let $a \in \mathrm{SU}(n)$. Choose an orthonormal basis for $\mathbb{C}^n$ consisting of eigenvectors of a. Let c be the change-of-basis matrix. Then, c is unitary and $c^{-1}ac$ is diagonal. Multiply c by a scalar so that its determinant in 1. Then $c \in \mathrm{SU}(n)$ as required.

(b) $\mathrm{SO}(n)$. Let $a \in \mathrm{SO}(n)$. Think of a as a unitary transformation of $\mathbb{C}^n$ satisfying $\bar{a} = a$. Let ϵ be an eigenvalue of a. Necessarily $\epsilon = e^{2\pi i\theta}$. If $\epsilon = \pm 1$ we can find a real eigenvector e for a, which we assume normalized: $(e, e) = 1$. Otherwise, we choose a complex normalized eigenvector v for ϵ. Its conjugate $\bar{v}$ is an eigenvector for $\bar{\epsilon}$. The space spanned by $e_+ = (1/2)(v + \bar{v})$ and $e_- = (1/2i)(v - \bar{v})$ is invariant under a and the matrix of a therein is a 2×2 matrix of the type indicated. Continuing in this way with the orthogonal complement one constructs an orthonormal basis with respect to which a is block-diagonal with 2×2 blocks of the required type and 1×1 blocks ± 1. The -1's come in pairs ($\det a = 1$), which may be combined to 2×2 blocks; pairs of $+1$'s may be combined as well, but if n is odd there will be one $+1$ left over. The change-of-basis matrix to the basis constructed is orthogonal; it may be assumed to have determinant $+1$, after an interchange of two basis vectors e_+, e_-, if necessary.

(c) $\mathrm{Sp}(n)$. Let $a \in \mathrm{Sp}(n)$. Consider $\mathbb{H}^n$ as right vector space over $\mathbb{C}$ and think of $a \in \mathrm{Sp}(n)$ as as a $\mathbb{C}$-linear, unitary transformation satisfying $a(xj) = (ax)j$. Let e be a normalized eigenvector of a with eigenvalue $\epsilon : ae = e\epsilon$. Then ej has eigenvalue $j^{-1}\epsilon j = \bar{\epsilon}$, has norm 1, and is orthogonal to $e : (ej, ej) = (e, e) = 1$ and $(e, ej) = 0$, because $j^* = j^{-1} = -j$. The complex subspace spanned by $\{e, ej\}$ is $\mathbb{H}$-stable with $\mathbb{H}$-basis $\{e\}$. Continuing in this way with its orthogonal complement one constructs an $\mathbb{H}$-basis for $\mathbb{H}^n$ with respect to which a is diagonal of the required type. The change-of-basis matrix belongs to $\mathrm{Sp}(n)$. QED

Corollary 2. *The exponential map for a compact classical groups* $\mathrm{SU}(n)$, $\mathrm{SO}(n)$, *and* $\mathrm{Sp}(n)$ *is surjective.*

Proof. The representatives of the conjugacy classes listed evidently belong to one-parameter subgroups. QED

Observe that in each case $K = \mathrm{SU}(n), \mathrm{SO}(n)$, or $\mathrm{Sp}(n)$ the representatives of the conjugacy classes listed in the theorem form in fact a subgroup T of K, isomorphic with a torus. T is called the (*standard*) *Cartan subgroup* of K. ('Standard' refers to the 'standard' basis on which depends the matrix realization. For another orthonormal basis for $\mathbb{C}^n, \mathbb{R}^n, \mathbb{H}^n$ one gets another Cartan subgroup, but these are all conjugate under K.)

The compact classical groups are connected, as one sees from the corollary. To discuss the connectedness properties of the other real classical groups we shall need a proposition from linear algebra:

Proposition 3 (Polar decomposition of a matrix). *Let E be a real or complex vector space with a positive-definite inner product. Then every $a \in \mathrm{GL}(E)$ can be uniquely written in the form*

$$a = k \exp Y,$$

where k is unitary and Y is Hermitian.

Let $K(E)$ denote the group of unitary transformations and $\mathsf{p}(E)$ the real vector space of Hermitian transformations of E. The mapping $K(E) \times \mathsf{p}(E) \to \mathrm{GL}(E), (k, Y) \to k \exp Y$, is an analytic bijection.

Explanation. Naturally, in the real case *unitary* = *orthogonal*, and *Hermitian* = *symmetric*.

Proof. Let momentarily $\mathsf{p} = \mathsf{p}(E)$ be the real vector space of Hermitian matrices and $P = P(E) = \{a \in \mathsf{p} : (x, ax) > 0 \text{ for all } x \neq 0\}$ the set of positive definite Hermitian matrices. P is evidently an open subset of p. We first prove:

$$\exp \text{ maps } \mathsf{p} \text{ bi-analytically onto } P. \tag{16}$$

First of all, since any Hermitian matrix is diagonal with real entries for a suitable orthonormal basis, it is clear that exp maps p onto P. Now suppose $p \in P$ is given and consider $\exp X = p$ as the equation for $X \in \mathsf{p}$. It suffices to consider this equation on a single eigenspace of p. If the eigenvalue is $\alpha > 0$ we get $p = \alpha 1$ and $\exp X = \alpha 1$. Now consider this equation on an eigenspace of X with eigenvalue λ. We get $e^\lambda = \alpha$, i.e. $\lambda = \log \alpha$. Thus $X = (\log \alpha)1$ is the unique solution. This shows that the map (16) is a bijection. We denote its inverse by $p \to \log p$. To show that this inverse is also analytic, it suffices to show that the map $\exp : \mathsf{p} \to P$ has an invertible differential $d\exp_X : \mathsf{p} \to \mathsf{p}$ for all $X \in \mathsf{p}$. This follows as in the proof of Proposition 7, §1.2.

We now turn to the map $(k, Y) \to k \exp Y$ in the theorem. We explicitly construct the inverse mapping by solving

$$a = k \exp Y \tag{17}$$

for k and Y. This is done as follows. Assume we have an eqn (17) with k unitary and Y Hermitian. Then

$$a^* a = \exp(2Y). \tag{18}$$

a^*a is Hermitian positive definite, hence from (16)–(18) we get

$$Y = \tfrac{1}{2} \log(a^* a), \quad k = a \exp -Y$$

as unique solution. It only remains to check that this k is unitary:

$$kk^* = a\exp(-2Y)a^* = a(a^*a)^{-1}a^* = 1.$$

QED

We now turn to the real classical groups. Recall that these are either the special linear groups $\mathrm{SL}(E)$ of a vector space E over $\mathbb{R}$, $\mathbb{C}$, or $\mathbb{H}$, or the subgroups thereof that preserve one of the forms listed after Table 1. If φ is one of these forms, write σ for the inverse-adjoint with respect to φ:

$$\sigma(a) = (a^{\varphi})^{-1}.$$

(The inverse is included to make σ an automorphism of $\mathrm{GL}(E) : \sigma(ab) = \sigma(a)\sigma(b)$. This automorphism is an *involution*, meaning $\sigma^2 = 1$.) Then the classical groups belonging to forms are all of the type

$$G = \{a \in \mathrm{SL}(E) \mid \sigma(a) = a\}. \tag{19}$$

Their Lie algebras are

$$\mathsf{g} = \{X \in \mathrm{SL}(E) \mid \sigma'(X) = X\}, \tag{20}$$

where σ' is the Lie map of $\sigma : \sigma'(X) = -X^{\varphi}$. For the sake of uniformity we now also admit the identity maps for σ and σ', and so that *all* real classical groups are of the type (19).

Write θ for the involution corresponding to the positive definite Hermitian form, denoted (x, y), obtained by taking all signs $+$ in the Hermitian form listed above. We use the same basis to define the positive definite form which was used to bring φ to one of the normal types listed. σ and θ commute:

$$\sigma\theta = \theta\sigma. \tag{21}$$

As a consequence, θ maps the group G defined by (19) into itself:

$$\theta(G) = G. \tag{22}$$

Set

$$K = G \cap K(E) \quad \text{and} \quad \mathsf{p} = \mathsf{g} \cap \mathsf{p}(E). \tag{23}$$

Proposition 4 (Polar decomposition of a classical group.). *Let G be a real classical group, K and p as defined above. Then the map*

$$K \times \mathsf{p} \to G, \quad (k, Y) \to k\exp Y$$

is an analytic bijection.

Proof. Apply Proposition 3 to an element $a \in G$. (In the quaternionic case, think of $\mathbb{H}$-linear transformations of E as $\mathbb{C}$-linear transformations that commute with multiplication by j.) Thus write

$$a = k\exp Y$$

as in Proposition 3. Equation (18) may be written as

$$\theta(a^{-1})a = \exp 2Y,$$

which shows that $\exp 2Y \in G$, i.e. $\sigma(\exp 2Y) = \exp(2Y)$, i.e. $\exp \sigma'(2Y) = \exp 2Y$, hence $\sigma'(Y) = Y$, since exp is $1-1$ on the Hermitian Y's. Furthermore, for Hermitian Y, $\det(\exp Y) = 1$ implies $\operatorname{tr} Y = 0$. (In the quaternionic case, $\operatorname{tr} Y$ is the trace of Y as $\mathbb{C}$-linear transformation. One should however observe that Y is in fact $\mathbb{H}$-linear: $j \exp(2Y) j^{-1} = \exp(2Y)$ implies $jYj^{-1} = Y$ for Hermitian Y.) From (20) one now sees that $Y \in \mathfrak{g}$, hence $\exp Y \in G$, and then $k = a \exp -Y \in G$ as well. QED

As a corollary, the identity component of G is

$$G_o = K_o \exp \mathfrak{p}, \tag{24}$$

where K_o is the identity component of K. Using this fact one verifies:

Proposition 5. *The real classical groups are all connected, except for* $\mathrm{SO}(p, q)$ *with* $p, q \neq 0$, *which has two connected components.*

Proof. We look at the exceptional case $G = \mathrm{SO}(p,q)$ to illustrate the method. In this case $K = S(\mathrm{O}(p) \times \mathrm{SO}(q)) = \{(a, b) \mid a \in \mathrm{O}(p), b \in \mathrm{O}(q), \det(a)\det(b) = +1\}$. Since $\mathrm{SO}(n)$ is connected, $K_o = \mathrm{SO}(p) \times \mathrm{SO}(q)$, and the two connected components of K are described by $\det a = \det b = +1$ and $= -1$, respectively.

As another sample we consider $\mathrm{Sp}(m, \mathbb{R})$. In this case K consists of the linear transformations a of $\mathbb{R}^{2m}$ which preserve the standard positive-definite bilinear form (x, y) as well as the skew-symmetric form (x, Jy), where J is the matrix

$$\begin{bmatrix} 0 & -1 \\ 1 & 0 \end{bmatrix}.$$

This means that

$$a^t = a^{-1}, \quad J^{-1}a^tJ = a^{-1}$$

or, equivalently,

$$a^t = a^{-1}, \quad aJ = Ja.$$

Since $J^2 = -1$, we can make $\mathbb{R}^{2m}$ into a complex vector space by defining $ix = Jx$. Then (x, y) may be realized as the real part of a positive definite Hermitian form on this complex vector space, and K consists of all $\mathbb{C}$-linear transformations preserving this form, i.e. $K = \mathrm{U}(m)$, which is connected.

We omit the remaining verifications, but list in Table 3.3 the various groups K at least for the *complex* classical groups G. The proposition follows from Table 3.3 since the compact classical groups are connected. QED

As an application one can determine the identity component of $\mathrm{SO}(p,q)$ very explicitly in general. Assume the form $\langle x, y \rangle$ on $\mathbb{R}^{p+q}$ defining $\mathrm{SO}(p,q)$ is chosen

Table 3.3

G	K
$\mathrm{SO}(p,q)$	$S(O(p)\times \mathrm{SO}(q))$
$\mathrm{SO}(n,\mathbb{C})$	$\mathrm{SO}(n)$
$\mathrm{Sp}(m,\mathbb{R})$	$\mathrm{U}(m)$
$\mathrm{Sp}(m,\mathbb{C})$	$\mathrm{Sp}(m)$
$\mathrm{SU}(p,q)$	$S(\mathrm{U}(p)\times \mathrm{SU}(q))$
$\mathrm{Sp}(p,q)$	$\mathrm{Sp}(p)\times \mathrm{Sp}(q)$
$\mathrm{SO}^*(2n)$	$\mathrm{SO}(2n)$

so that among the basis vectors $e_1,\dots,e_{p+q}$ the first p are positive and the remaining q are negative:

$$\langle x,y\rangle = \xi_1\eta_1+\xi_2\eta_2+\cdots+\xi_p\eta_p-\xi_{p+1}\eta_{p+1}+\cdots-\xi_{p+q}\eta_{p+q}.$$

We assume $p,q\geq 1$ and denote by $\mathbb{R}^p$ and $\mathbb{R}^q$ the subspaces of $\mathbb{R}^{p+q}$ spanned by the first p and last q basis vectors, respectively. Let $a\in \mathrm{O}(p,q)$. The subspace $a\mathbb{R}^p$ is again positive definite, hence the projection onto $\mathbb{R}^p$ along $\mathbb{R}^q$ must have rank p on $a\mathbb{R}^p$ (because $a\mathbb{R}^p\cap\mathbb{R}^q=0$). This means that the determinant $\det_p a=\det\langle ae_i,e_j\rangle, 1\leq i,j<p$, is non-zero. The same is true for $\det_q a=\det\langle ae_i,e_j\rangle, p+1\leq i,j<p+q$. By continuity considerations these determinants must keep the same sign on a connected component of $\mathrm{O}(p,q)$. Since we already know that $\mathrm{O}(p,q)$ has in all four components (and $\mathrm{SO}(p,q)$ has two), these components will be distinguished by the signs of $\det a$, $\det_p a$, and $\det_q a$ as soon as one can find four matrices a with distinct sign distributions. Considering diagonal matrices with entries ± 1 one sees that in fact two of the three determinants suffice to distinguish four such elements a. In particular:

$$\mathrm{SO}_o(p,q)=\{a\in\mathrm{SO}(p,q)\mid \det_p(a)>0\}=\{a\in\mathrm{SO}(p,q)\mid \det_q(a)>0\}.$$

Thus we recover Example 4, §2.4 as a special case; but the geometric discussion given there provides some additional insight into the connectivity properties of $\mathrm{SO}(2,1)$.

Problems for §3.1

1. Show that the following represent natural one-to-one correspondences:

   (a) $\{\mathbb{R}\text{-vector spaces } E\}\leftrightarrow \text{pairs}(F,C)\mid F$ a $\mathbb{C}$-vector space, $C:F\to F\mathbb{R}$-linear, $C(\alpha x)=\bar\alpha C(x), C^2=1\}$.

   (b) $\{(\text{right})\ \mathbb{H}\text{-vector spaces } E\}\leftrightarrow\{\text{pairs}(F,J)\mid F$ a (right) $\mathbb{C}$-vector space, $J:F\to F$, $\mathbb{R}$-linear, $J(\alpha x)=\bar\alpha J(x), J^2=-1\}$.

   [*Remark*: In this way vector spaces over $\mathbb{R}$ or $\mathbb{H}$ may be considered vector spaces over $\mathbb{C}$ with an additional piece of structure. This is sometimes useful to treat the three cases $\mathbb{R}$, $\mathbb{C}$, $\mathbb{H}$ in a uniform way.]

2. Verify the description (3) of $\mathrm{Aut}(\varphi)$. Check that $\mathrm{Aut}(\varphi)$ is a group. Verify the description of $\mathsf{aut}(\varphi)$.

3. *Cayley Transform.* Let $E \approx \mathbb{K}^n$ be an n-dimensional right vector space over $\mathbb{K} = \mathbb{R}, \mathbb{C}$, or $\mathbb{H}$, $M \approx M_n(\mathbb{K}^n)$ the matrix space. Define a map $M \cdots \to M, Z \to Z^c$ with domain $M' = \{Z \in M \mid 1 - Z \text{ invertible}\}$ by

$$Z^c = \frac{1+Z}{1-Z}.$$

(a) Show that $Z \to Z^c$ maps M' onto M' and $(Z^c)^c = Z$.

Let φ be a non-degenerate bilinear or sesquilinear form on E, Z^φ the adjoint of Z with respect to φ. Let $\mathrm{Aut}(\varphi)$ be its automorphism group, $\mathsf{aut}(\varphi)$ its Lie algebra, $\mathrm{Aut}'(\varphi)$ and $\mathsf{aut}'(\varphi)$ the intersections with M'.

(b) Show that $Z \to Z^c$ is a bi-analytic map $\mathrm{Aut}'(\varphi) \to \mathsf{aut}'(\varphi)$.

[This map may therefore be considered as a coordinate system on $\mathrm{Aut}(\varphi)$ with domain $\mathrm{Aut}'(\varphi)$.]

4. A real classical group leaves invariant a non-degenerate symmetric real-bilinear form on its Lie algebra. Prove this for

(a) $\mathrm{SL}(n, \mathbb{R})$, $\mathrm{SL}(n, \mathbb{C})$, and $\mathrm{SL}(n, \mathbb{H})$.

(b) $\mathrm{Aut}(\varphi)$, φ a non-degenerate bilinear or sesquilinear form on a vector-space over $\mathbb{R}$, $\mathbb{C}$, or $\mathbb{H}$.

[Suggestion: consider $\mathrm{Re}\,\mathrm{tr}(XY)$. To show that this form remains non-degenerate on $\mathsf{g} \subset M$, exhibit an orthogonal complement to g in M.]

5. Show that $\mathrm{SL}(2, \mathbb{R}) \approx \mathrm{Sp}(1, \mathbb{R})$ and $\mathrm{SL}(2, \mathbb{C}) \approx \mathrm{Sp}(1, \mathbb{C})$.

6. Show that $\mathrm{SL}(2, \mathbb{R})$ and $\mathrm{SU}(1, 1)$ are conjugate within $\mathrm{SL}(2, \mathbb{C})$. Find an explicit matrix $c \in \mathrm{SL}(2, \mathbb{C})$ so that $\mathrm{SU}(1,1) = c\mathrm{SL}(2, \mathbb{R})c^{-1}$. [Suggestion: in addition to a skew-symmetric form $\varphi(x, y)$, $\mathrm{SL}(2, \mathbb{R}) = \mathrm{Sp}(1, \mathbb{R})$ preserves the Hermitian form $i\varphi(\bar{x}, y)$ on $\mathbb{C}^2$.]

7. Show that $\mathrm{Sp}(p, q)$ is isomorphic with the subgroup of $\mathrm{Sp}(n, \mathbb{C})(n = p + q)$ which preserves the Hermitian form

$$\pm \bar{\xi}_1 \eta_1 \pm \cdots \pm \bar{\xi}_n \eta_n.$$

8. Show that $\mathrm{SO}^*(2n)$ is isomorphic with the subgroup of $\mathrm{SO}(2n, \mathbb{C})$ which preserves the skew-Hermitian form

$$(\bar{\xi}_1 \eta_{n+1} - \bar{\xi}_{n+1} \eta_1) + \cdots + (\bar{\xi}_n \eta_{2n} - \bar{\xi}_{2n} \eta_n),$$

or, equivalently, the Hermitian form obtained by multiplication by i. (This Hermitian form is of type $(p, q) = (n, n)$.)

9. Show that $\mathrm{Sp}(n,\mathbb{R}) \approx \mathrm{Sp}(n,\mathbb{C}) \cap \mathrm{SU}(n,n)$. [Suggestion: $\mathrm{Sp}(n,\mathbb{R})$ leaves invariant both a skew-symmetric form $\varphi(x,y)$ and a Hermitian form $i\varphi(\bar{x},y)$ on $\mathbb{C}^{2n}$.]

10. *Iwasawa decomposition in special linear groups.* Let $G = \mathrm{SL}(n,\mathbb{R})$,$\mathrm{SL}(n,\mathbb{C})$, or $\mathrm{SL}(n,\mathbb{H})$, $K = \mathrm{SU}(n)$, $\mathrm{SO}(n)$, $\mathrm{Sp}(n)$, A the group of diagonal matrices in G with positive real diagonal entries, N the group of unipotent-triangular matrices in G (upper triangular with 1's along the diagonal). Show that

$$G = KAN$$

in the sense that every element $g \in G$ can be *uniquely* written in the form $g = kan$ with $k \in K$, $a \in A$, $n \in N$. [Suggestion: Gram–Schmidt process; see problem 9 for §2.1. Deduce the existence of the decomposition $g = kan$. Prove uniqueness.]

11. The Lie algebras of the complex classical groups are *complex* subspaces of $M_n(\mathbb{C})$. Show that their complex dimensions are:

    (a) $\dim_{\mathbb{C}} \mathrm{SL}(n,\mathbb{C}) = n^2 - 1$,
    (b) $\dim_{\mathbb{C}} \mathrm{SO}(n,\mathbb{C}) = \frac{1}{2}n(n-1)$,
    (c) $\dim_{\mathbb{C}} \mathrm{Sp}(m,\mathbb{C}) = 2m^2 + m$.

12. Let $\mathfrak{g}$ be a *complex* Lie algebra. A *real* Lie subalgebra $\mathfrak{g}_o$ of $\mathfrak{g}$ is called a *real form of* $\mathfrak{g}$ if $\mathfrak{g} = \mathfrak{g}_o \oplus i\mathfrak{g}_o$ (direct sum of real vector spaces). Show that the Lie algebras of the compact classical groups are real forms of the complex classical groups:

    (a) $\mathfrak{su}(n)$ is a real form of $\mathfrak{sl}(n,\mathbb{C})$,
    (b) $\mathfrak{so}(n)$ is a real form of $\mathfrak{so}(n,\mathbb{C})$,
    (b) $\mathfrak{sp}(n)$ is a real form of $\mathfrak{sp}(n,\mathbb{C})$.

13. Verify the following explicit matrix realizations of the Lie algebras of the real classical groups. (Some have already been verified in the text. They are repeated here for reference.)

    (a) $\mathfrak{sl}(n,\mathbb{R})$: real $n \times n$ matrices X with $\operatorname{tr} X = 0$.
    (b) $\mathfrak{sl}(n,\mathbb{C})$: complex $n \times n$ matrices X with $\operatorname{tr} X = 0$.
    (c) $\mathfrak{sl}(n,\mathbb{H}) \approx \mathfrak{su}^*(2n)$: quaternion $n \times n$ matrices X with $\operatorname{Re}\operatorname{tr} X = 0$;

    $$\approx \text{ complex matrices } \begin{bmatrix} X & -\bar{Y} \\ Y & \bar{X} \end{bmatrix},$$

    X, Y of order $n \times n$, $\operatorname{Re}\operatorname{tr} X = 0$. (Re = real part)

    (d) $\mathfrak{so}(p,q)$: real matrices

    $$\begin{bmatrix} X & Z \\ Z^{\mathrm{t}} & Y \end{bmatrix},$$

    $X^{\mathrm{t}} = -X$ of order $p \times p$, $Y^{\mathrm{t}} = -Y$ of order $q \times q$, Z of order $p \times q$.

(e) $\mathsf{so}(n,\mathbb{C})$; complex $n \times n$ matrices X with $X^t = -X$.

(f) $\mathsf{sp}(m,\mathbb{R})$: real matrices

$$\begin{bmatrix} X & Y \\ Z & -X^t \end{bmatrix},$$

$Y^t = Y, Z^t = Z$, of order $m \times m$.

(g) $\mathsf{sp}(m,\mathbb{C})$: complex matrices

$$\begin{bmatrix} X & Y \\ Z & -X^t \end{bmatrix},$$

$Y^t = Y$, $Z^t = Z$, of order $m \times m$.

(h) $\mathsf{su}(p,q)$: complex matrices

$$\begin{bmatrix} X & Z \\ -\bar{Z}^t & Y \end{bmatrix},$$

$\bar{X}^t = -X$ of order $p \times p$, $\bar{Y}^t = -Y$ of order $q \times q$, Z of order $p \times q$, $\operatorname{tr} X + \operatorname{tr} Y = 0$.

(i) $\mathsf{Sp}(p,q)$ quaternion matrices:

$$\approx \text{complex matrices} \begin{bmatrix} X & Z \\ -\bar{Z}^t & Y \end{bmatrix},$$

$\bar{X}^t = -X$ of order $p \times p, \bar{Y}^t = -Y$ of order $q \times q$, Z of order $p \times q$, $\operatorname{tr} X + \operatorname{tr} Y = 0$;

$$\approx \text{ complex matrices } \begin{bmatrix} X & Z \\ -\bar{Z}^t & Y \end{bmatrix},$$

$$X = \begin{bmatrix} X_1 & -\bar{X}_2 \\ X_2 & \bar{X}_1 \end{bmatrix}, \quad Y = \begin{bmatrix} Y_1 & -\bar{Y}_2 \\ Y_2 & \bar{Y}_1 \end{bmatrix}, \quad Z = \begin{bmatrix} Z_1 & -\bar{Z}_2 \\ Z_2 & \bar{Z}_1 \end{bmatrix},$$

$\bar{X}_1^t = -X_1$ of order $p \times p$, $\bar{Y}_1^t = -Y_1$ of order $q \times q$.

(j) $\mathsf{so}^*(2n)$: quaternion matrices X with $\bar{X}^t = jXj$; $\approx$ complex matrices

$$\approx \text{complex matrices} \begin{bmatrix} X & -\bar{Y} \\ Y & \bar{X} \end{bmatrix},$$

X, Y of order $n, X^t = -X$, $\bar{Y}^t = Y$.

Appendix to §3.1: Bilinear and sesquilinear forms

We shall prove the facts about bilinear and sesquilinear forms over $\mathbb{R}$, $\mathbb{C}$, and $\mathbb{H}$ used in connection with the classical groups; the real and complex cases should be familiar. Treating all three cases together takes only minimal additional effort and should bring out better the similarities and differences between them.

Throughout this appendix $\mathbb{K}$ will denote $\mathbb{R}$, or $\mathbb{C}$, or $\mathbb{H}$, and E an n-dimensional right vector space over $\mathbb{K}$. We shall assume known the basic facts about vector spaces over possibly non-commutative fields, including the existence of bases
(already implicit in the 'dimension $= n$' condition) and the fact that the solution space of m linearly independent equations $\lambda_1(x) = 0, \dots, \lambda_m(x) = 0$, has dimension $n - m$. (The λ_k are m linearly independent $\mathbb{K}$-linear functions $E \to \mathbb{K}$).

A bi-additive function $\varphi : E \times E \to \mathbb{K}$ is a *bilinear form* if $\varphi(x\alpha, y\beta) = \alpha\varphi(x,y)\beta$ and is a *sesquilinear form* if $\varphi(x\alpha, y\beta) = \bar{\alpha}\varphi(x,y)\beta$. A form φ of either type is *non-degenerate* if $\varphi(x,y) = 0$ for all y implies $x = 0$. If a form φ on E is given, we say a subspace F of E is *non-degenerate* if the restriction of φ to F is non-degenerate. The subspace of E *orthogonal* to the subspace F is $F^\perp = \{x \in E \mid \varphi(x,y) = 0 \text{ for all } y \in F\}$.

The classification of forms is based on the following lemma.

Lemma A1. Let φ be a non-degenerate form on E, F a non-degenerate subspace of E. Then $F^\perp$ is also non-degenerate and $E = F \oplus F^\perp$, i.e each $x \in E$ is uniquely a sum $x = y + z$ with $y \in F$ and $z \in F^\perp$.

Proof. Let $e_1, e_2, \dots, e_m$ be a basis for F. Since φ is non-degenerate, the m linear equations $\varphi(x, e_1) = 0, \dots, \varphi(x, e_m) = 0$, which define $F^\perp$, are linearly independent and have therefore $n - m$ linearly independent solutions, say $e_{m+1}, \dots, e_n$. The n vectors $e_1, e_2, \dots, e_n$ are linearly independent: if $\sum_1^n \alpha_k e_k = 0$, then $\sum_1^m \alpha_k e_k = -\sum_{m+1}^n \alpha_k e_k$ lies both in F and in $F^\perp$, hence equal to 0, by non-degeneracy of F. Thus the $e_1, e_2, \dots, e_n$ form a basis for E, with the first m basis vector from F and the remaining $n - m$ from $F^\perp$. The assertion follows. QED

A bilinear form is *symmetric* if $\varphi(x,y) = \varphi(y,x)$, *skew-symmetric* if $\varphi(x,y) = -\varphi(y,x)$. A sesquilinear form φ is *Hermitian* if $\varphi(x,y) = \overline{\varphi(y,x)}$, *skew-Hermitian* if $\varphi(x,y) = -\overline{\varphi(y,x)}$.

In case $\mathbb{K} = \mathbb{R}$, bilinear = sesquilinear. In case $\mathbb{K} = \mathbb{C}$, skew-Hermitian forms are turned into Hermitian forms by multiplication by i, and vice versa. In case $\mathbb{K} = \mathbb{H}$, there are no non-zero bilinear forms: such would have to satisfy both $\varphi(x\alpha\beta, y) = (\alpha\beta)\varphi(x,y)$ and $\varphi(x\alpha\beta, y\alpha\beta) = \beta(\alpha\varphi(x,y))$.

We now fix a non-degenerate form φ on E. We shall repeatedly refer to the equation

$$\varphi(x+y, x+y) = \varphi(x,x) + \varphi(y,y) + \varphi(x,y) + \varphi(y,x). \tag{A.1}$$

We consider several cases.

(a) φ *symmetric*, $\mathbb{K} = \mathbb{R}$ or $\mathbb{C}$. In view of (A.1) there is a vector $x \in E$ with $\varphi(x,x) \neq 0$. Since $\varphi(\alpha x, \alpha x) = \alpha^2\varphi(x,x)$, we may choose $\alpha \in \mathbb{K}$ so that $e_1 = \alpha x$

satisfies

$$\varphi(e_1, e_1) = \begin{cases} \pm 1, & \text{if } \mathbb{K} = \mathbb{R} \\ 1, & \text{if } \mathbb{K} = \mathbb{C}. \end{cases}$$

The subspace F spanned by e_1 is non-degenerate. Appealing to the lemma we proceed in the same way with the subspace $F^{\perp}$, and successively construct a basis in terms of which φ is given by

$$\varphi(x, y) = \begin{cases} \pm\xi_1\eta_1 \pm \cdots \pm \xi_n\eta_n, & \text{if } \mathbb{K} = \mathbb{R}, \\ \xi_1\eta_1 + \cdots + \xi_n\eta_n, & \text{if } \mathbb{K} = \mathbb{C}, \end{cases}$$

as desired.

(b) φ *skew-symmetric*, $\mathbb{K} = \mathbb{R}$ or $\mathbb{C}$. Choose $x, y \in E$ so that $\varphi(x, y) \neq 0$. Since $\varphi(x\alpha, y) = \alpha\varphi(x, y)$, we may choose $\alpha \in \mathbb{K}$ so that $e_1 = \alpha x$ and $f_1 = y$ satisfy

$$\varphi(e_1, f_1) = 1.$$

The subspace F spanned by e_1 and f_1 is non-degenerate. So we may again appeal to Lemma A.1 to construct a basis $e_1, \ldots, e_m, e_{m+1} = f_1, \ldots, e_{2m} = f_m$ in terms of which the form φ is given by

$$\varphi(x, y) = (\xi_1\eta_{m+1} - \xi_{m+1}\eta_1) + \cdots + (\xi_m\eta_{2m} - \xi_{2m}\eta_m),$$

as desired.

(c) φ *Hermitian*, $\mathbb{K} = \mathbb{C}$ or $\mathbb{H}$. The equation (A.1) shows again that there is a vector x with $\varphi(x, x) \neq 0$: otherwise $\varphi(x, y) \equiv -\varphi(y, x)$ would imply that $\varphi(x, y)$ is linear as well as conjugate linear in both x and y. Since $\overline{\varphi(x, x)} = \varphi(x, x)$, one sees from the equation $\varphi(\alpha x, \alpha x) = \bar{\alpha}\varphi(x, x)\alpha$ that one can choose $\alpha \in \mathbb{K}$ so that $e_1 = \alpha x$ satisfies

$$\varphi(e_1, e_1) = \pm 1.$$

Proceeding in the familiar way we arrive at a basis in terms of which φ is given by

$$\varphi(x, y) = \bar{\xi}_1\eta_1 \pm \cdots \pm \bar{\xi}_n\eta_n,$$

as desired.

(d) φ *skew-Hermitian*, $\mathbb{K} = \mathbb{H}$. Equation (A.1) shows once more that there is a vector x with $\varphi(x, x) \neq 0$: otherwise $\varphi(x, y) \equiv -\varphi(y, x)$ would imply once more that $\varphi(x, y)$ is linear as well as conjugate linear in both x and y. Since $\overline{\varphi(x, x)} = -\varphi(x, x)$, one sees from the equation $\varphi(\alpha x, \alpha x) = \bar{\alpha}\varphi(x, x)\alpha$ one can choose $\alpha \in \mathbb{K}$ so that $e_1 = \alpha x$ satisfies

$$\varphi(e_1, e_1) = j,$$

for example. (See problem 5(b), §2.1.) Proceeding as before, one constructs a basis in terms of which φ is given by

$$\varphi(x, y) = \bar{\xi}_1 j\eta_1 + \cdots + \bar{\xi}_n j\eta_n,$$

This completes the classification.

It should also be noted that *the number p of $+$ signs in the symmetric case for $\mathbb{K}=\mathbb{R}$ or in the Hermitian case for $\mathbb{K}=\mathbb{C}$, $\mathbb{H}$ is independent of the basis chosen*: it is in fact the maximal number of mutually orthogonal vectors x on which $\varphi(x,x)>0$. To see this, suppose exactly the first p signs are $+$ in the normal form for φ belonging to the basis $\{e_k\}$. Suppose $f_1,\dots,f_p$ are any p mutually orthogonal vectors on which φ is positive and let f be a non-zero vector orthogonal to these f_k. Consider vectors of the form $x=\alpha f+\alpha_1 f_1+\cdots+\alpha_p f_p$ that satisfy

$$\varphi(x,e_1)=0, \quad \dots, \quad \varphi(x,e_p)=0. \tag{A.2}$$

This is a system of p homogeneous linear equations for the $p+1$ coefficients $\alpha,\alpha_1,\dots,\alpha_p$ in x, so there is a solution vector x for which not all coefficients are equal to 0, so that $x\neq 0$. Equation (A.2) says that the first p components of x with respect to the basis $\{e_k\}$ are equal to 0. Since these first p components correspond exactly to the $+$ signs in the normal form for φ it is clear that $\varphi(x,x)<0$. On the other hand,

$$\varphi(x,x)=|\alpha|^2\varphi(f,f)+|\alpha_1|^2\varphi(f_1,f_1)+\cdots+|\alpha_p|^2\varphi(f_p,f_p),$$

so

$$|\alpha|^2\varphi(f,f)=\varphi(x,x)-|\alpha_1|^2\varphi(f_1,f_1)-\cdots-|\alpha_p|^2\varphi(f_p,f_p).$$

The right side of this equation is <0, hence so is the left side, and therefore so is $\varphi(f,f)$. This shows that indeed no further f can be adjoined to the f_k.

3.2 Cartan subgroups

G shall denote a complex classical group, $\mathrm{SL}(E)$, $\mathrm{SO}(E)$, or $\mathrm{Sp}(E)$, E an N-dimensional complex vector space, equipped with the appropriate form φ in the latter two cases. We choose a basis $e_1,\dots,e_N$ for E as follows. For $\mathrm{SL}(E)$ the basis is arbitrary. For $\mathrm{SO}(E)$ the basis is assumed to reduce the symmetric form $\varphi(x,y)$ to

$$(\xi_1\eta_{n+1}+\xi_{n+1}\eta_1)+\cdots+(\xi_n\eta_{2n}+\xi_{2n}\eta_n) \quad \text{if } N=2n \text{ is even,} \tag{1}$$

$$(\xi_1\eta_{n+1}+\xi_{n+1}\eta_1)+\cdots+(\xi_n\eta_{2n}+\xi_{2n}\eta_n)+\xi_{2n+1}\eta_{2n+1} \quad \text{if } N=2n+1 \text{ is odd.} \tag{2}$$

(This normal form for $\varphi(x,y)$ will be more convenient here than the one discussed in §3.1. Problem 1 explains the relation between the two.) For $\mathrm{Sp}(E)$ the basis is assumed to reduce the skew-symmetric form to

$$(\xi_1\eta_{n+1}-\xi_{n+1}\eta_1)+\cdots+(\xi_n\eta_{2n}-\xi_{2n}\eta_n), \quad \text{with } N=2n. \tag{3}$$

Elements of G will be identified with matrices by means of the basis chosen.

Let H be the diagonal subgroup of G, $\mathfrak{h}$ its Lie algebra. We use the letter H both for the group and for elements of its Lie algebra, hoping that its meaning will be clear from the context. Table 3.1 lists the various possibilities, diagonal matrices being written as N-tuples.

The labelling A–D of the types is customary; the subscript denotes the dimension of h, called the *rank* of G. In type A we have set $n = N$.

We shall consider the diagonal entries $\lambda_k = \lambda_k(H)$ as functions of $H \in \mathsf{h}$. They provide a (complex) linear coordinate system on h; in type A they satisfy $\lambda_1 + \cdots + \lambda_n = 0$, so that the coordinate map $H \to (\lambda_1(H), \ldots, \lambda_n(H))$ should be considered as taking on values in the $(n-1)$-dimensional subspace of $\mathbb{C}^n$ described by this equation, rather than in $\mathbb{C}^n$ itself. The diagonal entries $\epsilon_k = \epsilon_k(h)$ of $h \in H$ provide coordinates on H in an analogous way. These ϵ_k are related to the λ_k by $\epsilon_k(h) = e^{\lambda_k(H)}$ if $h = \exp H$.

A complex linear functional $\lambda \in \mathsf{h}^*$ is of the form

$$\lambda = l_1\lambda_1 + \cdots + l_n\lambda_n$$

with $(l_1, \ldots, l_n) \in \mathbb{C}^n$. In type A the n-tuple is determined only modulo n-tuples of the form $(l, \ldots, l)$. Any such λ gives a function $e^\lambda : \mathsf{h} \to \mathbb{C}$, $H \to e^{\lambda(H)}$. When all l_k are integers, this function on h *lifts* to a function on H, also denoted e^λ, defined by

$$e^\lambda(\exp H) = e^{\lambda(H)}.$$

Thus as function on H,

$$e^\lambda = \epsilon_1^{l_1}\epsilon_2^{l_2}\cdots\epsilon_n^{l_n}.$$

$e^\lambda : H \to \mathbb{C}^\times$ is a holomorphic homomorphism and $\lambda : \mathsf{h} \to \mathbb{C}$ is its Lie map. Any holomorphic homomorphism $H \to \mathbb{C}^\times$ is of the form e^λ for some such λ with $l_k \in \mathbb{Z}$ for all k. (Example 4, §2.6) These $\lambda \in \mathsf{h}^*$ with integral coefficient n-tuples $(l_1, \ldots, l_n) \in \mathbb{Z}^n$ will be called *weights* of H, a term also used for the homomorphisms e^λ themselves.

The subgroup H of G is maximal abelian: any matrix which commutes with a diagonal matrix with distinct eigenvalues is itself diagonal, since it leaves the one-dimensional eigenspaces invariant. It can be shown that any maximal abelian subgroup of G consisting of semisimple (= diagonalizable) elements is conjugate to H (Problem 6(c)), but this will not be needed here. We simply define a *Cartan subgroup* of G to be any subgroup conjugate to H; it is the diagonal subgroup of G for a basis of the same kind as the one defining H.

The normalizer $N_G(H) = \{n \in G \mid nHn^{-1} = H\}$ acts on H and on its Lie algebra h by conjugation. The subgroup of $N_G(H)$ leaving H or (equivalently)

Table 3.4 Cartan subgroups and Cartan subalgebras

Type	G	$h \in H$	$H \in \mathsf{h}$
A_{n-1}	$\mathrm{SL}(n, \mathbb{C})$	$(\epsilon_1, \ldots, \epsilon_n), \prod \epsilon_j = 1$	$(\lambda_1, \ldots, \lambda_n), \sum \lambda_j = 0$
B_n	$\mathrm{SO}(2n+1, \mathbb{C})$	$(\epsilon_1, \ldots \epsilon_n, \epsilon_1^{-1}, \ldots, \epsilon_n^{-1}, 1)$	$(\lambda_1, \ldots, \lambda_n, -\lambda_1, \ldots, -\lambda_n, 0)$
C_n	$\mathrm{Sp}(n, \mathbb{C})$	$(\epsilon_1, \ldots, \epsilon_n, \epsilon_1^{-1}, \cdots, \epsilon_n^{-1})$	$(\lambda_1, \cdots, \lambda_n, -\lambda_1, \ldots, -\lambda_n)$
D_n	$\mathrm{SO}(2n, \mathbb{C})$	$(\epsilon_1, \ldots, \epsilon_n, \epsilon_1^{-1}, \ldots, \epsilon_n^{-1})$	$(\lambda_1, \cdots, \lambda_n, -\lambda_1, \cdots, -\lambda_n)$

h pointwise fixed is precisely H, so that the quotient group $N_G(H)/H$ may be considered a group of transformations of H or of h. This group $W = N_G(H)/H$ is called the *Weyl group* of the pair (G, H). Table 3.5 shows the action of a typical element $s \in W$ on h in the coordinates λ_k, along with some additional data to be discussed later.

The third column lists a general element of $W : j \to j'$ is a permutation of $(1, 2, \ldots, n)$ and the sign $\pm$ depends on j (all possibilities occur).

As a sample, we verify the description of W in type D. A matrix $s \in N_G(H)$ permutes the coordinate lines $\langle e_j \rangle$, because these are the joint eigenspaces of H. (We use the same letter s to denote elements of $N_G(H)$ and of W.) The pairs $\langle e_j \rangle, \langle e_{n+j} \rangle$, $j = 1, \ldots, n$, of non-orthogonal coordinate lines retain this property under such a permutation, and the value $\varphi(e_j, e_{n+j}) = 1$ remains the same.

Thus either

$$se_j = \sigma_j e_{j'} \quad \text{and} \quad se_{j+k} = \sigma_j^{-1} e_{j'+k},$$

or

$$se_j = \sigma_j e_{j'+k} \quad \text{and} \quad se_{j+k} = \sigma_j^{-1} e_{j'} \tag{4}$$

for some permutation $j \to j'$ of $(1, 2, \ldots, n)$ and some $\sigma_j \in \mathbb{C}$. This matrix s acts on h by the rule $\lambda_j \to \pm\lambda_{j'}$; the sign $+$ or $-$ depends on whether the first or second possibility prevails. Furthermore, the condition '$\det s = 1$' implies that the second possibility occurs an even number of times, indicated by $\Pi(\pm 1) = 1$ in Table 3.2. Conversely, every transformation $\lambda_j \to \pm\lambda_{j'}$ with $\Pi(\pm 1) = 1$ comes from a matrix $s \in N_G(H)$, as one sees from the above argument. (One may even choose all $\sigma_j = 1$.). This completes the verification of the description of W in type D. The other types are similar. (The restriction $\Pi(\pm 1) = 1$ does not appear in type B, because one can always satisfy '$\det s = 1$' by setting $se_{2n+1} = \pm e_{2n+1}$ with the appropriate sign. The restriction $\Pi(\pm 1) = 1$ does not appear in type C either, because there σ_j^{-1} is replaced by $-\sigma_j^{-1}$ when the second case in (4) prevails.)

It remains to explain the last column of the table. Consider the decomposition of g into joint eigenspaces of $\operatorname{ad}(\mathsf{h})$. For $\mathsf{g} = \mathsf{sl}(n, \mathbb{C})$ this decomposition is evident from the formula

$$\operatorname{ad}(H)E_{jk} = (\lambda_j - \lambda_k)E_{jk} \tag{5}$$

where $\lambda_k = \lambda_k(H)$ and the E_{jk} are the matrix units for the basis $e_1, \ldots, e_n$:

$$E_{jk}e_k = e_j, \qquad E_{jk}e_l = 0 \quad \text{for } l \neq k.$$

Table 3.5 Weyl group, roots

Type	G	$s \in W$	$\lvert W \rvert$	Φ $(j < k)$
A_{n-1}	$\mathrm{SL}(n, \mathbb{C})$	$\lambda_j \to \lambda_{j'}$	$n!$	$\pm(\lambda_j - \lambda_k)$
B_n	$\mathrm{SO}(2n+1, \mathbb{C})$	$\lambda_j \to \pm\lambda_{j'}$	$2^n n!$	$\pm(\lambda_j \pm \lambda_k), \pm\lambda_k$
C_n	$\mathrm{Sp}(n, \mathbb{C})$	$\lambda_j \to \pm\lambda_{j'}$	$2^n n!$	$\pm(\lambda_j \pm \lambda_k), \pm 2\lambda_k$
D_n	$\mathrm{SO}(2n, \mathbb{C})$ $\prod(\pm 1) = 1$	$\lambda_j \to \pm\lambda_{j'}$	$2^{n-1} n!$	$\pm(\lambda_j \pm \lambda_k)$

Table 3.6 Root vectors

Type	$\alpha \in \Phi$	E_α
A_{n-1}	$\lambda_j - \lambda_k$	E_{jk}
B_n	$\lambda_j - \lambda_k$	$E_{jk} - E_{n+k,n+j}$
	$\lambda_j + \lambda_k$	$E_{j,n+k} - E_{k,n+j}$
	$-\lambda_j - \lambda_k$	$E_{n+j,k} - E_{n+k,j}$
	λ_j	$E_{j,2n+1} - E_{2n+1,j+n}$
	$-\lambda_j$	$E_{2n+1,j} - E_{n+j,2n+1}$
C_n	$\lambda_j - \lambda_k$	$E_{jk} - E_{n+k,n+j}$
	$\lambda_j + \lambda_k$	$E_{j,n+k} + E_{k,n+j}$
	$-\lambda_j - \lambda_k$	$E_{n+j,k} + E_{n+k,j}$
	$2\lambda_j$	$E_{j,n+j}$
	$-2\lambda_j$	$E_{n+j,j}$
D_n	$\lambda_j - \lambda_k$	$E_{jk} - E_{n+k,n+j}$
	$\lambda_j + \lambda_k$	$E_{j,n+k} - E_{k,n+j}$
	$-\lambda_j - \lambda_k$	$E_{n+j,k} - E_{n+k,j}$

Equation (5) shows that in type A the eigenvalues of ad(h) are the $\alpha = \lambda_j - \lambda_k$ as functions of $H \in \mathsf{h}$. The non-zero ones among these α's are listed under the heading $\alpha \in \Phi$ ($j < k$) in the table; $j < k$ indicates the restriction on index pairs jk.

For the remaining types B–D, we have

$$\mathsf{g} = \{X \in M_N(\mathbb{C}) \mid X^\varphi = -X\},$$

where X^φ is the adjoint of X with respect to the form φ defining g. Since $H \in \mathsf{h}$ is diagonal, the eigenvectors of ad(H) on all of $M_N(\mathbb{C})$ are still the E_{jk} and the eigenvalues are still the $\lambda_j - \lambda_k$, except that the diagonal entries $(\lambda_1, \ldots, \lambda_N)$ of $H \in \mathsf{h}$ are subject to the restriction $H^\varphi = -H$. The eigenvectors of ad(H) in g belonging to a given eigenvalue α (as function of H) are then

$$E_\alpha = E_{jk} - (E_{jk})^\varphi$$

for which $\lambda_j - \lambda_k = \alpha$ on h ($j, k = 1, 2, \ldots, N$). Considering now the types B–D case-by-case one finds the non-zero eigenvalues α and the corresponding eigenvectors E_α as listed in Table 3.6. (Type A is also listed for reference.) The index pairs jk are restricted by $j \neq k$.

The non-zero eigenvalues $\alpha = \alpha(H)$, considered as linear functions $\alpha \in \mathsf{h}^*$, are called the *roots* of the pair g, h. These α form the *root system* Φ of g, h, listed in the previous table.

The joint eigenspace of ad(h) in g with eigenvalue 0 is precisely h itself (again because a matrix commuting with a diagonal matrix $H \in \mathsf{h}$ with distinct eigenvalues must itself belong to h). Thus the *root-space decomposition* of g into eigenspaces of h looks like

$$\mathsf{g} = \mathsf{h} + \sum_{\alpha \in \Phi} \mathbb{C} E_\alpha,$$

each *root-vector* $E_\alpha \in \mathsf{g}$ occurs with multiplicity 1, as one sees from Table 3.6.

Two types of bracket relations in g are used frequently:

$$[H, E_\alpha] = \alpha(H)E_\alpha, \tag{6}$$

$$[E_\alpha, E_\beta] \begin{cases} \in \mathbb{C}E_{\alpha+\beta} & \text{if } \alpha+\beta \text{ is a root,} \\ \in \mathsf{h} & \text{if } \beta = -\alpha, \\ = 0 & \text{otherwise.} \end{cases} \tag{7}$$

Equation (6) follows from the definition of E_α; eqn (7) from the Jacobi Identity:

$$\operatorname{ad}(H)[E_\alpha, E_\beta] = [\operatorname{ad}(H)E_\alpha, E_\beta] + [E_\alpha, \operatorname{ad}(H)E_\beta].$$

We note:

Proposition 1. *The Weyl group permutes the roots.*

Explanation. We consider W as a group of linear transformations of h through the adjoint representation (here denoted by a dot, for short):

$$s \cdot H = sHs^{-1}.$$

Then W can also be considered a group of linear transformations of h^*, operating by the rule

$$(s \cdot \lambda)(H) = \lambda(s^{-1} \cdot H).$$

The assertion is that for any root α, $s\alpha$ is again a root.

Proof. This is evident from Table 3.5, but is better understood as follows. Let $s \in N_G(H)$, $\alpha \in \Phi$. Then

$$\begin{aligned} \operatorname{ad}(H)\operatorname{Ad}(s)E_\alpha &= \operatorname{Ad}(s)\operatorname{ad}(s^{-1}Hs)E_\alpha \\ &= \alpha(s^{-1}Hs)\operatorname{Ad}(s)E_\alpha. \end{aligned}$$

Thus $\operatorname{Ad}(s)E_\alpha$ is an eigenvector for $\operatorname{Ad}(H)$ with eigenvalue $\alpha(s^{-1} \cdot H)$. QED

Remark 2. W preserves the inner product $(\lambda, \mu) = l_1m_1 + \cdots + l_nm_n$ of real n-tuples λ, μ. In types A and D all roots have the same length; in types B and C there are *short* and *long* roots, the two root-lengths having the ratio $1 : \sqrt{2}$. In each type, the roots of the same length are conjugate under W, with the exception of D_2.

We shall now prove that the complex classical Lie algebras are simple except for D_1 and D_2. (Recall that a Lie algebra g is simple if it has dimension >1 and no ideals except $\{0\}$ and g.) This fact is of intrinsic interest, but will not be needed later.

Theorem 3. *The complex classical Lie algebras are simple except for* $D_1 = \mathsf{so}(2, \mathbb{C})$, *which is one-dimensional, and* $D_2 = \mathsf{so}(4, \mathbb{C})$, which is isomorphic with $A_1 \times A_1 = \mathsf{sl}(2, \mathbb{C}) \times \mathsf{sl}(2, \mathbb{C})$.

Table 3.7 The complex and compact classical groups

Type	G	K	$\dim_{\mathbb{C}} G = \dim_{\mathbb{R}} K$
A_{n-1}	$\mathrm{SL}(n, \mathbb{C})$	$\mathrm{SU}(n)$	$n^2 - 1$
B_n	$\mathrm{SO}(2n+1, \mathbb{C})$	$\mathrm{SO}(2n+1)$	$2n^2 + n$
C_n	$\mathrm{Sp}(n, \mathbb{C})$	$\mathrm{Sp}(n)$	$2n^2 + n$
D_n	$\mathrm{SO}(2n, \mathbb{C})$	$\mathrm{SO}(2n)$	$2n^2 - n$

Proof. Let g be a complex classical Lie algebra, G the corresponding group. The ideals of g are precisely the $\mathrm{ad}(\mathsf{g})$-stable subspaces, (i.e. stable under $\mathrm{ad}(X)$ for all $X \in \mathsf{g}$). We shall show that there are no such subspaces except $\{0\}$ and all of g.

Suppose z is an ideal of g. Since z is in particular $\mathrm{ad}(\mathsf{h})$-stable, it must decompose into a sum of joint eigenvectors of $\mathrm{ad}(\mathsf{h})$. The eigenvectors with non-zero eigenvalues are root vectors E_α, and if $\mathsf{z} \neq \{0\}$, as we shall assume, then z must contain some such root vectors; otherwise z is contained in h, which is impossible except in case D_1.

Since z is also stable under $N_G(H)$, $E_{s\alpha}$ occurs along with E_α in z for any $s \in W$. In view of Remark 2 above this shows that, in types A and D, z contains all E_α. (In type D_n we require $n \geq 3$). In types B and C, z contains along with E_α all E_β for which β has the same length as α. From Table 3.6 one can check that in types B and C there are bracket relations of the form $[E_\alpha, E_\beta] = \text{const.}\ E_{\alpha+\beta}$ with α and $\alpha + \beta$ of different length and const. $\neq 0$. This implies that also in type B and C, z contains all E_α. Finally, one may verify that the elements of the form $[E_\alpha, E_{-\alpha}]$ span h, so that h is contained in z as well. Thus $\mathsf{z} = \mathsf{g}$.

The isomorphism $\mathsf{so}(4, \mathbb{C}) \approx \mathsf{sl}(2, \mathbb{C}) \times \mathsf{sl}(2, \mathbb{C})$ may be seen from problem 8, §2.6.

QED

We now consider the unitary subgroup $K = G \cap \mathrm{U}(E)$ of a complex classical group G. (The Hermitian inner product on E is the one defined in exercise 1(b) below.) The groups K are the compact classical groups, listed in Table 3.7.

(It should be noted that for the basis of E chosen above $\mathrm{SO}(2n + 1)$ or $\mathrm{SO}(2n)$ is not the subgroup of $\mathrm{SO}(E)$ leaving invariant the real subspace of E spanned by the basis vectors e_k. See problem 1.) The compact classical groups are connected, as we know. Furthermore, k is a real form of g, meaning

$$\mathsf{g} = \mathsf{k} + i\mathsf{k}$$

as direct sum of real vector spaces. This decomposition amounts to writing a complex matrix $Z \in M_n(\mathbb{C})$ as $Z = X + iY$ with X and Y skew-Hermitian.

The diagonal subgroup $H \cap K$ of K is denoted T; T or any subgroup of K conjugate to it is called a Cartan subgroup of K. T is the subgroup of H on which the coordinates ϵ_k have modulus 1; t is the subalgebra of h on which the coordinates λ_k are purely imaginary. We shall apply to T without further comment the notation introduced for H: h, H, e_k, λ_k, e^λ, $\lambda = (l_1, \ldots, l_n)$.

The Weyl group was defined as $W = N_G(H)/H$. The action of W on h by conjugation leaves t invariant, since t consists precisely of matrices in h with imaginary eigenvalues. Since the only element of W that fixes all of t (or one element with distinct eigenvalues) is the identity, W may also be considered a group of linear transformations of t. *Every such transformation from* $W = N_G(H)/H$ may in fact be represented by an element from K, as is evident from the discussion following Table 3.5. Thus $W = N_K(T)/T$ as transformation group of t (T, h, or H). We shall use this alternative realization of W without further comment.

Theorem 1, §3.1 may now be restated and sharpened as follows.

Theorem 4. *Every element of K is conjugate to an element of T, unique up to conjugation by W.*

Proof. The conjugacy statement is Theorem 1, §3.1. The uniqueness statement is proved as follows. Assume two elements of T are conjugate by an element $c \in K$. Then c maps the eigenspace of one with given eigenvalue to the eigenspace of the other with the same eigenvalue. Since these eigenspaces are spanned by certain coordinate vectors e_j, one can modify c by an element of K so that it permutes the coordinate lines $\langle e_j \rangle$ themselves. This element then represents the required element of W. QED

Problems for §3.2

1. Let E be an N-dimensional complex vector space with a non-degenerate, symmetric bilinear form φ, $e_1, \dots, e_N$ a basis for E that reduces φ to the form (1) or (2).

(a) Show that the basis $\{e'_k\}$ of E defined by

$$e'_j = \frac{1}{\sqrt{2}}(e_j + e_{n+j}) \quad e'_{n+j} = \frac{i}{\sqrt{2}}(e_j - e_{n+j}) \quad [e'_{2n+1} = e_{2n+1}]$$

($j = 1, \dots, n$) reduces the form φ on E given by (1) or (2) to

$$\xi'_1\eta'_1 + \dots + \xi'_{2n}\eta'_{2n}[+\xi'_{2n+1}\eta'_{2n+1}].$$

Conclude that, with respect to the basis $e'_1, \dots, e'_N$, $\mathrm{SO}(E)$ is represented by matrices $a \in M_N(\mathbb{C})$, $aa^t = 1$, $\det a = +1$.

(b) Provide E with the positive definite Hermitian form given by

$$\xi_1\bar{\eta}_1 + \dots + \xi_{2n}\bar{\eta}_{2n}[+\xi_{2n+1}\bar{\eta}_{2n+1}] = \xi'_1\bar{\eta}'_1 + \dots + \xi'_{2n}\bar{\eta}'_{2n}[+\xi'_{2n+1}\bar{\eta}'_{2n+1}]$$

in terms of the coordinates with respect to $\{e_k\}$ and $\{e'_k\}$. (Check the equality). Show that with respect to the basis $\{e'_k\}$ $\mathrm{SO}(E) \cap \mathrm{U}(E)$ is represented by matrices $a \in M_N(\mathbb{R})$, $aa^t = 1$, $\det a = +1$.

(c) Order the basis $\{e'_k\}$ as $e'_1, e'_{n+1}, \ldots, e'_n, e'_{2n}, [e'_{2n+1}]$. Show that with respect to this ordered basis the subgroup H of $\mathrm{SO}(E)$ is represented by block-diagonal matrices $(t_1, t_2, \ldots, t_n, [1])$ where each t_k is a 2×2 block

$$\begin{bmatrix} \cos 2\pi\theta & -\sin 2\pi\theta \\ \sin 2\pi\theta & \cos 2\pi\theta \end{bmatrix}, \quad \theta \in \mathbb{C}.$$

(The 1 for $\dim E = 2n+1$ is a 1×1 block.) $T = H \cap \mathrm{U}(E)$ consists of such matrices with $\theta \in \mathbb{R}$.

2. Simplify the relations (7):

$$[E_\alpha, E_\beta] = \begin{cases} N_{\alpha\beta} E_{\alpha+\beta}, N_{\alpha\beta} \neq 0 & \text{if } \alpha + \beta \text{ is a root,} \\ H_\alpha, & \text{if } \beta = -\alpha, \\ 0 & \text{otherwise.} \end{cases}$$

Here $H_\alpha \in \mathsf{h}$, $\alpha \in \Phi$, is a non-zero multiple of α if h is identified with h^* by the bilinear form $\mathrm{tr}(XY)$. [Suggestion: use Table 3.6; it suffices to give the details in type D, for example.]

3. Verify the list of roots α and root vectors E_α in Table 3.6; (a) in type B. (b) in type C, (c) in type D.

4. Verify the description of $s \in W$ given in Table 3.5 in types B and C.

5. In types B_n, C_n, D_n, show that the sign-changes in W form a normal subgroup isomorphic to $(\mathbb{Z}_2)^n$ or, in type D_n, to $(\mathbb{Z}_2)^{n-1}$. Show that every element $s \in W$ may be uniquely written as $s = \sigma\epsilon$ where σ is a permutation $(l_1, \ldots, l_n) \to (l_{1'}, \ldots, l_{n'})$ and ϵ a sign-change $(l_1, \ldots, l_n) \to (\pm l_1, \ldots, \pm l_n)$, even ($\prod(\pm 1) = 1$) in type D_n.

These facts are summarized in the Table 3.8.

6. With the notation of this section, show:

(a) Every *semisimple* element of G is conjugate (by G) to an element of H. [Suggestion: construct an appropriate basis consisting of eigenvectors.]

(b) Every semisimple element of $\mathbf{g}$ is conjugate (by G) to an element of h. [Same suggestion.]

Table 3.8 Weyl group

Type	W
A_{n-1}	S_n
B_n, C_n	$S_n \rtimes (\mathbb{Z}_2)^n$ (semidirect product)
D_n	$S_n \rtimes (\mathbb{Z}_2)^{n-1}$ (semidirect product)

(c) Every abelian subalgebra a of g consisting of semisimple elements is conjugate (by G) to a subalgebra of h. [Suggestion: decompose E into joint eigenspaces E_λ, $\lambda \in \mathsf{a}^*$, of a. Find an element $X \in \mathsf{a}$ so that distinct λ have distinct values $\lambda(X)$ on X. Apply (b) to X to find an element $\mathsf{a} \in G$ with $\mathrm{Ad}(\mathsf{a})X \in \mathsf{h}$. Show that $\mathrm{Ad}(a)\mathsf{a} \subset \mathsf{h}$.]

(d) Every *connected* abelian subgroup A of G consisting of semisimple elements is conjugate (by G) to a subgroup of H. [Suggestion: use (b).]

(e) Every maximal abelian subalgebra a of g consisting of semisimple elements is conjugate (by G) to h.

(f) Every *connected* maximal abelian subgroup A of G consisting of semisimple elements is conjugate (by G) to H.

(g) The statement (f) is false if 'connected' is omitted. [Consider the subgroup of $\mathrm{SO}(E)$ consisting of diagonal matrices $(\pm 1, \ldots, \pm 1)$ for the basis $\{e'_k\}$ in problem 1.]

(h) The statements (e) and (f) are false if 'consisting of semisimple elements' is omitted. [Suggestion: consider the subgroup of $\mathrm{SL}(E)$ consisting of matrices

$$\begin{bmatrix} 1 & x \\ 0 & 1 \end{bmatrix},$$

where $x = (\xi_1, \ldots, \xi_{n-1})$ and the lower right-hand 1 is the $(n-1) \times (n-1)$ identity matrix. (Any basis for E.)]

7. Prove the assertions of Remark 2.

8. Complete the verifications required for the proof of Theorem 3.

9. Recall that the n-torus $\mathbb{T}^n$ consists of all $n \times n$ diagonal matrices with complex diagonal entries of modulus $= 1$. More generally, any linear group which is isomorphic to some $\mathbb{T}^n$ by an analytic group isomorphism is called a *torus*.

   Every maximal torus in a compact classical group is a Cartan subgroup.

   Prove this assertion for: (a) $\mathrm{SO}(n)$, (b) $\mathrm{SU}(n)$, (c) $\mathrm{Sp}(n)$ [Suggestion: Show that the Lie algebra t of a maximal torus T is contained in the Lie algebra of a Cartan subgroup by constructing a basis with respect to which the elements of t have (simultaneously) the appropriate matrix form.]

3.3 Roots, weights, reflections

In this section we are concerned with geometric and combinatorial aspects of the root systems of the classical groups. The groups themselves, or their Lie algebras, are no longer needed explicitly.

Roots and, more generally weights are integral linear combinations of the λ_k. In particular, they lie in the real subspace L of h^* spanned by the λ_k. An element $\lambda = l_1\lambda_1 + \cdots + l_n\lambda_n$ of L will also be written as an n-tuple $(l_1, \ldots, l_n)$ of real numbers. In type A_{n-1} two such n-tuples are identified if they differ by an n-tuple of the form $(l, \ldots, l)$; the n-tuples may be normalized by the condition $l_1 + \cdots + l_n = 0$ (or in other ways).

We introduce a positive-definite inner product in L by setting

$$(\lambda, \mu) = l_1 m_1 + \cdots + l_n m_n$$

in terms of the components l_k of λ and m_k of μ.

The root systems of types A–D are listed in Table 3.5, §3.2. The *rank* of a root system is the dimension of the space L in which it lives (in agreement with the definition of *rank* for the groups). The roots in Φ, as listed in Table 3.5, §3.2 occur in pairs $\pm\alpha$; those with the plus sign will be called *positive roots*, those with the minus sign *negative*. The collection of positive roots is denoted Φ_+. The root systems of rank 2 are shown in Figure 1 at the end of this section.

An element λ of L is called *dominant weight* if $(\lambda, \alpha) \geq 0$ for all positive roots α, *strictly dominant weight* if the inequalities are all strict. In the rank 2 case they form the shaded regions in Figure 1. The dominant elements of L are described as n-tuples $(l_1, \ldots, l_n)$ in Table 3.9, where the positive roots are also listed for reference.

The prettiest aspect of a root system has to do with reflections. The *reflection* along a root α (or along any element α of a Euclidean space) is the linear transformation s_α of L that sends α to $-\alpha$ and leaves the hyperplane orthogonal to α pointwise fixed. These two properties lead to the formula

$$s_\alpha(\lambda) = \lambda - 2\frac{(\alpha, \lambda)}{(\alpha, \alpha)}\alpha. \tag{1}$$

(The term 'reflection in the hyperplane orthogonal to α' would be more appropriate for s_α, but is too cumbersome.) The reflections s_α for the relevant α are listed in Table 3.10, which shows the effect of s_α on the basis vectors λ_j. It is understood that basis vectors not explicitly listed remain fixed.

Reflections enter the picture through the following crucial observation.

Proposition 1. *The reflections s_α along the roots $\alpha \in \Phi$ belong to the Weyl group W and generate W.*

Proof. The transpositions $\lambda_j \leftrightarrow \lambda_k$ by themselves generate all permutations

$$(l_1, \ldots, l_n) \to (l_{1'}, \ldots, l_{n'});$$

Table 3.9 Dominant weights

Type	$\Phi_+ (j < k)$	Dominant λ
A_{n-1}	$\lambda_j - \lambda_k$	$l_1 \geq \cdots \geq l_n$
B_n	$\lambda_j \pm \lambda_k$, λ_k	$l_1 \geq \cdots \geq l_n \geq 0$
C_n	$\lambda_j \pm \lambda_k$, $2\lambda_k$	$l_1 \geq \cdots \geq l_n \geq 0$
D_n	$\lambda_j \pm \lambda_k$	$l_1 \geq \cdots \geq l_{n-1} \geq \lvert l_n \rvert$

Table 3.10 Reflections

α	s_α
$\lambda_j - \lambda_k$	$\lambda_j \leftrightarrow \lambda_k$
$\lambda_j + \lambda_k$	$\lambda_j \leftrightarrow -\lambda_k$
λ_j, $2\lambda_j$	$\lambda_j \leftrightarrow -\lambda_j$

together with the transpositions with sign switches $\lambda_j \leftrightarrow -\lambda_k$ they generate all permutations with an even number of sign changes

$$(l_1, \ldots, l_n) \to (\pm l_{1'}, \ldots, \pm l_{n'}), \quad \prod(\pm 1) = 1;$$

with the further addition of one single switch $\lambda_j \leftrightarrow -\lambda_j$ they generate all permutations with sign changes. Examining Table 3.6, §3.2 and Table 3.10 above in the light of these facts one sees what is being asserted. QED

We pause momentarily for a bit of history. In 1926 Weyl *defined* the groups now bearing his name as the groups generated by the s_α. For the classical groups, which alone concern us here, it is evident that these are exactly the groups $N_G(H)/H$. But in the abstract setting of semisimple groups it is not evident that the two definitions of W coincide. Indeed, in 1926 Weyl had only indirect evidence that the two definitions must be equivalent (as he mentions on p. 361 of his paper). The proof of their equivalence is due to Cartan (1925), who at that point was already influenced by Weyl's work.

We continue. There are three properties of the action of W on L that will be essential for the proof of Weyl's Character Formula in §6.7. To state them we adopt the following terminology. An element λ of L is a *regular weight* if $(\lambda, \alpha) \neq 0$ for all roots α; $\lambda \in L$ is *higher* than $\mu \in L$ (μ is *lower* than λ) if

$$\lambda = \mu + (\text{a linear combination of positive roots with real coefficients } \geq 0).$$

We admit the possibility that $\lambda = \mu$ (empty sum) and use the term *strictly higher* (or *strictly lower*) to exclude that case. (It is *not* required that *all* coefficients be > 0.)

The relation 'higher than' is a partial order on L. There is also a total order on L, the *lexicographic order*, defined as follows. Consider elements of L as n-tuples $\lambda = (l_1, \ldots, l_n) \in \mathbb{R}^n$ (in type A_{n-1} the n-tuples are here normalized by the condition $l_1 + \cdots + l_n = 0$.) λ precedes μ lexicographically if the first component l_j of λ that differs from the corresponding component m_j of μ satisfies $l_j > m_j$.

If λ is higher than μ then λ precedes μ lexicographically: this follows from the observation that $\mu + \alpha$ precedes μ lexicographically for every positive root, as one sees from the explicit expressions for the roots. In particular there are only finitely many dominant elements lower than a given λ, as there are only finitely many dominant elements lexicographically later than a given λ. (But 'dominant' is certainly *not* equivalent to 'higher than 0'; for example, the positive roots themselves are generally *not* dominant. Rather than being equivalent the two notions are in a sense *dual* to each other; see problem 8.)

The three properties of W referred to above may be stated thus:

Proposition 2.

(a) If $\lambda \in L$ is regular, then the only $s \in W$ with $s\lambda = \lambda$ is $s = 1$.

(b) Every $\lambda \in L$ is W-conjugate to a unique dominant element.

(c) If $\lambda \in L$ is dominant, then $s\lambda$ is lower than λ for all $s \in W$, strictly lower if λ is regular and $s \neq 1$.

Proof. Assertions (a) and (b) may be checked in each case A–D by representing the elements of L as n-tuples and by using the description of the $s \in W$ in Table 3.5, §3.2 together with the description of the dominant λ in Table 3.9 above. We consider the type D_n as example. There an element $\lambda = (l_1, \ldots, l_n)$ is regular if

$$l_j \neq \pm l_k \quad \text{for all } j \neq k \tag{2}$$

and dominant if

$$l_1 \geq \cdots \geq l_{n-1} \geq |l_n|. \tag{3}$$

An element $s \in W$ is of the form

$$(l_1, \ldots, l_n) \to (\pm l_{1'}, \ldots, \pm l_{n'}), \quad \prod(\pm 1) = 1. \tag{4}$$

From (2) and (4) it is evident that $s\lambda = \lambda$ for regular λ implies $s = 1$.

Next, the entries of any n-tuple λ may be ordered decreasingly in absolute value by a permutation without sign changes. With sign changes the entries may be made ≥ 0, and if an odd number of sign changes is required for this purpose the last entry l_n may be changed to $-l_n$ so that an even number of sign changes is used. This proves (a) and (b) in type D_n. The other types are similar.

(c) It may be proved simultaneously in all types, which in this case is simpler than a type-by-type verification. Start with any element μ of L. If μ is not dominant there is a positive root α for which $(\mu, \alpha) < 0$. Then $s_\alpha \mu = \mu - (2(\mu, \alpha)/(\alpha, \alpha))\alpha$ is higher than μ. If $s_\alpha \mu$ is still not dominant, we apply the same procedure to $s_\alpha \mu$, and continue until we arrive at a λ that is dominant. (The process must terminate in a finite number of steps, because the value of the inner product of $\mu, s_\alpha \mu, \ldots$ with a fixed dominant regular element is strictly increasing.) The final λ is W-conjugate to μ (being obtained from μ by a sequence of reflections along roots), so $\mu = s\lambda$ with $s \in W$ and is lower than λ, by construction, strictly lower for regular λ, by part (a).

To finish the proof of (c) this argument may now be applied to $\mu = s\lambda$ for a given dominant λ (which is then the unique dominant λ W-conjugate to $s\lambda$). QED

Note, incidentally, that the preceding proof shows once more that every $s \in W$ may be written as a product of reflections along roots.

A *highest weight* element of a given subset of L has the property that no other element of the set is higher than it. The proposition says that every W-orbit $W \cdot \lambda$ in L has a unique highest element, namely the dominant element in the orbit. (For arbitrary subsets of L, there will be several highest elements, however.)

We close this section with a few remarks about *abstract root systems*. It may already be apparent that the essential properties of root systems and Weyl groups, expressed in Propositions 1 and 2, should follow from a few axioms, quite independently of the origins of the root systems in group theory. This is the case, and the appropriate *abstract* definition of 'root system' reads as follows.

A *root system* is a subset Φ of a Euclidean space L satisfying:

R1. Φ is finite, spans E, and does not contain 0.

R2. For $\alpha \in \Phi$, the only multiples of α in Φ are $\pm\alpha$.

R3. For $\alpha \in \Phi$, the reflection s_α, $s_\alpha \lambda = \lambda - 2((\lambda, \alpha)/(\alpha, \alpha))\alpha$, leaves Φ invariant.

R4. $2((\beta, \alpha)/(\alpha, \alpha))$ is an integer for all $\alpha, \beta \in \Phi$.

The basic properties of abstract root systems may be found in the book by Humphreys (1972), for example; the standard reference is Bourbaki (1968) where a wealth of further information may be found.

Any root system (in the abstract sense) may be decomposed into *irreducible* root systems. It turns out that in addition to the four classical families A_n, B_n, C_n, D_n there are only five exceptional irreducible root systems, denoted G_2, F_4, E_6, E_7, E_8. The classification of root systems leads to the classification of complex simple Lie algebras: the irreducible root systems correspond one-to-one to the complex simple Lie algebras (up to isomorphism), of which there are therefore also only five exceptional ones, beyond the four classical families.

The root systems of rank 2 are shown in Figure 1, including the exceptional G_2. The positive roots are the roots that are linear combinations with coefficients ≥ 0 of the two roots α_1, α_2 shown; the shaded area consists of the dominant elements of L.

Problems for §3.3

1. Verify Table 3.9.

2. Verify Table 3.10.

3. Verify parts (a) and (b) of Proposition 2 in types A–C.

4. Verify that the classical root systems satisfy the axioms R1–R4.

   The following problems refer to the *classical* root systems A–D. The tables provided in the text may be used.

5. A positive root is called *simple* root if it cannot be written as the sum of two positive roots.

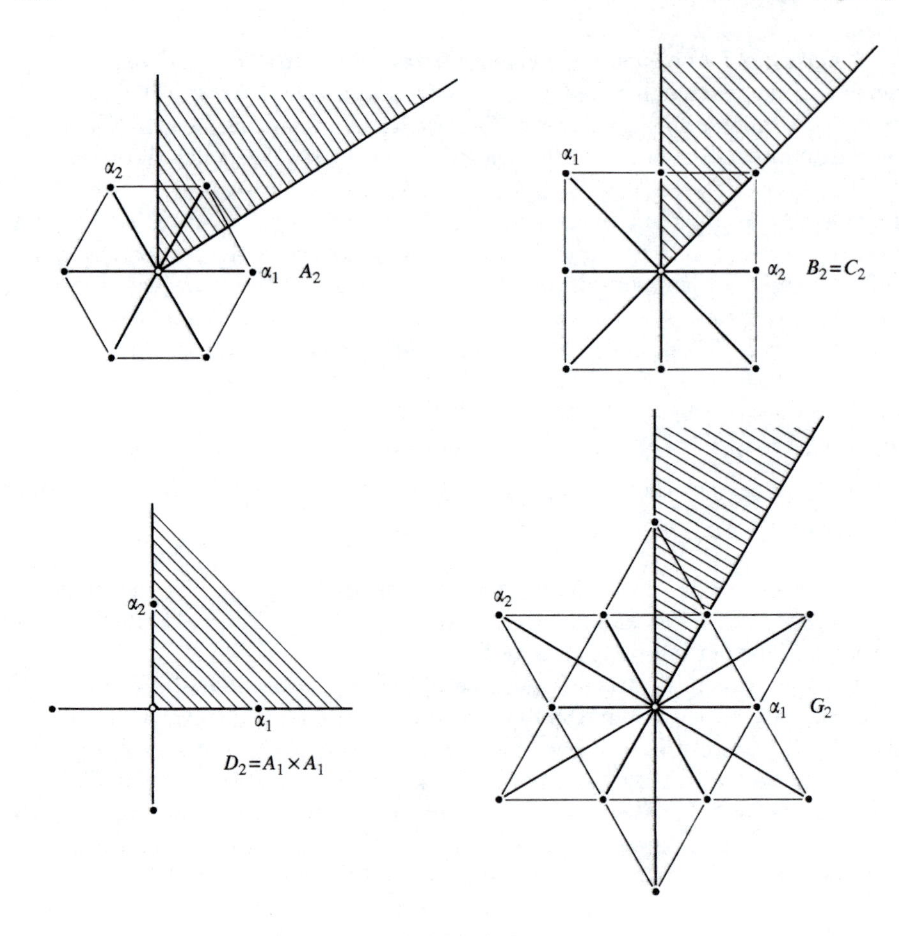

Fig. 1 The root systems of rank 2.

(a) Verify the expressions for the simple roots listed in Table 3.11. (The positive roots are also listed for reference.)

(b) Verify that the simple roots form a basis for L and show that the components of the positive roots with respect to this basis are integers ≥ 0. Find the explicit expressions for the positive roots in terms of the simple roots in each case A–D.

6. Show that W is already generated by the reflections along the *simple* roots.

7. (a) Show that the reflection s_α along a *simple* root α permutes the positive roots other than α.

   (b) Let $\rho = \frac{1}{2}\sum_{\alpha\in\Phi_+}\alpha$ be half the sum of the positive roots. Use part (a) to show that $(2(\rho,\alpha)/(\alpha,\alpha)) = 1$ for every simple root α.

8. Let $\{\alpha_j\}$ be the basis of L consisting of the simple roots. Its *dual basis* $\{\alpha^j\}$ (for the inner product (λ,μ) on L) is defined by $(\alpha^j,\alpha_k)=\delta_{jk}$

Table 3.11 Simple roots

Type	$\Phi_+(j < k)$	Simple roots $(1 \leq j \leq n-1)$
A_{n-1}	$\lambda_j - \lambda_k$	$\lambda_j - \lambda_{j+1}$
B_n	$\lambda_j \pm \lambda_k,\ \lambda_j$	$\lambda_j - \lambda_{j+1},\ \lambda_n$
C_n	$\lambda_j \pm \lambda_k,\ 2\lambda_j$	$\lambda_j - \lambda_{j+1},\ 2\lambda_n$
D_n	$\lambda_j \pm \lambda_k$	$\lambda_j - \lambda_{j+1},\ \lambda_{n-1} + \lambda_n$

Table 3.12 Highest root

Type	Highest root
A_{n-1}	$\lambda_1 + \lambda_n$
B_n	$\lambda_1 + \lambda_2$
C_n	$2\lambda_1$
D_n	$\lambda_1 + \lambda_2$

(a) $\lambda \in L$ is dominant weight if and only if $\lambda = \sum_j a_j \alpha^j,\ a_j \geq 0$.

(b) $\lambda \in L$ is higher than 0 if and only if $\lambda = \sum_j a^j \alpha_j,\ a^j \geq 0$.

9. Verify that each classical root system has a *unique* highest root, except D_2, which has two highest roots (Table 3.12).

3.4 Fundamental groups of the classical groups[2]

The connectivity properties of the compact classical groups were determined by Weyl (1925–26) as an essential part of his investigations of the representations of these groups. His methods and results were extended by Cartan (1927) to all compact semisimple groups. We shall retrace the historical development, following first Weyl then Cartan: there is something to be learnt from both. Some elementary facts from homotopy theory will be needed, but the essential points of the discussion should be understandable from the explanations given here. (Details may be found in any book on algebraic topology, for example in Massey (1967).)

Two continuous paths $a_o(\tau), a_1(\tau), \tau \in [0,1]$, in a linear group, or in any other topological space, with the same endpoints a_o and $a_1\,[a_o(0) = a_1(0) = a_o,\ a_o(1) = a_1(1) = a_1]$ are said to be *homotopic* if there is a one-parameter family $\{a_\sigma(\tau)\}_\sigma$ of such paths, with $a_\sigma(\tau)$ depending continuously on $(\sigma, \tau) \in [0,1] \times [0,1]$, so that $a_\sigma(\tau)\mid_{\sigma=0} = a_o(\tau)$ and $a_\sigma(\tau)\mid_{\sigma=1} = a_1(\tau)$. It is understood that the endpoints

[2]The material of this section will not be used in this book.

remain fixed: $a_s(0) \equiv a_o, a_s(1) \equiv a_1$. The homotopy classes of loops [continuous closed paths: $a(0) = a(1)$] through a fixed based point form a group, the group operation being induced by concatenation of paths. This group is called the *fundamental group* of the space, usually denoted π_1. The choice of base-point is of no consequence in a connected space: loops at one base-point may be transformed to loops at another base-point by a round trip along a fixed path connecting the two points, and this provides an isomorphism of the corresponding fundamental groups.

The space is *simply connected* if $\pi_1 = \{1\}$, which means that every loop through a fixed-base point a_o is homotopic to a point. If $\pi_1 \approx \mathbb{Z}_2$ we call the space *doubly connected*, which means that there is a loop $a(\tau)$ through a_o not homotopic to a point, and that every loop through a_o is homotopic either to the point a_o or else to $a(\tau)$.

Theorem 1. SU(n) *and* Sp(n) *are simply connected;* SO(n) *is doubly connected.*

Proof. Let K denote SU(n), SO($2n+1$), Sp(n), or SO($2n$), realized as in §3.2, T its Cartan subgroup. Every $a \in K$ may be written as $a = ctc^{-1}$ with $t \in T$. We write $t = t(\epsilon_1, \ldots, \epsilon_n)$, as in Table 3.4, §3.2 with $|\epsilon_k| = 1$, say $\epsilon_k = e^{2\pi i\theta_k}$, (subject to $\Pi\epsilon_k = 1$ for $K = \mathrm{SU}(n)$). We say a is *regular* if $e^\alpha(t) \neq 1$ for all roots α; otherwise a is *singular*.

The element $t \in T$ is unique up to conjugation by W. And for fixed t, the element c can only be replaced by cb where b commutes with t. When a is regular, only such b's belong again to T; but if a is singular, say $e^\alpha(t) = 1$, α a root, then c can in addition be modified by any element of K that belongs to the connected group with Lie algebra $\mathbb{R}H_\alpha + \mathbb{R}E_\alpha + \mathbb{R}E_{-\alpha}$, where $H_\alpha = [E_\alpha, E_{-\alpha}]$. This subgroup of K is three-dimensional (isomorphic with SU(2) or SO(3), see problem 1). *The singular elements form therefore a subset of K of at least three dimensions less*, because of the collapsing of these three-dimensional groups to single points. A continuous path that happens to meet the singular set can be made to miss the singular set by a slight displacement; there is room to spare: two free dimensions would suffice. The extra dimension is put to use as follows. A continuous one-parameter family $a_\sigma(\tau)$ of paths may be considered a surface parametrized by (σ, τ). Since such a surface requires only two dimensions, it too can be made to miss the singular set by a slight displacement. (Weyl considered this to be evident. The idea is that two 'reasonable' subsets whose dimensions add up to less than the dimension of the surrounding space can be made disjoint

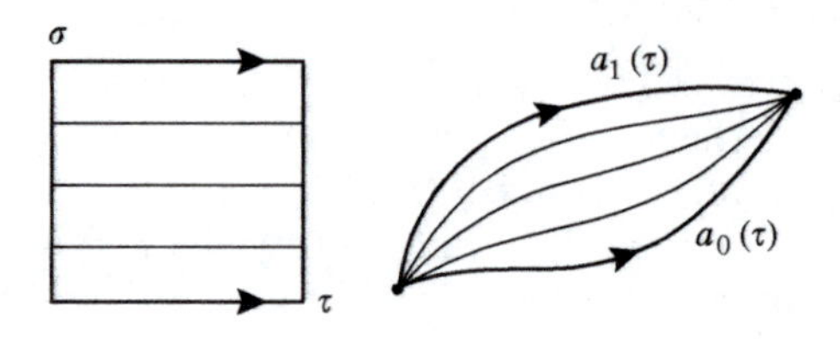

Fig. 1

by a slight displacement; the surrounding space is assumed to look locally like $\mathbb{R}^n$. A proof can be found in Freudental and de Vries (1969), for example.) As a consequence, if we choose a base-point in the regular set, then every loop in K is homotopic to a loop that lies entirely in K_{reg}; and if two such loops are homotopic in K then they are already homotopic in K_{reg}.

Choose as base-point a regular element t_o in T. Let $a(\tau)$, $0 \leq \tau \leq 1$, be a continuous loop through $t_o[a(0) = a(1) = t_o]$ that lies in K_{reg}. Then we may write

$$a(\tau) = c(\tau)t(\tau)c(\tau)^{-1}$$

with $c(\tau) \in K$ and $t(\tau) \in T_{\text{reg}}$ depending continuously on τ. (The continuity property will be taken for granted here; actually, we only require that the coset $c(\tau)T \in K/T$ depend continuously on τ, and that will follow from Proposition 3, §5.3.)

Write $t(\tau) = \exp H(\tau)$ with $H(\tau)$ depending continuously on τ. (Since $H \to \exp H$ is locally bijective, this may done by subdividing the path $t(\tau)$ into sufficiently small pieces.) The initial point $H(0)$ may be assumed to be a given element H_o of t with $\exp H_o = t_o$.

Assume first that $H(\tau)$ returns to its starting point: $H(1) = H_o$. Introduce a real parameter σ and consider

$$a_\sigma(\tau) = \exp(\sigma H_o)c(\tau)\exp\left((1-\sigma)H(\tau)\right)c(\tau)^{-1}.$$

As σ varies from 0 to 1 the initial loop $a(\tau)$ shrinks to the point t_o, while $a_\sigma(0) = a_\sigma(1) = t_o$ remains fixed. Such a loop $a(\tau)$ is therefore homotopic to a point. It follows that two loops that lead to the same endpoint $H(1)$ are homotopic: the composite of one with the other traversed in reverse is a loop whose $H(\tau)$ returns to H_o and which is therefore homotopic to a point. The endpoint $H(1)$ of the $H(\tau)$ belonging to a loop $a(\tau)$ depends only on the homotopy class of $a(\tau)$: $H(1)$ is subject to the condition

$$c(1)\exp H(1)c(1)^{-1} = \exp H_o.$$

Since $\exp H_o = t_o$ is regular $c(1)$ must normalize T, say $c(1)$ represents $s \in W$, and $H_1 = H(1)$ satisfies

$$H_1 = s^{-1}H_o + Z,$$

where $Z \in t$ satisfies $\exp Z = 1$. These H_1 from a discrete set in t and a continuous function that takes on values in this set, such as the $H_\sigma(1)$ belonging to a homotopy $a_\sigma(\tau)$, must be constant.

It remains to determine the possible $H_1 = H(1)$. For this purpose we follow the motion of the ϵ_k on the unit circle as a travels around the loop $a(\tau)$ in K. As functions of τ, the ϵ_k are given by $\epsilon_k = e^{2\pi i\theta_k(\tau)}$ where $2\pi i\theta_k(\tau) = \lambda_k(H(\tau))$ in terms of the (imaginary) coordinates λ_k on t. Thus the possible end positions $H(1)$ of the $H(\tau)$ correspond to the possible end positions $\theta_k(1)$ of the angles $\theta_k(\tau)$. It must be noted that the motion of the ϵ_k is subject to the condition that $e^\alpha(t(\tau)) \neq 1$ for all τ, because $t(\tau)$ remains in the regular set. We now determine case-by-case the few possibilities that remain.

$K = \mathrm{SU}(n)$. The conditions are $\epsilon_j \epsilon_k^{-1} \neq 1$ $(j \neq k)$, $\Pi \epsilon_k = 1$. The ϵ_k cannot pass each other on the circle ($\epsilon_j \epsilon_k^{-1} \neq 1$) and must therefore all return to their initial positions (rather than undergoing some permutation). If they have travelled m times around the circle before returning, then each angle θ_k will have changed by the same value m.

On the other hand the angle sum $\theta_1 + \cdots + \theta_n$ has remained constant, because it is capable only of the discrete values $0, \pm 1, \pm 2, \ldots$ ($\Pi \epsilon_k = 1$) and cannot pass continuously from one to the other. Therefore $m = 0$ and all θ_k return to their initial values. It follows that $\mathrm{SU}(n)$ is simply connected.

Comment. The condition $\Pi \epsilon_k = 1$ is crucial: $\mathrm{U}(n)$ is not simply connected (already for $n = 1$). $K = \mathrm{Sp}(n)$. The conditions are $\epsilon_j \epsilon_k^{\pm 1} \neq 1 (j \neq k), \epsilon_k^2 \neq 1$. We mark the ϵ_k and their inverses on the unit circle. Again the ϵ_k cannot pass each other ($\epsilon_j \epsilon_k^{-1} \neq 1$) and in addition they cannot pass through $\pm 1 (\epsilon_k^2 \neq \pm 1)$. Consequently, the ϵ_k can only oscillate about their initial positions and their angles θ_k must return to their initial values. $\mathrm{Sp}(n)$ is therefore also simply connected.

$K = \mathrm{SO}(2n+1)$. The conditions are $\epsilon_j \epsilon_k^{\pm 1} \neq 1$ $(j \neq k)$, $\epsilon_k \neq 1$. In this case, it is possible that the leftmost pair of points on the unit circle, labelled $\epsilon_n, \epsilon_n^{-1}$ in Figure 1, returns interchanged, the points having passed through -1. This means that $\theta_n(1) = 1 - \theta_n(0)$. All other θ_k must return to their initial values. This may indeed happen because one can take for $a(\tau)$ a path $c(\tau) \exp H(\tau) c(\tau)^{-1}$ for which $H(\tau) \in t$ is defined by such a motion of the θ_k, say

$$\begin{aligned} \theta_1 &= \theta_1(0), \ldots, \theta_{n-1} = \theta_{n-1}(0), \\ \theta_n &= (1-\tau)\theta_n(0) + \tau(1 - \theta_n(0)). \end{aligned}$$

The $\theta_k(0)$ are the values of the θ_k for the given $H(0) = H_o$; $c(\tau)$ is a path from $c(0) = 1$ to an element $c(1) = s$ that normalizes T and acts on the θ_k by $\theta_n \to -\theta_n$ (all other θ_k stay fixed). Thus there are two possible end positions of the θ_k. It follows that $\mathrm{SO}(2n+1)$ is doubly connected.

$K = \mathrm{SO}(2n)$. The conditions are $\epsilon_j \epsilon_k^{\pm 1} \neq 1$ $(j \neq k)$. The situation is similar to that of $\mathrm{SO}(2n+1)$, except it is also conceivable that the rightmost pair, labelled $\epsilon_1, \epsilon_1^{-1}$ in Figure 1, returns interchanged, the points having passed through $+1$. This means that $\theta_1(1) = -\theta_1(0)$. If this happens, then the leftmost pair returns interchanged as well. (Only an *even* number of such interchanges may occur because of the relation $t_o = st(1)s^{-1}$.) Thus there are again two possible end positions of the θ_k. It follows that $\mathrm{SO}(2n)$ is also doubly connected.

This finishes the proof of Theorem 1. QED

It is possible to determine the fundamental groups strictly from the geometry of roots and weights, without following the dance of the ϵ_k on the unit circle. To illustrate this method, due to E. Cartan, we now determine the fundamental groups of the adjoint groups (the most interesting case), even though the result could also be derived from the above theorem by consideration of Ad as a covering.

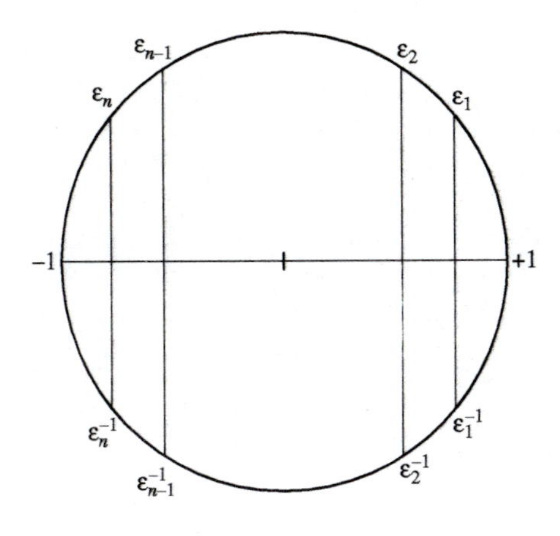

Fig. 2

We change notation: K will now denote Ad of the previous K and T the image of the previous T in K. To determine the fundamental group of the new K we proceed exactly as before: the homotopy class of $a(\tau) = c(\tau)\exp(H(\tau))c(\tau)^{-1}$ is uniquely characterized by the element $H_1 = H(1)$, which must be of the form

$$H_1 = s^{-1}H_o + Z \tag{1}$$

for some $s \in W$, and some $Z \in t$ with $\exp Z = 1$ in T.

It remains to find the possible H_1 that can occur as $H(1)$ for some loop $a(\tau)$. In addition to (1), H_1 must lie in the connected component C_o of H_o in $\{H \in t \mid \alpha(H) \notin 2\pi i\mathbb{Z} \text{ for all } \alpha \in \Phi\}$, so that $t(\tau) = \exp H(\tau)$ may stay in $T_{\text{reg}} = \{t \in T \mid e^{\alpha}(t) \neq 1 \text{ for all } \alpha \in \Phi\}$. These conditions are also sufficient. For given such an H_1 we may take $H(\tau) = H_o + \tau(H_1 - H_o)$, the line from H_o to H_1, which lies in C_o because C_o is bounded by some planes of the form $\alpha(H) = 2\pi i k, k \in \mathbb{Z}$. For $c(\tau)$ we take a path from $c(0) = 1$ to $c(1) = s$ (the s belonging to H_1). Then

$$a(\tau) = c(\tau)\exp(H(\tau))c(\tau)^{-1}$$

has all the required properties.

If we wish to take 1 as base-point for the loops rather than $t_o = \exp H_o$, then the $a(\tau)$ above must be replaced by

$$a(\tau) = \exp(-H_o)c(\tau)\exp(H(\tau))c(\tau)^{-1}.$$

This may also be written as

$$a(\tau) = \exp(-H_o)c(\tau)\exp(H_o + \tau(s^{-1}H_o - H_o + Z))c(\tau)^{-1} \tag{2}$$

because of the formula (1) for H_o and the definition of $H(\tau)$.

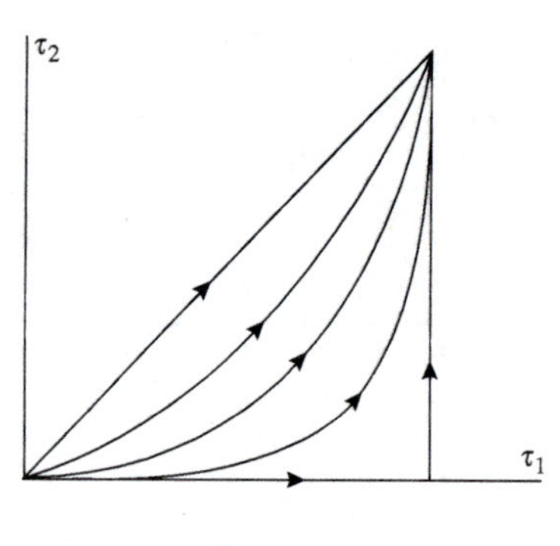

Fig. 3

In (2), replace H_o by σH_o for any real σ. The result is still a loop at 1, say $a_\sigma(\tau)$, and as σ varies from 1 to 0 the path $a_\sigma(\tau)$ varies from the initial $a(\tau)$ to the loop

$$c(\tau)\exp(\tau Z)c(\tau)^{-1}$$

Now replace $c(\tau)$ by $c(\sigma\tau)$. The result is still a loop at 1 and for $\sigma = 0$ becomes

$$\exp \tau Z. \tag{3}$$

Summary. Every homotopy class at 1 in K is represented by a loop $\exp\tau Z$, $\tau \in [0,1]$, where $Z \in t$ satisfies $\exp Z = 1$ in T. Fix a regular element $H_o \in t$. Then $Z \in t$ may be chosen so that for some $s \in W$ the element $s^{-1}H_o + Z$ lies in the same connected component C_o of H_o in $\{\alpha(H) \notin 2\pi i\mathbb{Z}\}$ and is then uniquely determined.

The elements $Z \in t$ satisfying $\exp Z = 1$ form a group $\Gamma(T)$ under addition:

$$\Gamma(T) = \{Z \in t \mid \exp Z = 1\};$$

$\Gamma(T)$ may be described as $\Gamma(T) = \{H \in t \mid \lambda(H) \in 2\pi i\mathbb{Z}$ for all weights λ of $T\}$. Of course not *all* weights of T are required in the definition of $\Gamma(T)$; just enough to insure that $\exp Z = 1$ in T. In the present situation, when K is the adjoint group, the roots suffice: $\Gamma(T) = \{H \in t \mid \alpha(H) \in 2\pi i\mathbb{Z}$ for all roots $\alpha \in \Phi\}$.

The map which sends $Z \in \Gamma(T)$ into the homotopy class of the path $\exp\tau Z$, $\tau \in [0,1]$, is actually a group homomorphism $\Gamma(T) \to \pi_1(K)$. This is seen by considering $\exp(\tau_1 Z_1)\exp(\tau_2 Z_2)$: as (τ_1,τ_2) runs along the two legs of the triangle in the $\tau_1\tau_2$-plane of Figure 3 (parametrized by a parameter τ running form 0 to 1), $\exp(\tau_1 Z_1)\exp(\tau_2 Z_2)$ runs through the composite path; as (τ_1,τ_2) runs along the base of the triangle (again parametrized by $[0,1]$), $\exp(\tau_1 Z_1)\exp(\tau_2 Z_2)$ runs through the pointwise-product path. And a homotopy between the two paths in the $\tau_1\tau_2$-plane gives a homotopy of the corresponding loops on T.

We have shown that the homomorphism $\Gamma(T) \to \pi_1(T)$, $Z \to$ class of the loop $\exp\tau Z, \tau \in [0,1]\}$ is surjective and admits a section $\pi_1(K) \to \Gamma(T)$ (a left inverse, not a group homomorphism) depending on a regular element $H_o \in t$, as described in the summary. This last property will allow us to determine the kernel of $\Gamma(T) \to \pi_1(K)$, as follows.

We note first of all that for any $Z \in \Gamma(T)$, the loop $\exp(\tau Z)$ is homotopic to $\exp(\tau s Z)$ for any $s \in W$: one only has to connect s to 1 by a path $c(\sigma)$ on K; then $c(\sigma)\exp(\tau Z)c(\sigma)^{-1}$ provides the desired homotopy. This means that $sZ - Z$ lies in the kernel of the homomorphism $[\exp\tau(\ )] : \Gamma(T) \to \pi_1(K)$. When s is the reflection s_α in a root $\alpha \in \Phi$ one can write

$$s_\alpha(H) = H - \alpha(H)H_\alpha$$

where $H_\alpha \in it$ is defined by the equation

$$H_\alpha = \frac{2\alpha}{(\alpha, \alpha)}$$

when one identifies h and h^* by the bilinear from $(\,,)$ introduced in §3.3 to define the reflections s_α. (The form is extended $\mathbb{C}$-linearly from L to h^*). Thus $\alpha(Z)H_\alpha = Z - s_\alpha Z$ lies in the kernel of $\Gamma(T) \to \pi_1(K)$ for all $Z \in \Gamma(T)$ and all $\alpha \in \Phi$. If one chooses $Z \in \Gamma(T)$ so that $\alpha(Z) = 2\pi i$ for a particular $\alpha \in \Phi$ (as is possible) one sees that $2\pi i H_\alpha$ lies in the kernel for all $\alpha \in \Phi$. Let

$$\Gamma(\Phi^\vee) = \sum_\alpha \mathbb{Z}H_\alpha$$

denote the subgroup of it generated by the H_α. (The $^\vee$ is included as a reminder that one has to pass from the α to the H_α). We have just shown that $2\pi i\Gamma(\Phi^\vee)$ is contained in the kernel of $\Gamma(T) \to \pi_1(K)$, so that we have a surjective homomorphism

$$[\exp\tau(\ )] : \Gamma(T)/2\pi i\Gamma(\Phi^\vee) \to \pi_1(K). \tag{4}$$

We shall show that this map is one-to-one, hence an isomorphism.

We know that every element of $\pi_1(K)$ is represented by a loop $\exp\tau Z$, $\tau \in [0,1]$, for a *unique* $Z \in \Gamma(T)$ satisfying

$$s^{-1}H_o + Z \in C_o. \tag{5}$$

for some $s \in W$. H_o is a fixed element of t with $\alpha(H_o) \notin 2\pi i\mathbb{Z}$ for all $\alpha \in \Phi$ and C_o is the component of H_o in $\{\alpha(H) \notin 2\pi i\mathbb{Z}\}$. Since both $H \to s^{-1}H$ and $H \to H + Z$ permute the hyperplanes $\alpha = 2\pi i k$, $k \in \mathbb{Z}$, these transformations also permute the connected components and (5) is equivalent to

$$s^{-1}C_o + Z = C_o. \tag{6}$$

To prove that (4) is one-to-one it therefore suffices to show that every element of $\Gamma(T)/2\pi i\Gamma(\Phi^\vee)$ has a representative $Z \in \Gamma(T)$ for which (6) holds with some $s \in W$: because of its uniqueness mentioned above, Z can be uniquely recognized from its image in $\pi_1(K)$. Equivalently, for every $Z \in \Gamma(T)$ there should be a $Y \in 2\pi i\Gamma(\Phi^\vee)$ so that

$$C_o + Z = sC_o + Y. \tag{7}$$

Both sides of this equation are components of $\{H \in t \mid \alpha(H) \notin 2\pi i\mathbb{Z}\}$; thus it suffices to show that any two components of $\{H \in t \mid \alpha(H) \notin 2\pi i\mathbb{Z}\}$ are related by a transformation of the form $H \to sH + Y$ with $s \in W$ and $Y \in 2\pi i\Gamma(\Phi^\vee)$. These (affine) transformations form a group (because W transforms $2\pi i\Gamma(\Phi^\vee)$ into itself). In fact:

Lemma 2. *The group of transformations of t of the form $H \to sH + Y$ $(s \in W, Y \in 2\pi i\Gamma(\Phi^\vee)$ is exactly the group generated by the reflections $s_{\alpha,\nu}$ in the hyperplanes $\alpha(H) = \nu(\alpha \in \Phi, \nu \in 2\pi i\mathbb{Z})$:*

$$\begin{aligned}\sigma_{\alpha,\nu}H &= H - ((\alpha, H) - \nu)H_\alpha \\ &= s_\alpha H + \nu H_\alpha.\end{aligned}$$

Proof. The $s_{\alpha,\nu}$ are evidently of the form $H \to sH + Z$. The equation

$$H - \nu H_\alpha = s_\alpha s_{\alpha,\nu} H$$

shows that the translations by the νH_α belong to the group generated by the $s_{\alpha,\nu}$. Since W is generated by the reflections $s_\alpha = s_{\alpha,0}$ and $2\pi i\Gamma(\Phi^\vee)$ is generated by the νH_α, the lemma follows. QED

We now place ourselves in the following situation. E is an n-dimensional Euclidean space. We assume given a *discrete* family $\mathcal{H}$ of hyperplanes in E (of the form $\alpha(x) = \text{const.}, \alpha \in E^* = E$) which is permuted by the reflection in any one of its members. The discreteness means that only finitely many hyperplanes from $\mathcal{H}$ can intersect a given compact set; the reflection condition means that the group A of (affine) transformations of E generated by these reflections permutes the hyperplanes in $\mathcal{H}$.

We introduce some geometric terminology. The connected components of complement of these hyperplanes will be called *chambers*. The boundary of a chamber C lies in certain hyperplanes form $\mathcal{H}$ the have an $(n-1)$-dimensional intersection with the closure of C. Such a hyperplane is called a *wall* of C and its part in the closure of C is called a *face* of C. Two chambers are *adjacent* if they share a face. The reflection in the their common wall interchanges two adjacent chambers.

In case $\mathcal{H}$ is the family of hyperplanes $\{\alpha = m \mid \alpha \in \Phi, m \in \mathbb{Z}\}$ corresponding to a root system Φ of rank 2, the chambers are sketched in Figure 4.

Lemma 3. *Any two chambers are related by a transformation from A.*

Proof. Any two chambers C, C' may be joined by a sequence of adjacent ones: $C = C_o, C_1, \ldots, C_k = C'$. To see this one considers a continuous path joining a point in C to a point in C'. This path meets only finitely hyperplanes from $\mathcal{H}$ (because of the discreteness assumption), and may be assumed to meet each such hyperplane in the interior of a face between two adjacent walls (after a small displacement, if necessary). The sequence of adjacent chambers met by the path then leads form C to C'. And the product of the corresponding sequence of reflections in the walls encountered sends C to C'. QED

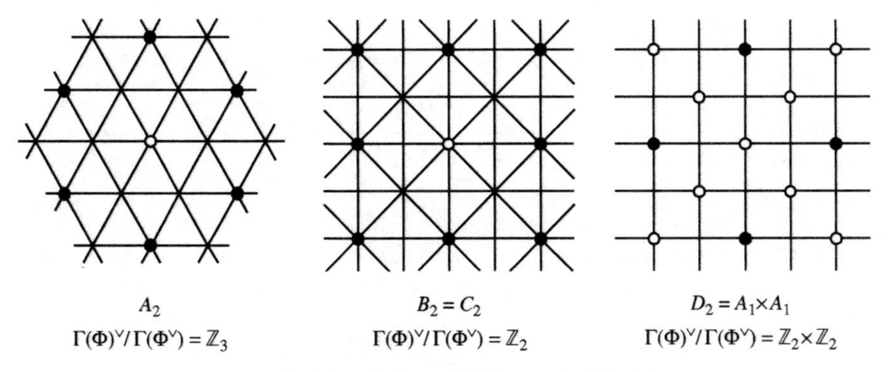

$\Gamma(\Phi)^\vee$ = {lattice points}, $\Gamma(\Phi^\vee) = \{\circ, \bullet\}$, $\Phi^\vee = \{\bullet\}$

Fig. 4

Applied to the case at hand, the lemma says that any two components of $\{H \in \mid \alpha(H) \notin 2\pi i\mathbb{Z}\}$ are related by a transformation of the form $H \to sH + Y$ with $s \in W$ and $H \in 2\pi i\Gamma(\Phi^\vee)$, as required. This finishes the proof that the map $[\exp \tau(\ )] : \Gamma(T)/2\pi i\Gamma(\Phi^\vee) \to \pi_1(K)$ is an isomorphism.

Remark 4. In the case of a root system Φ, Lemma 3 may be applied to the hyperplanes $\{\alpha = 0, \alpha \in \Phi\}$ in $L(L = \mathfrak{t}^* \approx \mathfrak{t})$. The corresponding group is then the Weyl group W and the lemma says that W acts transitively on the connected components of $\{\alpha \neq 0\}$, called *chambers* of the root system. Or the lemma may be applied to the hyperplanes $\{\alpha = m \mid \alpha \in \Phi, m \in \mathbb{Z}\}$ (essentially the case above, without the factor $2\pi i$) and then says that the *affine Weyl group* (generated by the reflection in these hyperplanes) acts transitively on the *affine chambers* (also called *alcoves*). In both of these cases any two chambers are in fact related by a *unique* transformation from the corresponding reflection group. We know this for the Weyl group; for the affine Weyl group see Bourbaki (1968) or Gutkin (1986).

To state the final result we get rid of some factors $2\pi i$ by setting $\Gamma(\Phi)^\vee = \{H \in i\mathfrak{t} \mid \alpha(H) \in \mathbb{Z} \text{ for all } \alpha \in \Phi\}$. Then $\Gamma(T) = 2\pi i\Gamma(\Phi)^\vee$ and the result may be stated thus:

Theorem 5. *Let K be the adjoint group of a compact group. Then the map*

$$[\exp 2\pi i\tau -] : \Gamma(\Phi)^\vee/\Gamma(\Phi^\vee) \to \pi_1(K)$$

which sends the class of $Z \in \Gamma(\Phi)^\vee$ in $\Gamma(\Phi)^\vee/\Gamma(\Phi^\vee)$ into the class of the loop $\exp(2\pi i\tau Z)$*, $\tau \in [0, 1]$, in $\pi_1(K)$ is an isomorphism.*

It remains to explicitly list $\Gamma(\Phi)^\vee$, $\Gamma(\Phi^\vee)$, and $\Gamma(\Phi)^\vee/\Gamma(\Phi^\vee)$ in each case. This we do in the Table 3.13.

The λ_j in the table are the coordinates of a general element $H(\lambda_1, \ldots, \lambda_n)$ in $\mathfrak{h}$ as in §3.2. We omit the verifications.

Table 3.13 Fundamental groups

Type	$\Gamma(\Phi)^\vee$	$\Gamma(\Phi^\vee)$	$\Gamma(\Phi)^\vee/\Gamma(\Phi^\vee)$
A_{n-1}	$\lambda_j = k_j - k_o/n$	$\lambda_j \in \mathbb{Z}$	$\mathbb{Z}_n$
	$k_j \in \mathbb{Z}$,	$\sum k_j = k_o$	$\sum \lambda_j = 0$
B_n	$\lambda_j \in \mathbb{Z}$	$\sum \lambda_j \in 2\mathbb{Z}$	$\mathbb{Z}_2$
C_n	$\lambda_j = k_j + k_o/2$	$\lambda_j \in \mathbb{Z}$	$\mathbb{Z}_2$
	$k_j, k_o \in \mathbb{Z}$		
D_n	$\lambda_j = k_j + k_o/2$	$\lambda_j \in \mathbb{Z}$	$\mathbb{Z}_2 \times \mathbb{Z}_2$ (n even),
	$k_j, k_o \in \mathbb{Z}$	$\sum \lambda_j \in 2\mathbb{Z}$	$\mathbb{Z}_4$ (n odd)

The somewhat disappointing fundamental groups listed in the last column of the table are nevertheless the largest of *any* connected group with one of these Lie algebras. For any connected group K whose Lie algebra is isomorphic with the Lie algebra of a compact classical group is a covering of its adjoint group by $\mathrm{Ad} : K \to \mathrm{Ad}(K)$. $\pi_1(K)$ is then isomorphic with the subgroup of $\pi_1(\mathrm{Ad}(K))$ consisting of those classes of loops that arise as images of loops in K. This happens for the class of the loop $\mathrm{Ad}(\exp 2\pi i\tau Z)$ in $\mathrm{Ad}(K)$ if and only if $\exp(2\pi i\tau Z)$ closes to a loop in K, i.e. if and only if $\exp(2\pi i Z) = 1$ in K, i.e. $\lambda(Z) \in \mathbb{Z}$ for all weights λ of the Cartan subgroup T of K. Thus *Theorem 5 remains valid for any such K provided $\Gamma(\Phi)^\vee$ is replaced by $\{Z \in T \mid \lambda(Z) \in \mathbb{Z}$ for all weights λ of $T\}$*. In particular, the fundamental groups of the compact classical groups themselves may be found once more in this way.

The fundamental groups of the compact classical groups may also be determined by an inductive argument, which may be found in Weyl's book (1938), VII.12, and is attributed there to Hurewicz. But Weyl's original method and Cartan's extension thereof provide deeper insight into the structure of these groups.

A final comment. Theorem 1 implies that there must be a double covering of $\mathrm{SO}(n)$, which is then simply connected. This group is denoted $\mathrm{Spin}(n)$; an explicit construction of $\mathrm{Spin}(n)$ may be found again in Weyl's book (1938), VII.13.

Problems for §3.4

1. Let G be a complex classical group. (a) Show that the the connected subalgebra $g_\alpha = \mathbb{R}H_\alpha + \mathbb{R}E_\alpha + \mathbb{R}E_{-\alpha}$, $H_\alpha = [E_\alpha, E_{-\alpha}]$, of $\mathfrak{g}$ generated by two root vectors $E_\alpha, E_{-\alpha}$ is isomorphic with $\mathsf{sl}(2,\mathbb{C})$ by the map

$$\begin{bmatrix}1 & 0\\ 0 & -1\end{bmatrix} \to H_\alpha, \quad \begin{bmatrix}0 & 1\\ 0 & 0\end{bmatrix} \to E_\alpha, \quad \begin{bmatrix}0 & 0\\ 1 & 0\end{bmatrix} \to E_{-\alpha}.$$

Deduce that the corresponding connected subgroup of G is isomorphic with $\mathrm{SL}(2,\mathbb{C})$ or with $\mathrm{SL}(2,\mathbb{C})/\{\pm 1\}$.

(b) Let K be the unitary subgroup of G. Show that $K \cap G_\alpha$ is isomorphic with SU(2) or with SU(2)/$\{\pm 1\}$.

2. Show that the non-trivial homotopy class of SO(n) is represented by a uniform rotation in any 2-dimensional plane. [Explain what this means.]

3. Verify the data in Table 1 for (a) A_{n-1}, (b) B_n, (c) C_n, and (d) D_n.

4. Let G be any connected linear group (or even any connected topological group, if this concept is familiar). Show that $\pi_1(G)$ is abelian. [Suggestion: think about Figure 3.]

4

Manifolds, homogeneous spaces, Lie groups

4.1 Manifolds

In a *vector space* one has coordinate systems any two of which are related by a linear transformation (and this property could be used to define the notion 'vector space'). Somewhat analogously, an *analytic manifold* is a point-set together with coordinate systems any two of which are related by an analytic transformation; but the coordinate systems are not required to be defined on the whole point-set. Formalizing this notion one arrives at:

Definition. An n-dimensional *analytic manifold* is a set M (whose elements are called *points*) together with a collection of partially defined maps $M \cdots \to \mathbb{R}^n$, called *coordinate systems*, subject to the axioms MFLD 1–4 below.

Notation and terminology. A coordinate system is typically denoted (U, x), $U \subset M$ is the *coordinate domain* (not always explicitly mentioned),

$$x : U \to \mathbb{R}^m, \quad p \to x(p) = (\xi_1(p), \ldots, \xi_n(p))$$

the *coordinate map*. The components $\xi_k(p)$ of the *coordinate point* $x(p)$ are the *coordinates* of p.

MFLD 1. Each coordinate system (U, x) provides a one-to-one correspondence between the points p in its domain $U \subset M$ and the points $x(p)$ in an *open* subset of $\mathbb{R}^n$.

MFLD 2. The *coordinate transformation* $\tilde{x}(p) = f(x(p))$ relating the coordinates of $p \in U \cap \tilde{U}$ in two different coordinate systems (U, x) and $(\tilde{U}, \tilde{x})$ is given by an analytic map $f : \mathbb{R}^n \cdots \to \mathbb{R}^n$. The domain $x(U \cap \tilde{U})$ of f is required to be open in $\mathbb{R}^n$.

MFLD 3. Every point of M lies in the domain of some coordinate system.

We shall sometimes use the notation $x \to p(x)$ for the inverse of a coordinate map $p \to x(p)$. The domains need not be explicitly mentioned. Use of the same

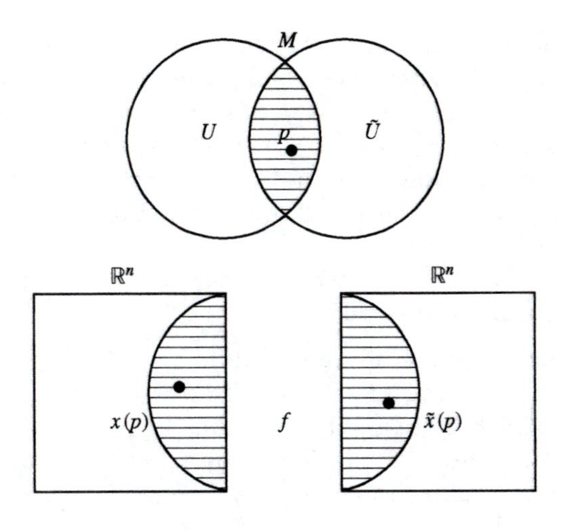

Fig. 1

symbol $x = (\xi_1, \ldots, \xi_n)$ for both the coordinate map and coordinate points is customary and convenient, just as in calculus. Figure 1 illustrates the second axiom, which is evidently the main point.

Coordinates may be used to introduce a *topology* on M in the same way as has been done for linear groups using the exponential maps. A *neighborhood* of a point p_o in M is any subset of M containing all p whose coordinate points $x(p)$ in some coordinate system satisfy $\|x(p) - x(p_o)\| < \epsilon$ for some $\epsilon > 0$. A subset of M is *open* if it contains a neighborhood of each of its points, and *closed* if its complement is open.

As a further axiom we add the *Hausdorff Separation Axiom*:

MFLD 4. Any two points of M have disjoint neighborhoods.

We now extend the meaning of *coordinate system* to include any one-to-one correspondence $p \to \tilde{x}(p)$ of an open subset of M with an open subset of $\mathbb{R}^n$ for which, in a neighborhood of each point, the transformation $\tilde{x}(p) = f(x(p))$ to some original coordinate system $p \to x(p)$ is analytic in both directions (as in MFLD 2). These new coordinate systems are admitted on the same basis as the original coordinate systems with which M comes equipped in virtue of the definition. We shall take the expression 'a coordinate system around p_o' to mean 'a coordinate system defined in some neighborhood of p_o'. In view of the Inverse Function Theorem, any analytic transformation $f : \mathbb{R}^n \cdots \to \mathbb{R}^n$ with an invertible differential at $x(p_o)$ allows one to introduce a coordinate system $p \to \tilde{x}(p)$ around p_o in this extended sense in terms of a given coordinate system $p \to x(p)$ defined at p_o by setting $\tilde{x}(p) = f(x(p))$.

Comments

(a) Coordinate systems are sometimes called *charts*. Any collection of coordinate systems satisfying the axioms MFLD 1–4 is called an *atlas*. Starting from the atlas on a manifold with which it comes equipped in virtue of the axioms, the coordinate systems in the extended sense form another atlas. The latter is called *maximal*, evidently with reason. Two atlases that give rise to the same maximal atlas are considered to define the *same* manifold structure on a point-set.

(b) Instead of analytic functions one can use C^k functions for any $0 \leq k \leq \infty$ in the axioms MFLD 1–4 and in the subsequent extension of the meaning of 'coordinate system'. One then gets C^k manifolds. If one replaces $\mathbb{R}^n$ by $\mathbb{C}^n$ and '(real) analytic' by 'holomorphic' one gets *complex manifolds*. For us 'manifold' will always mean 'analytic manifold'.

Example 1 (Vector spaces as manifolds). Let E be an n-dimensional real vector space. Given an ordered basis $e_1, e_2, \ldots, e_n$ for E one can introduce a coordinate system on E by taking for $x(p)$ the n-tuple of components of $p \in E$ with respect to this basis. This single coordinate system makes E into a manifold. The resulting manifold structure is independent of the basis chosen: the coordinates defined similarly in terms of other bases appear among the coordinate systems in the generalized sense. Such coordinate systems on a vector space are called *linear*. In addition there are all sorts of non-linear coordinate systems, among them the spherical and cylindrical coordinates on $\mathbb{R}^3$ familiar from calculus and the various less familiar curvilinear coordinate systems from physics and analytic geometry. (Some care needs to be taken to restrict these coordinates to an *open* domain on which they give a one-to-one correspondence with an *open* set, in accordance with the definition of coordinates in the extended sense.)

Example 2 (Linear groups as manifolds). Let G be a linear group. We check in detail that the exponential coordinate systems defined in §2.3 satisfy the axioms MFLD 1–4. Recall that the domain of such a coordinate system is

$$a_0 U = \{a_0 \exp X \mid X \in \mathbf{g},\ \|X\| < R\},$$

with $R > 0$ chosen so that the log series converges on $\{a \in M \mid a = \exp X, X \in M, \|X\| < R\}$. The coordinate map $a_0 U \to \mathbf{g},\ a \to X = X(a)$, is then defined by

$$a = a_0 \exp X,\ \|X\| < R.$$

(That the coordinate map takes on values in $\mathbf{g}$ rather than in $\mathbb{R}^m$ is of course immaterial.)

MFLD 1. $a_0 U \to \mathbf{g}$, maps its domain bijectively onto $\{X \in \mathbf{g} \mid \|X\| < R\}$, which is open in $\mathbf{g}$.

MFLD 2. The coordinate transformation $X \to \tilde{X}$ between two such coordinate systems $(a_0 U, X)$ and $(\tilde{a}_0 U, \tilde{X})$ is defined by

$$a_0 \exp X = \tilde{a}_0 \exp \tilde{X}, \quad \|X\|, \|\tilde{X}\| < R.$$

Its domain consists of those $X \in \mathbf{g}$ for which this equation has a solution for X ($\|X\|, \|\tilde{X}\| < R$); it is open in $\mathbf{g}$ and (if non-empty) the coordinate transformation defined thereon is given by $\tilde{X} = \log(\tilde{a}_0^{-1} a_0 \exp X)$, hence is analytic.

(The log of $a_0^{-1}a_0 \exp X$ necessarily exists if the above equation has a solution with $\|X\| < R$.)

MFLD 3. This is clear.

MFLD 4. For any two matrices $a_0 \neq b_0$ in M one can find $\epsilon > 0$ so that $\{a = a_0 \exp X \mid X \in M, \|X\| < \epsilon\}$ and $\{b = b_0 \exp Y \mid Y \in M, \|Y\| < \epsilon\}$ are disjoint. When $a_0 \neq b_0$ lie in G, then $\{a = a_0 \exp X \mid X \in \mathbf{g}, \|X\| < \epsilon\}$ and $\{b = b_0 \exp Y \mid Y \in g, \|Y\| < \epsilon\}$ are disjoint neighborhoods of a_0 and b_0 in G. This completes the verification that G together with its exponential coordinate systems is an analytic manifold.

A map $f : M \to N$ between manifolds is *of class* $C^k, 0 \leq k \leq \infty, \omega$, if f maps any sufficiently small neighborhood of a point of M into the domain of a coordinate system on N and the equation $q = f(p)$ defines a C^k map when p and q are locally expressed in terms of coordinates. This definition is independent of the local coordinates chosen and applies equally to a map $f : M \cdots \to N$ defined only on an open subset of N.

A *curve* in M is an M-valued function $p = p(\tau)$ of a real variable τ, defined on some interval. Exceptionally, the domain of a curve is not always required to be open. The curve is of class C^k if it extends to such a function also in a neighborhood of any endpoint of the interval that belongs to its domain.

For each τ in its domain, a C^1 curve $p(\tau)$ has a *coordinate tangent vector* $x'(\tau)$ relative to a coordinate system x:

$$x'(\tau) = \frac{d}{d\tau} x(\tau),$$

where $x(\tau) = x(p(\tau))$ is the coordinate point of $p(\tau)$. $x'(\tau) \in \mathbb{R}^n$ depends of course on the coordinate system: if $\tilde{x} = \tilde{x}(p)$ is another coordinate system defined around $p(\tau)$, then

$$\tilde{x}'(\tau) = df_{x(\tau)}(x'(\tau)),$$

where $\tilde{x} = f(x)$ is the coordinate transformation of MFLD 2. This leads to the definition of a *tangent vector* v at $p \in M$ as a rule that associates to each coordinate system x around p a coordinate vector $y \in \mathbb{R}^n$ subject to the transformation law

$$\tilde{y} = df_{x(p)}(y).$$

Here y and $\tilde{y}$ are the coordinate vectors associated with v by the coordinate systems x and $\tilde{x}$, and $\tilde{x} = f(x)$ is the coordinate transformation.

With this definition, every C^1 curve $p(\tau)$ has a *tangent vector* for each τ, denoted $p'(\tau)$ or $dp/d\tau$ (with $p = p(\tau)$ understood), and every tangent at p_o is of the form $p'(0)$ for some analytic curve with $p(0) = p_o$. (The tangent vector at p_o represented by $y \in \mathbb{R}^n$ in a coordinate system x around $p_o = p(x_0)$ is, for example, the tangent vector at $\tau = 0$ of the curve $p(\tau)$ represented by the line $x_0 + \tau y$ in $\mathbb{R}^n$ in these coordinates.)

Tangent vectors at the same point p may be added and scalar multiplied by adding and scalar multiplying their coordinate vectors relative to coordinate systems. In this way, the tangent vectors at $p \in M$ form a vector space, called the *tangent space* of M at p, denoted T_pM. Relative to a coordinate system around $p \in M$, elements of T_pM are represented by their coordinate vectors, leading to an identification $T_pM \approx \mathbb{R}^n$ depending on the coordinate system. If the coordinate system is written as $x = (\xi_1, \xi_2, \ldots, \xi_n)$, then the vector at p represented by the k-th standard basis vector of $\mathbb{R}^n$ is denoted by

$$\left(\frac{\partial}{\partial \xi_k}\right)_p.$$

It is the tangent vector

$$\left.\frac{\partial p(x)}{\partial \xi_k}\right|_{x=x(p)}$$

of the k-th *coordinate line* through p, i.e. of the curve $p(\xi_1, \xi_2, \ldots, \xi_n)$ parametrized by ξ_k, the remaining coordinates being given their value at p. The $(\partial/\partial\xi_k)_p$ form a basis for T_pM. The components of a vector at p with respect to this basis are just the components of its coordinate vector relative to x; they are called its *components relative to the coordinate system* x.

The definition of 'tangent vector' as rule associating to a coordinate system an n-vector subject to a transformation law is a bit awkward, since one naturally thinks of a tangent vector as a geometric object in its own right, *represented* by a coordinate n-vector relative to a coordinate system, just as points are represented by coordinate n-tuples. For this reason other definitions of 'tangent vector' are frequently given (as equivalence classes of curves, or as point derivations); but the definition given above is the one most directly related to the definition of manifolds in terms of coordinate systems, and should be sufficiently intuitive in view of its relation to tangent vectors of curves. It is also in agreement with the principle that 'everything local about manifolds comes down to $\mathbb{R}^n$ via coordinates'.

Example 1 (Continued). If $M = E$ is a vector space, then tangent vectors at a point $p \in E$ may be represented by their n-tuples of components relative to a *linear* coordinate system defined using a basis $\{e_k\}$, and elements of E themselves may be represented by their n-tuples of components relative to the same basis $\{e_k\}$. The resulting correspondence $T_pM \approx E$ is in fact independent of the basis $\{e_k\}$, so that one may identify $T_pM = E$ for any $p \in M$.

Example 2 (Continued). The tangent space at 1 to a linear group G considered as manifold may be identified with its Lie algebra $\mathbf{g}$ using exp as coordinate system around 1. The tangent space at any point $a \in G$ may also be identified with $\mathbf{g}$, using either $a\exp(X) \to X$ or $\exp(X)a \to X$ as coordinate system around a on G. These two identifications differ by the transformation $\mathrm{Ad}(a)$ of $\mathbf{g}$. The tangent space to G at $a \in G$ may also be identified with the subspace $a\mathbf{g} = \mathbf{g}a$ of the matrix space by associating with the tangent vector to a curve $a(\tau)$ in G its tangent vector in the matrix space.

Let $f : M \cdots \to N$ be a C^1 map defined in a neighborhood of $p \in M$. Then one can define a linear map $df_p : T_pM \to T_{f(p)}N$ by the requirement that the equation

$$w = df_p(v)$$

holds when $v \in T_pM$ is represented by its coordinate vector relative to a coordinate system x around p, its image $w \in T_{f(p)}N$ is represented by its coordinate vector relative to a coordinate system y around $f(p)$ on N, and df_p is taken as the differential at $x(p)$ of the map $x \to y$ representing f in the coordinates x and y. In short, df_p is the usual differential when points as well as tangent vectors are represented in coordinates. The tangent vector of a C^1 curve $p(\tau)$ in M is related to the tangent vector to the image curve $f(p(\tau))$ in N by

$$\frac{df(p)}{d\tau} = df_p\left(\frac{dp}{d\tau}\right),$$

where $p = p(\tau)$ is understood. This is the formula usually used to calculate differentials.

The definition of df_p applies in particular to a scalar-valued function $f : M \cdots \to \mathbb{R}$. In this case, df_p is a linear functional on T_pM. It follows from the definitions that the value of df_p on a basis vector $(\partial/\partial\xi_k)_p$ is $\partial f(p)/\partial\xi_k$. The partial is taken by thinking of $f(p)$ as a function of the coordinates of p with respect to the coordinate system $x = (\xi_1, \xi_2, \ldots, \xi_n)$.

Of particular importance are the differentials of the coordinate functions $\xi_k = \xi_k(p)$ of a coordinate system. $(d\xi_k)_p$ associates to a vector v at p the k-th component of the coordinate vector representing v, i.e. the k-th component of v with respect to the basis $(\partial/\partial\xi_1)_p, \ldots, (\partial/\partial\xi_n)_p$.

A *vector field* X on M associates to each point p in M a tangent vector $X(p) \in T_pM$. We also admit vector fields defined only on open subsets of M. Examples are the *basis vector fields* $\partial/\partial\xi_k$ of a coordinate system. On the coordinate domain, every vector field can be written as

$$X = \sum_k X_k \frac{\partial}{\partial\xi_k}$$

for certain scalar functions X_k. X is said to be of class C^k if the X_k have this property.

Dually, *a differential 1-form* φ on an open subset of M associates to each point p in its domain a linear functional φ_p on T_pM. Examples are the *coordinate differentials* $d\xi_k$. On the coordinate domain, every differential 1-form φ can be written as

$$\varphi = \sum_k \varphi_k \, d\xi_k$$

for certain scalar functions φ_k. φ is said to be of class C^k if the φ_k have this property.

An analytic vector field X may be considered as an operator on real-valued analytic functions φ defined on open subsets of M by

$$X\varphi(p) = d\varphi_p(X(p)).$$

If one writes in coordinates $X = \sum_k X_k(\partial/\partial\xi_k)$, then X is the differential operator given by this symbol, providing one more reason for the notation.

Example 3 (Spherical coordinates). Let $M = \mathbb{R}^3$ with standard coordinates x, y, z. Spherical coordinates ρ, θ, ϕ are defined by

$$x = \rho \sin\phi \cos\theta, \quad y = \rho \sin\phi \sin\theta, \quad z = \rho\cos\phi.$$

The coordinates ρ, θ, ϕ are restricted by

$$\rho > 0, \quad 0 < \theta < \pi, \quad 0 < \phi < 2\pi$$

(strict inequalities to insure openness of domain). The relations between the basis vector fields and the coordinate differentials are found by routine differentiation:

$$dx = \frac{\partial x}{\partial \rho}d\rho + \frac{\partial x}{\partial \theta}d\theta + \frac{\partial x}{\partial \phi}d\phi = \sin\phi\cos\theta\, d\rho - \rho\sin\phi\sin\theta\, d\theta + \rho\cos\phi\cos\theta\, d\phi$$

$$dy = \frac{\partial y}{\partial \rho}d\rho + \frac{\partial y}{\partial \theta}d\theta + \frac{\partial y}{\partial \phi}d\phi = \sin\phi\sin\theta\, d\rho + \rho\sin\phi\cos\theta\, d\theta + \rho\cos\phi\sin\theta\, d\phi$$

$$dz = \frac{\partial z}{\partial \rho}d\rho + \frac{\partial z}{\partial \theta}d\theta + \frac{\partial z}{\partial \phi}d\phi = \cos\phi\, d\rho - \rho\sin\phi\, d\phi$$

$$\frac{\partial}{\partial\rho} = \frac{\partial x}{\partial\rho}\frac{\partial}{\partial x} + \frac{\partial y}{\partial\rho}\frac{\partial}{\partial y} + \frac{\partial z}{\partial\rho}\frac{\partial}{\partial z} = \sin\phi\cos\theta\frac{\partial}{\partial x} + \sin\phi\sin\theta\frac{\partial}{\partial y} + \cos\phi\frac{\partial}{\partial z}$$

$$\frac{\partial}{\partial\theta} = \frac{\partial x}{\partial\theta}\frac{\partial}{\partial x} + \frac{\partial y}{\partial\theta}\frac{\partial}{\partial y} + \frac{\partial z}{\partial\theta}\frac{\partial}{\partial z} = -\rho\sin\phi\sin\theta\frac{\partial}{\partial x} + \rho\sin\phi\cos\theta\frac{\partial}{\partial y}$$

$$\frac{\partial}{\partial\phi} = \frac{\partial x}{\partial\phi}\frac{\partial}{\partial x} + \frac{\partial y}{\partial\phi}\frac{\partial}{\partial y} + \frac{\partial z}{\partial\phi}\frac{\partial}{\partial z} = \rho\cos\phi\cos\theta\frac{\partial}{\partial x} + \rho\cos\phi\sin\theta\frac{\partial}{\partial \phi} - \rho\sin\phi\frac{\partial}{\partial z}.$$

The *product* $M \times N$ of an m-dimensional manifold M and an n-dimensional manifold N is an $m+n$-dimensional manifold in an obvious way: the mappings of the form $(p, q) \to (x(p), y(q))$, where x, y are coordinate systems on M, N satisfy the axioms MFLD 1–4. (But the coordinate systems on $M \times N$ in the extended sense are generally not of this form.)

Less obvious is the notion of *submanifold*. Let M be an n-dimensional manifold. A subset S of M is called an m-dimensional *(regular) submanifold* of M if every point $p \in S$ has a neighborhood U in M that lies in the domain of some coordinate system $\xi_1, \ldots, \xi_n$ on M with the property that the points of S in U are precisely those points in U whose coordinates satisfy $\xi_{m+1} = 0, \ldots, \xi_n = 0$. We shall summarize this definition by saying that a submanifold is *given by the equations* $\xi_{m+1} = 0, \ldots, \xi_n = 0$ locally around p. We also admit the possibility that no equations are present ($m = n$), in which case the definition simply says that S must be open in M. At the other extreme, when $m = 0$, S is discrete, i.e. every point in S has a neighborhood in M which contains only this one point of S.

By definition, a *coordinate system on a submanifold* S consists of the restrictions $\xi_1|_S, \ldots, \xi_m|_S$ on S of the first m coordinates of a coordinate system $\xi_1, \ldots, \xi_n$ on M of the type mentioned above (for some p in S); as their domain we take the points of S in U, U being chosen open in M. (We omit the straightforward verification that these coordinate systems make S into a manifold.)

The Closed Subgroup Theorem (or its Corollary 2, §2.7) may now be stated as follows.

Theorem 4 (Closed Subgroup Theorem (Second version)). *A closed subgroup of a linear group is a submanifold.*

Corollary 5. *A closed subgroup of* $\mathrm{GL}(E)$ *is a submanifold of the matrix space* M.

Let S be a submanifold on M. Suppose $p(\tau)$ is a differentiable curve in S. Then $p(\tau)$ is also a differentiable curve in M and its tangent vector $p'(\tau)$ can be considered a tangent vector on S as well as a tangent vector on M. Choose a coordinate system $\xi_1, \ldots, \xi_n$ on M so that the part of S in U consists of those points p whose last $n-m$ coordinates are zero. The last $n-m$ components of the coordinate tangent vector representing $p'(\tau)$ are therefore zero as well. One sees from this that the tangent vectors to S at $p \in S$ may be identified with those tangent vectors to M at p whose last $n-m$ components are zero in a coordinate system $\xi_1, \ldots, \xi_n$ as in the definition of 'submanifold'. We may summarize these observations as follows.

Lemma 6. *Let S be a submanifold of M. Suppose S is given by the equations*

$$\xi_{m+1} = 0, \quad \ldots, \quad \xi_n = 0$$

locally around p. Then T_pS may be identified with the subspace of T_pM given by the linear equations

$$(d\xi_{m+1})_p = 0, \quad \ldots, \quad (d\xi_n)_p = 0.$$

This subspace consists of tangent vectors at p of differentiable curves in M that lie in S.

This lemma may be generalized:

Proposition 7. *Let M be a manifold, $f_1, \ldots, f_k$ analytic functions on M. Let S be the set of points $p \in M$ satisfying*

$$f_1(p) = 0, \quad \ldots, \quad f_k(p) = 0.$$

Suppose the differentials $df_1, \ldots, df_k$ are linearly independent at every point of S. Then S is an $(n-k)$-dimensional submanifold of M. Its tangent space at $p \in S$ is the subspace of vectors $v \in T_pM$ satisfying the linear equations

$$(df_1)_p(v) = 0, \quad \ldots, \quad (df_k)_p(v) = 0.$$

Proof. Fix $p \in S$. Choose a coordinate system $\xi_1, \ldots, \xi_n$ on M around p. Since the differentials $df_i = \sum_j (\partial f_i / \partial \xi_j) d\xi_j$ are linearly independent at p, their coefficient matrix $(\partial f_i / \partial \xi_j)$ has rank k at p. By renumbering the coordinates we may assume that

$$\left. \frac{\partial(f_1, \ldots, f_k)}{\partial(\xi_1, \ldots, \xi_k)} \right|_p \neq 0.$$

If we set $f_{k+1} = \xi_{k+1}, \ldots, f_n = \xi_n$, then this Jacobian determinant is that same as

$$\left.\frac{\partial(f_1,\ldots,f_n)}{\partial(\xi_1 \ldots,\xi_n)}\right|_p,$$

which is therefore also $\neq 0$. By the Inverse Function Theorem, the equations

$$\tilde{\xi}_1 = f_1(\xi_1,\ldots,\xi_n), \quad \ldots, \quad \tilde{\xi}_n = f_n(\xi_1,\ldots,\xi_n)$$

define coordinates $\tilde{\xi}_1,\ldots,\tilde{\xi}_n$ around p on M. In these coordinates S is locally around p given by the equations

$$\tilde{\xi}_1 = 0, \quad \ldots, \quad \tilde{\xi}_k = 0$$

as required by the definition of 'submanifold'. According to Lemma 6, in these coordinates the tangent space T_pS is given by the equations

$$(\tilde{\xi}_1)_p = 0, \quad \ldots, \quad (d\tilde{\xi}_k)_p = 0, \text{ i.e. } (df_1)_p = 0, \quad \ldots, \quad (df_k)_p = 0.$$

QED

Comment. Even when the differentials of the f_j are linearly dependent at some points of S, the subset of S where they are linearly independent is still a submanifold.

Example 8 (The sphere as submanifold). Let $S = \{p \in \mathbb{R}^3 \mid x^2+y^2+z^2 = 1\}$. The differential $d(x^2+y^2+z^2) = 2(x\,dx + y\,dy + z\,dz)$ is everywhere nonzero on S, hence S is a submanifold, and its tangent space at a general point $p = (x,y,z)$ is given by $x\,dx + y\,dy + z\,dz = 0$. This means that, as a subspace of $\mathbb{R}^3$, the tangent space at the point $p = (x,y,z)$ on S consists of all vectors $v = (a,b,c)$ satisfying $xa + yb + zc = 0$, as one would expect.

In spherical coordinates ρ, θ, ϕ on $\mathbb{R}^3$ (Example 3) the sphere S^2 is given by $\rho = 1$ on the coordinate domain, as required by the definition of 'submanifold'. According to the definition, θ and ϕ provide coordinates on S^2 (where defined). If one identifies the tangent spaces to S^2 with subspaces of $\mathbb{R}^3$ then the coordinate vector fields $\partial/\partial\theta$ and $\partial/\partial\phi$ are given by the formulas of Example 3 with $\rho = 1$. The coordinate differentials $d\theta$ and $d\phi$ on S^2 are obtained from the formulas of Example 3 by setting $\rho = 1$ and $d\rho = 0$.

Example 9 (The real classical groups as manifolds). The real classical group defined by a non-degenerate form φ on a vector space E over $\mathbb{R}$, $\mathbb{C}$, or $\mathbb{H}$ is

$$G = \{a \in M \mid a^\varphi a = 1, \det a = 1\}.$$

G is evidently a closed subgroup of $\mathrm{GL}(E)$ and therefore a submanifold, by the Closed Subgroup Theorem. As an illustration of Proposition 7, we verify this fact directly.

Consider the function $f(a) = a^\varphi a$ from M to the real subspace $\{b \in M \mid b^\varphi = b\}$. Its differential at $a \in M$ is $Xa \to a^\varphi(X^\varphi + X)a$, hence is surjective for invertible $a \in M$. So, the components $f_1, \ldots, f_k$ of f with respect to a basis for $\{b \in M \mid b^\varphi = b\}$ have linearly independent differentials at invertible $a \in M$. This shows that $\mathrm{Aut}(\varphi) = \{a \in M \mid a^\varphi a = 1\}$ is an analytic submanifold of M and furthermore that the tangent space at 1 of $\mathrm{Aut}(\varphi)$ is $\{X \in M \mid X^\varphi + X = 0\}$, which is its Lie algebra. The manifold structure on $\mathrm{Aut}(\varphi)$ as a submanifold of M may be seen to be the same as its manifold structure as a linear group.

Within $\mathrm{Aut}(\varphi)$ we now consider $G = \{a \in \mathrm{Aut}(\varphi) \mid \det a = 1\}$. To see that G is a submanifold of $\mathrm{Aut}(\varphi)$, and therefore also of M, we appeal to a general fact:

Proposition 10. *Let $f : G \to H$ be a homomorphism of linear groups. Its kernel $K = \{a \in G \mid f(a) = 1\}$ is a submanifold of G.*

Proof. This follows from the Closed Subgroup Theorem, but is easily seen directly: it suffices to consider a neighborhood of 1 in G. In exponential coordinates K is given locally at 1 by the linear equation $Lf(X) = 0$, from which its submanifold property is evident. QED

A final remark. There is another (inequivalent) definition of 'submanifold', namely a subset S of M that has itself *some* manifold structure for which the inclusion $S \to M$ has an injective differential everywhere. Such submanifolds may be called *embedded submanifolds* to distinguish them from the *(regular) submanifolds* defined above. In general, the manifold structure on an embedded submanifold is not uniquely determined by the requirement that the differential $S \to M$ be everywhere injective, and there may be no natural way of choosing a particular one. The definition has the formal advantage that any subgroup of a linear group (with its manifold structure defined in Example 2) is an embedded submanifold in this sense, but not necessarily a regular submanifold, as one sees from the irrational line on the torus.

Problems for §4.1

1. Show that the manifold structure on a linear group $G \subset M$ has the following property, which characterizes it uniquely: A map from an analytic manifold into G is analytic if and only if it is analytic as a map into M. [See Proposition 1, §2.3; for uniqueness consider the identity map $G \to G$.]

2. Verify that the manifold structure on a closed subgroup G of $\mathrm{GL}(E)$ as submanifold of the matrix space $M(E)$ coincides with its manifold structure as linear group. [See Example 9 and Theorem 4.]

3. Show that the cone in $\mathbb{R}^3$ given by $x^2 + y^2 - z^2 = 0$ in $\mathbb{R}^3$ is *not* submanifold if the origin is included, but *is* a manifold if the origin is excluded. [Suggestion: consider tangent vectors at 0 of curves on the cone.]

4. Show that the irrational line on the torus $\mathbb{T}^2$ is not a submanifold of $\mathbb{T}^2$.

5. Suppose G is a linear group that is a submanifold of the matrix space M.
   (a) Show that a real-valued f on G is of class C^k if and only if $f(a)$ extends to a C^k function of the matrix coefficients of a in a neighborhood of any point in the domain of f.
   (b) Show that tangent space to G (considered as submanifold of M) at $a \in G$ is the subspace $a\mathfrak{g} = \mathfrak{g}a$ of M.
6. Justify the formulas in Examples 3 and 8 from the definitions.
7. Show that addition and scalar multiplication of tangent vectors at a point p on a manifold, as defined in the text, does indeed produce tangent vectors.
8. Verify that the coordinate systems on a submanifold S of M, as defined in the text, satisfy MFLD 1–4.
9. (a) Verify that the coordinate systems $(p, q) \to (x(p), y(q))$ on a product manifold $M \times N$ satisfy MFLD 1–4.
   (b) Show that $T_{(p,q)}(M \times N) \approx T_pM \times T_qN$ in a natural way.
10. Verify that the identification $T_pE = E$ for a real vector space E is indeed independent of the basis of E chosen. [See Example 1, continued.]
11. Let G be a linear group. Let $\mathfrak{g} = \mathfrak{s}_1 \oplus \cdots \oplus \mathfrak{s}_k$ be a decomposition of its Lie algebra $\mathfrak{g}$ into a direct sum of subspaces. Show the equation

$$a = \exp(X_1)\exp(X_2)\cdots\exp(X_k), \quad X_j \in \mathfrak{s}_j, \quad \|X\| < \epsilon,$$

defines a coordinate system around 1 in G. (Some $\epsilon > 0$; $(X_1, \ldots, X_k)$ is the coordinate point corresponding to $a \in G$).

12. Let X be a linear transformation of a real, n-dimensional vector space E, considered as a vector field on E. Show that in terms of a linear coordinate system $\xi_1, \xi_2, \ldots, \xi_n$ on E corresponding to a basis $e_1, e_2, \ldots, e_n$,

$$X = \sum_{jk} X_{jk}\xi_k \frac{\partial}{\partial \xi_j},$$

where (X_{jk}) is the matrix of X with respect to the given basis.

13. Fix a basis $(e_1, \cdots, e_n)$ for E to represent its elements as column n vectors. Represent $P \in \mathrm{Gr}_m(E)$ by an $n \times m$ matrix $p = [p_1, \cdots, p_m]$ whose columns form a basis for P, unique up to right multiplication $p \mapsto pa$ by an invertible $m \times m$ matrix a. Show:
   (a) Given $P \in \mathrm{Gr}(E)$ there is a permutation matrix $se_i = e_{s(i)}$ so that first m rows of sp are linearly independent. Then sp can be uniquely written in the form $sp = \begin{bmatrix} 1 \\ x \end{bmatrix} a$ where $x = x(P)$ is an $(n-m) \times m$ matrix, which is independent of the basis p chosen.

(b) As s runs over runs over all permutation of $(e_1, \cdots, e_n)$, the partially defined maps $P \mapsto x(P)$ from $\mathrm{Gr}(E)$ to $\mathbb{R}^{(n-m)\times m}$ or $\mathbb{C}^{(n-m)\times m}$ satisfy the axioms MFLD 1-4. (Specify the coordinate domains. Explain why it suffices to take permutations of the form $(e_{i_1}, \cdots, e_{i_m}, \cdots)$ where $i_1 < \cdots < i_m$ and the dots indicate the remaining e_is in their proper order as well.)

14. Let M be a manifold.

(a) Fix $p_o \in M$. Show that the set M_0 of points $p \in M$ that can be joined to p_o by a continuous curve $p(\tau)$ $(p(\tau_0) = p_o, p(\tau_1) = p)$ is open in M.

(b) Show that the following conditions are equivalent:

(i) Any two points of M can be joined by a continuous curve.

(ii) M is not the disjoint union of two non-empty open sets.

Under these conditions M is said to be connected. The set M_0 in (a) is called the connected component of M containing p_o.

(c) Show that M_0 is the largest connected open subset containing p_o. Show that M_0 is the disjoint union of its (distinct) connected components.

15. *Immersion Theorem.* Let $f : M \to N$ be an analytic map between two manifolds. Suppose f has an injective differential at $p \in M$. Show that, in suitable coordinates around p and $f(p)$, f looks like the inclusion $\mathbb{R}^m \to \mathbb{R}^n$,

$$(\xi_1, \dots, \xi_m) \to (\xi_1, \dots, \xi_m, 0, \dots, 0),$$

$m = \dim M$, $n = \dim N$. [Suggestion: review the proof of Proposition 7.]

16. *Submersion Theorem.* Let $f : M \to N$ be an analytic map between two manifolds. Suppose f has a surjective differential at $p \in M$. Show that, in suitable coordinates around p and $f(p)$, f looks like the projection $\mathbb{R}^m \to \mathbb{R}^n$,

$$(\xi_1, \dots, \xi_n, \xi_{n+1}, \dots, \xi_m) \to (\xi_1, \cdots, \xi_n)$$

$m = \dim M$, $n = \dim N$. [Suggestion: start with coordinates on M and N so that the differential of f at the given point p looks like such a projection. Then adjust the coordinates as in the proof of Proposition 7.]

4.2 Homogeneous spaces

An *action* of a group G on a set M associates to each $a \in G$ an invertible transformation $\tau(a)$ of M so that

$$\tau(ab) = \tau(a)\tau(b),$$

i.e. τ is a homomorphism of G into the group of all invertible transformations of M.

When G is a linear group, we also require that M be an analytic manifold and that the action be *analytic*, meaning the map $G \times M \to M, (a,p) \to \tau(a)p$, be analytic. $\tau(a)p$ is often also denoted $a \cdot p$, or even simply ap. The definition requires that G acts on M on the left. One may also require that G acts on M on the right: $p\tau(a) = p \cdot a = pa$. A left action may be turned into right action by setting $p \cdot a = a^{-1} \cdot p$, and *vice versa.*

Example 1.

(a) $G = \mathrm{SO}(3)$, $M = \mathbb{R}^3$, obvious action.

(b) $G = \mathrm{SO}(3)$, $M = S^2$, obvious action.

(c) G any group, $M = G$. Action:

(i) left translation: $l(a)p = ap$,

(ii) right translation: $r(a)p = pa^{-1}$,

(iii) conjugation: $c(a)p = apa^{-1}$.

(d) G any linear group, $M = \mathbf{g}$ its Lie algebra. Action: $\mathrm{Ad}(a)p = apa^{-1}$.

The *orbit* of a point $p \in M$ under G is $G \cdot p = \{a \cdot p \mid a \in G\}$. The action is *transitive* if there is only one orbit, i.e. any point of M can be transformed into any other point by an element of G. The *stabilizer* of a point $p \in M$ is the subgroup $G_p = \{a \in G \mid a \cdot p = p\}$.

Lemma 2.

(a) M is the disjoint union of the distinct G-orbits.

(b) The stabilizer of $p \in M$ is a subgroup of G.

(c) Suppose G acts transitively on M. Fix a base point $o \in M$, and let H be its stabilizer. Then the map $G/H \to M, aH \to o$, is a bijection from the set $G/H = \{aH \mid a \in G\}$ of left cosets of H to M. (QED)

From now on G will denote a linear group acting on an analytic manifold M. The action of G on M induces an action of G on tangent vectors: if v is a tangent vector at $p \in M$, then $a \cdot v = d\tau(a)_p v$ is a tangent vector at $a \cdot p \in M$. $(a, v) \to a \cdot v$ is an action, in the sense that $(ab) \cdot v = a \cdot (b \cdot v)$, but we have not made the set of all tangent vectors on M into a manifold. (This is easily done, but will not be needed.) If we fix a point $o \in M$, then its stabilizer H acts on the tangent space T_oM at o by linear transformations, and this action is again analytic.

Example 1 (Continued). Refer to the examples above.

(a) The orbits of SO(3) in $\mathbb{R}^3$ are spheres, except for the origin. The action on tangent vectors is again by matrix multiplication, if the tangent vectors at any point of $\mathbb{R}^3$ are identified with vectors in $\mathbb{R}^3$. If one writes a tangent vector at $p \in \mathbb{R}^3$ as a pair (p, v) with $v \in \mathbb{R}^3$, then the action is $a \cdot (p, v) = (ap, av)$. The action of the stabilizer H at a point $o \in \mathbb{R}^3$ on the tangent space $T_oR^3 = \mathbb{R}^3$ is therefore again by matrix multiplication.

(b) The action of SO(3) on S^2 is transitive. If one identifies the tangent space at $p \in S^2$ with the subspace of R^3 orthogonal to p, then the action on tangent vectors is also by matrix multiplication as in (a). The stabilizer $H \approx \mathrm{SO}(2)$ at a point $o \in S^2$ acts on $T_oS^2 \approx \mathbb{R}^2$ in the obvious way. Generally, when G acts on a vector space by linear transformations, the action on tangent vectors is the same linear action when tangent vectors are identified with vectors in the vector space. This observation applies also to the action on submanifolds of the vector space that are stable under the action, such as the action of $\mathrm{SO}(3)$ on S^2.

(c) The left and right translation action of a group on itself are transitive. The tangent space at $a \in G$ may be identified with $a\mathbf{g} = \mathbf{g}a$ and the tangent action is by matrix multiplication on the left or on the right. The orbits of the conjugation action of a group on itself are the conjugacy classes. The stabilizer of $a \in G$ under the conjugation action is the centralizer $Z_G(a) = \{c \in G \mid cac^{-1} = a\}$ of a in G. For $a = 1$, this stabilizer is all of G, and the action of G on its tangent space g at 1 is the adjoint representation $\mathrm{Ad}(a)v = ava^{-1}$.

(d) The orbits of the adjoint action of G on g are called *adjoint orbits* of G. Under the exponential map $\exp \mathbf{g} \to G$, an adjoint orbit in $\mathbf{g}$ gets mapped into a conjugacy class in G, because exp *intertwines* the adjoint action on $\mathbf{g}$ with the conjugation action on $G : \exp(cXc^{-1}) = c(\exp X)c^{-1}$.

Example 3 (The complex projective line $\mathbb{CP}^1$). The points of $\mathbb{CP}^1$ are the lines through the origin in $\mathbb{C}^2$; the symbol $[z_1, z_2]$ with $(z_1, z_2) \neq (0,0)$ in $\mathbb{C}^2$, represents the line $\{(tz_1, tz_2) \mid t \in \mathbb{C}\}$ considered as point element of $\mathbb{CP}^1$. z_1, z_2 are called *homogeneous coordinates* of the point $p = [z_1, z_2]$; they are defined only up to a non-zero common factor $t \in \mathbb{C}^\times$, and are therefore not coordinates in the usual sense. Coordinates in the usual sense may be defined as follows. As long as $z_2 \neq 0$ the point $p = [z_1, z_2]$ may be uniquely written in the form $p = [z, 1]$: $z = z_1/z_2$. This map $p \to z, \mathbb{CP}^1 \to \mathbb{C}$, with domain $\{[z_1, z_2] \mid z_2 \neq 0\}$, defines *inhomogeneous coordinates* on $\mathbb{CP}^1$. Another system of coordinates is defined similarly on the domain $\{[z_1, z_2] \mid z_1 \neq 0\}$. Together these two coordinate systems make $\mathbb{CP}^1$ into a one-dimensional *complex* manifold.

The inhomogeneous coordinates extend to a bijection $\mathbb{CP}^1 \to \mathbb{C} \cup \{\infty\}$ with $[1,0] \to \infty$. $\mathbb{C} \cup \{\infty\}$ inherits in this way the structure of a one-dimensional complex manifold, the *Riemann sphere*. It may be identified with the usual sphere $S^2 = \{\xi_1^2 + \xi_2^2 + \xi_3^2 = 1\}$ in $\mathbb{R}^3$ through *stereographic projection* as shown in Figure 1.

The formula for the stereographic projection $S^2 \to \mathbb{C} \cup \{\infty\}, (\xi_1, \xi_2, \xi_3) \to z$ is

$$z = \frac{\xi_1 + i\xi_2}{1 - \xi_3}. \tag{1}$$

(a) **The action of $\mathbf{SL(2, \mathbb{C})}$.** Notation:

$$\mathrm{SL}(2, \mathbb{C}) = \left\{ \begin{bmatrix} \alpha & \beta \\ \gamma & \delta \end{bmatrix} \mid \alpha\beta, \gamma, \delta \in \mathbb{C}, \alpha\delta - \beta\gamma = 1 \right\}.$$

The natural action of $\mathrm{GL}(2, \mathbb{C})$ on $\mathbb{C}^2$ induces an action on $\mathbb{CP}^1$, which is transitive. The scalar matrices fix all of $\mathbb{CP}^1$, so that one actually has an action of the quotient group $\mathrm{GL}(2, \mathbb{C})/\mathbb{C}^\times$, called the projective general linear group in one dimension, denoted $\mathrm{PGL}(1, \mathbb{C})$. Since $\mathrm{PGL}(1, \mathbb{C})$ may also be realized as

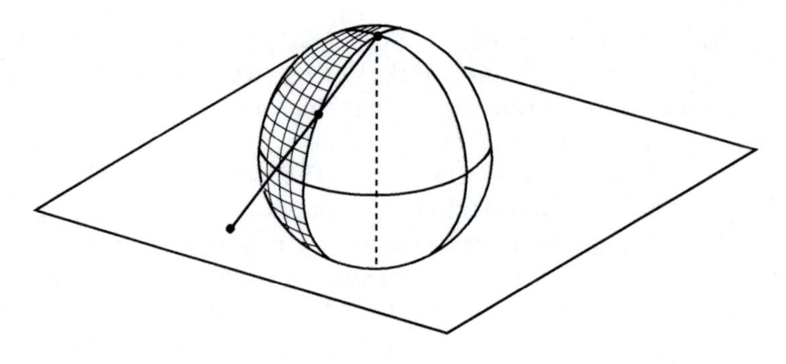

Fig. 1

$\mathrm{SL}(2,\mathbb{C})/\{\pm 1\}$, we may equivalently consider the action of $\mathrm{SL}(2,\mathbb{C})$ on $\mathbb{CP}^1$, as is customary.

In homogeneous coordinates $[z_1, z_2]$ the action is simply matrix multiplication:

$$\begin{bmatrix} \alpha & \beta \\ \gamma & \delta \end{bmatrix} \begin{bmatrix} z_1 \\ z_2 \end{bmatrix} = \begin{bmatrix} \alpha z_1 + \beta z_2 \\ \gamma z_1 + \delta z_2 \end{bmatrix}$$

In inhomogeneous coordinates the action is by *linear fractional transformations*:

$$z \to \frac{\alpha z + \beta}{\gamma z + \delta}.$$

It may be shown that $\mathbb{CP}^1$ admits no other biholomorphic transformations (see H. Cartan (1963), for example).

(b) **The action of SU(2).** Notation:

$$\mathrm{SU}(2) = \left\{ \begin{bmatrix} \alpha & \beta \\ -\bar{\beta} & \bar{\alpha} \end{bmatrix} \mid \alpha, \beta \in \mathbb{C},\ \alpha\bar{\alpha} + \beta\bar{\beta} = 1 \right\}.$$

The action of SU(2) on $\mathbb{CP}^1$ is still transitive. This action has a geometric interpretation in terms of the bijection $\mathbb{CP}^1 \approx S^2$:

Proposition 3A. *Under the stereographic projection* $\mathbb{CP}^1 \approx S^2$, SU(2) *acts on* S^2 *through the double covering* $\mathrm{SU}(2) \to \mathrm{SO}(3)$ *in §2.1.*

Proof. We construct the inverse of stereographic projection $S^2 \to \mathbb{CP}^1$ defined by eqn (1) as follows. With $(\xi_1, \xi_2, \xi_3) \in \mathbb{R}^3$ associate the matrix $X \in \mathsf{su}(2)$ defined by

$$X = i \begin{bmatrix} \xi_3 & \xi_1 + i\xi_2 \\ \xi_1 - i\xi_2 & -\xi_3 \end{bmatrix}.$$

Then the sphere S^2 is realized as the surface in $\mathsf{su}(2)$ with equation

$$\tfrac{1}{2}\,\mathrm{tr}(XX^*) = 1. \tag{2}$$

(b) **The action of** $SU(2)$. Notation:
Write the homogeneous coordinates of a point $[z_1, z_2] \in \mathbb{CP}^1$ as a 2×1 matrix:

$$Z = \begin{bmatrix} z_1 \\ z_2 \end{bmatrix}.$$

One checks immediately that the equation

$$X = i\left(\frac{2ZZ^*}{Z^*Z} - 1\right).$$

defines a map $\mathbb{C}P^1 \to S^2 \subset \mathsf{su}(2)$ which satisfies (1) for $[z_1, z_2] = [z, 1]$, and hence represents the inverse of the stereographic projection $S^2 \to \mathbb{CP}^1$. Furthermore, under this map the action $Z \to aZ$ of $a \in \mathrm{SU}(2)$ on $\mathbb{CP}^1$ evidently corresponds to the restriction to S^2 of the adjoint action $X \to aXa^{-1}$ on $\mathrm{SU}(2)$. The latter corresponds to the covering $\mathrm{SU}(2) \to \mathrm{SO}(3)$ under the identification $\mathsf{su}(2) \approx \mathbb{R}^3, X \leftrightarrow (\xi_1, \xi_2, \xi_3)$. QED

Comment. The proposition gives in fact an alternative *construction* of the covering $\mathrm{SU}(2) \to \mathrm{SO}(3)$. This is the point of view taken by Weyl (1931, p. 144).

The action of

$$\mathrm{SL}(2,\mathbb{R}) = \left\{ \begin{bmatrix} \alpha & \beta \\ \gamma & \delta \end{bmatrix} \mid \alpha, \beta, \gamma, \delta \in \mathbb{R}, \alpha\delta - \beta\gamma = 1 \right\}.$$

For real $\alpha, \beta, \gamma, \delta$ with $\alpha\delta - \beta\gamma = 1$,

$$\mathrm{Im}\,\frac{\alpha z + \beta}{\gamma z + \delta} = \frac{\mathrm{Im} z}{|\gamma z + \delta|^2}.$$

One sees from this that $\mathrm{SL}(2,\mathbb{R})$ has three orbits on $\mathbb{CP}^1 = \mathbb{C} \cup \{\infty\}$: the upper half-plane $\{\mathrm{Im} z > 0\}$, the lower half-plane $\{\mathrm{Im} z < 0\}$, and the extended real line $\{\mathrm{Im} z = 0\} \cup \{\infty\}$. It may be shown that the upper (or lower) half-plane admits no other biholomorphic transformations (see again H. Cartan (1963), for example).

The action of

$$\mathrm{SU}(1,1) = \left\{ \begin{bmatrix} \alpha & \beta \\ \bar{\beta} & \bar{\alpha} \end{bmatrix} \mid \alpha, \beta \in \mathbb{C}, \alpha\bar{\alpha} - \beta\bar{\beta} = 1 \right\}.$$

The element of $\mathrm{PGL}(1,\mathbb{C})$ represented by $\left[\begin{smallmatrix} 1 & -i \\ 1 & i \end{smallmatrix}\right]$ carries the upper half-plane $\{\mathrm{Im} z > 0\}$ onto the disk and transforms the action of $\mathrm{SL}(2,\mathbb{C})$ into the action of $\mathrm{SU}(1,1)$:

$$\mathrm{SU}(1,1) = \begin{bmatrix} 1 & -i \\ 1 & i \end{bmatrix} \mathrm{SL}(2,\mathbb{R}) \begin{bmatrix} 1 & -i \\ 1 & i \end{bmatrix}^{-1}$$

The actions of $\mathrm{SL}(2,\mathbb{R})$ and of $\mathrm{SU}(1,1)$ on $\mathbb{CP}^1$ are in this sense equivalent.

The stabilizer H of a point $p \in M$ under the action of the linear group G is always a closed subgroup of G, being the inverse image of the point $p \in M$ under the continuous map $G \to M$, $a \to ap$. Lemma 2(c) says that any manifold with transitive G-action is in one-to-one correspondence with G/H for such a subgroup H of G. Suppose now we *start* with a closed subgroup H of G. We shall define on G/H a manifold structure so that the action of G on G/H by left translation is analytic.

Recall the *Closed Subgroup Theorem* (Theorem 1, §2.7): Let G be a linear group, H a closed subgroup of G. Let s be a vector space complement for the Lie algebra h of H in the Lie algebra g of G: $\mathsf{g} = \mathsf{s} \oplus \mathsf{h}$. There is an open neighborhood U of 0 in s so that the map

$$U \times H \to G, \quad (X, b) \to \exp(X)b$$

is an analytic bijection onto an open neighborhood of H in G.

With this notation, introduce coordinates around the point $1H$ on G/H by taking $X \in U$ as coordinate of the point $\exp(X)H$ on G/H. Coordinates around an arbitrary point $a_o H$ on G/H are defined by the equation

$$aH = a_o \exp(X)H,$$

with $X \in U$ serving as coordinate of aH. (The coordinates depend on the choice of the representative a_o for $a_o H$.)

Proposition 4. *With these coordinates G/H becomes an analytic manifold on which G acts analytically by left translation $a \cdot (bH) = (ab)H$.*

Proof. Consider a coordinate system on G/H as defined above: $\mathsf{g} = \mathsf{s} \oplus \mathsf{h}$, $a_o \in G$, U an open neighborhood of 0 in s for which $U \times H \to \exp(U)H$ is bi-analytic onto the open subset $\exp(U)H$ of G. The coordinate domain in G/H is $a_o \exp(U) \cdot H$ and the coordinate map is $a_o \exp(U) \cdot H \to U$, $a_o \exp(X) \cdot H \to X$. The axiom MFLD 1 is satisfied by the choice of U. To verify MFLD 2, consider another such coordinate system $\tilde{a}_o \exp(\tilde{U}) \cdot H \to \tilde{U}$, corresponding to $\tilde{\mathsf{s}}$, $\tilde{U}$, and $\tilde{a}_o$. The coordinate transformation

$$U \to \tilde{U}, \quad X \to \tilde{X},$$

is defined by $\tilde{a}_o \exp(\tilde{X}) \cdot H = a_o \exp(X) \cdot H$, or equivalently

$$\exp(\tilde{X}) \in \tilde{a}_o^{-1} a_o \exp(X)H.$$

Its domain is $\{X \in U \mid \tilde{a}_o^{-1} a_o \exp(X) \in \exp(\tilde{U})H\}$, which is open in s. On this domain, the coordinate transformation $X \to \tilde{X}$ is the composite of $X \to \tilde{a}_o^{-1} a_o \exp(X)$ with the map $\exp(\tilde{U})H \to \tilde{U}$, hence is analytic. This shows that MFLD 2 is satisfied. MFLD 3 is evident. To verify MFLD 4, assume $a_1 \cdot H \neq a_2 \cdot H$ in G/H. Then $a_1^{-1} a_2$ lies outside of H, and since H is a submanifold of G it is clear that $a_1^{-1} a_2$ has a neighborhood that lies outside of H. We may assume this neighborhood to be of the form $(a_1 A_1)^{-1}(a_2 A_2)$, where A_1 and A_2 are neighborhoods of 1 in G, which may be chosen arbitrarily small. $a_1 A_1 \cdot H$ is a neighborhood of $a_1 \cdot H$ in G/H, $a_2 A_2 \cdot H$ a neighborhood of $a_2 \cdot H$ (as one may check from the definitions). Their intersection is empty, as otherwise one would get an equation $a_1 b_1 H = a_2 b_2 H$ with $b_1 \in A_1$, $b_2 \in A_2$, leading to $(a_1b_1)^{-1}a_2b_2 \in H$, which is impossible. This completes the verification that G/H is an analytic manifold with the coordinate systems indicated. To see that the action $G \times G/H \to G/H$ is analytic, one remarks that in coordinate systems on G/H of the above type this map becomes locally a map $G \times U \to \tilde{U}$, obtained as a composite of $(a, X) \to aa_o \exp X$ and $\exp(\tilde{U})H \to \tilde{U}$. QED

We can now prove an analytic version of Lemma 2(c):

Proposition 5. *Let G be a linear group, M be an analytic manifold with a transitive analytic action $G \times M \to M$. Fix $o \in M$ and let H be its stabilizer. If G has countably many connected components, then the bijection $G/H \to M$, $aH \to a \cdot o$ is bi-analytic.*

Proof. It suffices to show that the bijection $G/H \to M$ is locally bi-analytic at $1H$. Choose a subspace s of g complementary to h and a submanifold N of a neighborhood of o in M whose tangent space at o is complementary to the image of differential at 0 of $\mathsf{s} \to M$, $X \to \exp(X) \cdot o$: if this differential is written $X \to X \cdot o$, then we require that $T_oM = \mathsf{s} \cdot o \oplus T_oN$. Then the map $\mathsf{s} \times N \to M, (X,p) \to \exp(X)p$ has an invertible differential at $(0, o)$, namely $(X, v) \to X \cdot o + v$. Hence this map is locally bi-analytic. We choose $U = \exp\{X \in \mathsf{s} \mid \|X\| < \epsilon\}$ and a coordinate ball around o in N, which we may as well denote by N again, so that the map $U \times N \to M$ is bi-analytic onto a neighborhood of o in M. The set UH contains a neighborhood of 1 in G. As in the proof of the Lie correspondence (eqn (1), §2.5) we may write G_o as a countable union $\cup_{j=1}^{\infty} a_jUH$. Since G has countably many components, it is expressible as such a union as well. Then $M = \cup_{j=1}^{\infty} a_jU \cdot o$. By Baire's Covering Lemma (Lemma 2, §2.5, which applies to manifolds as well as to groups), the closure of some $a_jU \cdot o$ contains an open subset of M. Hence so does the closure of $U \cdot o$, and by shrinking the radius of the ball U we may assume that $U \cdot o$ itself is open in M. On the other hand, the neighborhood $UN \cdot o$ of o in M looks like $U \times N$, from which one sees that the ball N must be zero-dimensional, i.e. the point $\{o\}$. This proves that the map $U \to M$ is analytic onto a neighborhood of o in M, from which it follows that $G/H \to M$ is bi-analytic at $1H$. QED

A *homogeneous space* for a linear group G is by definition an analytic manifold M with a transitive analytic action of G so that for every (equivalently: some) $o \in M$ the bijection $G/H \to M$, $aH \to a \cdot o$ is bi-analytic. By what has just been shown, the last condition is superfluous if G has countably many connected components.

Remark 6. The map $G \to G/H$, $a \to aH$, is analytic and its differential at 1 maps g onto the tangent space of G/H at $1H$ with kernel h. The tangent space of G/H at $1H$ may therefore be identified with the quotient space g/h. Under this identification, the tangent vector at $1H$ of a curve on G/H of the form $c(\tau)H$, $c(\tau) \in G$ of class C^1 in τ with $c(0) = 1$, gets identified with the image of $c'(0)$ in g/h. The action of H on the tangent space of G/H at $1H$ is the adjoint action on g/h. (QED)

Problems for §4.2

1. Prove Lemma 2.

2. Prove Remark 6.

3. Let S^2 be the unit sphere in $\mathbb{R}^2$, o a point of S^2. The stabilizer of o in SO(3) may be identified with SO(2). Show that the map $\mathrm{SO}(3)/\mathrm{SO}(2) \to S^2$, $a\mathrm{SO}(2) \to ao$, is an analytic bijection.

4. Show that the tangent space at $aH \in G/H$ can naturally be identified with $\mathfrak{g}/\operatorname{Ad}(a)\mathfrak{h}$. [Suggestion: given $a(t)H$ in G/H consider $a(t)a(0)^{-1}$ in G.]

5. Let G be a linear group, H a closed subgroup of G. Show that the map $G \to G/H$, $a \to aH$, is analytic and has a surjective differential everywhere. Show that this property characterizes the manifold structure of G/H.

6. (a) Show that the Grassmannian $\mathrm{Gr}_m(E)$ (problem 13, §4.1) is a homogeneous space for $\mathrm{GL}(E)$ under the natural action. Show that the stabilizer in $\mathrm{GL}(E)$ of a point $P_o \in \mathrm{Gr}_m(E)$ consists of all matrices in $\mathrm{GL}(E)$ of the form
$$\begin{bmatrix} * & * \\ 0 & * \end{bmatrix}$$
if one uses a basis for E whose first m vectors from a basis for P_o.

   (b) Suppose E has a positive definite inner product (Hermitian if E is complex), and let $K(E)$ be the group of linear transformations preserving this inner product. ($O(n)$ in the real case, $U(n)$ in the complex case.) Show that $\mathrm{Gr}_m(E)$ is also a homogeneous space for $K(E)$. Show that the stabilizer of a point $P_o \in \mathrm{Gr}_m(E)$ consists of all matrices in $K(E)$ of the form
$$\begin{bmatrix} * & 0 \\ 0 & * \end{bmatrix}$$
if one uses an orthonormal basis for E whose first m vectors form a basis for P_o, and is isomorphic with $O(m) \times O(n-m)$ in the real case and with $U(m) \times U(n-m)$ in the complex case.

7. Let G be a linear group, H, L closed subgroups of G so that $G = HL$. Show that the map $G/(H \cap L) \to (G/H) \times (G/L), a(H \cap L) \to (aH, aL)$, is bi-analytic.

8. Let G be a linear group, H a closed *normal* subgroup of G. Show that the group operations on G/H
$$(G/H) \times (G/H) \to G/H, \quad (aH)(bH) \to (ab)H$$
$$G/H \to G/H, \quad aH \to a^{-1}H$$
are analytic.

9. Let H be a *connected* closed subgroup of G. Suppose G/H is connected. Show that G is connected. [Suggestion: consider G_oH in G/H.]

10. Use the previous problem and induction to show that the following groups are connected.

    (a) $\mathrm{SO}(n)$ [Suggestion: consider the action on S^{n-1}].

    (b) $\mathrm{SL}(n, \mathbb{R})$ [Suggestion: consider the action on $\mathbb{R}^n - \{0\}$].

11. Let H be a *connected* closed subgroup of G. Suppose G/H is simply connected. Show that G is simply connected. [This problem assumes some familiarity with homotopy theory. Suggestion: Let $a(\tau)$ be a closed path in G with $a(0) = a(1) = 1$. Deduce from the simple-connectedness of G/H that there is a homotopy $a_\sigma(\tau)$, $0 < \sigma < 1$, with $a_0(\tau) = a(\tau)$ and $a_1(\tau) \in H$. Use the connectedness of H to conclude the proof.]

12. Use the previous problem and induction to show that $\mathrm{SU}(n)$ is simply connected. [Suggestion: consider the action on $S^{2n-1} \subset \mathbb{C}^n$. Assume known that S^N is simply connected for $N > 2$. Why does this method fail for $\mathrm{SO}(n)$?]

13. Let K be a compact linear group acting on a manifold M. Let o be a fixed-point of $K (k \cdot o = o$ for all $k \in K)$. Show:

    (a) There exists arbitrarily small K-stable neighborhoods of o in M. [Suggestion: Let be a neighborhood U of o. For each $k \in K$ there is a neighborhood V_k of k in K and a neighborhood U_k of o in M so that $V_k \cdot U_k \subset U$.]

    (b) There is a bi-analytic map $f : M \cdots \to T_oM$ from a K-stable neighborhood of o in M onto a K-stable neighborhood of 0 in T_oM satisfying $f(k \cdot p) = k \cdot f(p)$. Deduce that in suitable coordinates around o in M the action of K on M (locally) looks like its linear action on T_oM. [Suggestion: Start with any map $f : M \cdots \to T_oM$ with $df_o = 1$. Consider

    $$f^o(p) = \frac{1}{|K|} \int\limits_K k \cdot f(k^{-1}p)\, dk.$$

    Calculate $(df^o)_o$ by differentiation under the integral sign. [This problem anticipates the invariant integral on K. In the meantime assume K is finite and interpret the integral over K as a sum.]

    (c) Find an explicit such map f when

       (i) $M = S^2, K \subset \mathrm{SO}(3)$, the stabilizer of $o = e_3$.

       (ii) $M = \{\text{upper half-plane } \mathrm{Im} z > 0\}$, $K \subset \mathrm{SL}(2, \mathbb{R})$ the stabilizer of $o = i$. (See Example 3.)

       (iii) $M = K$ with adjoint action, $o = 1$.

(iv) $M = \{$real symmetric $n \times n$ matrices$\}$, $K = \mathrm{SO}(n)$ with action

$$k \cdot p = kpk^*, \quad o = 1.$$

14. Let G be a linear group acting transitively on a manifold M, H the stabilizer of $o \in M$. Show that the bijection $G/H \to M$, $aH \to a \cdot o$, is bi-analytic if and only if its tangent map at $1H$ is surjective. [No conditions on the number of components.]

15. Let G be a linear group acting transitively on a connected manifold M. Assume G has countably many components. Show that the connected subgroup G_o already acts transitively on M.

16. Let G be a linear group acting on a manifold M. Show:

    (a) If for every $p \in M$ the map $G \to M$, $a \to a \cdot p$, has a surjective differential at p, then G acts transitively on M.

    (b) The converse of (a) holds provided G has countably many components.

17. Let G be a linear group, H a closed subgroup of H. Show that the analytic structure on G/H is uniquely characterized by the property that the natural map $G \to G/H$ is analytic and its differential at 1 is surjective. [Suggestion: Show first that the kernel of the differential is exactly h; for this purpose consider $d/d\tau \exp(\tau X)H$.]

4.3 Lie groups

A *Lie group* is a group G with the structure of an analytic manifold so that the group operations are analytic, i.e. the maps

$$G \times G \to G, \ (a, b) \to ab, \ \text{and} \ G \to G, \ a \to a^{-1}$$

are analytic.

Linear groups are Lie groups, as was shown in Example 2, §4.1. The passage from linear groups to general Lie groups is to a large extent a matter of language and technique. The purpose of this section is to explain in what sense this is the case and to serve as a bridge to the extensive literature on Lie groups. The material will not be used later. We shall divide the discussion into three subsections.

Vector fields and one-parameter groups on manifolds. Returning to our starting point, we consider once more the differential equation

$$p'(\tau) = X(p(\tau)) \tag{1}$$

with the initial condition

$$p(0) = p_o. \tag{2}$$

But this time we take for X an analytic vector field on an arbitrary analytic manifold, rather than a linear vector field on a vector space. The basic facts about the solution of this differential equation may be stated as follows.

Theorem 1. *Let M be an analytic manifold, X an analytic vector field on M.* *(a) For any $p_o \in M$ the differential equation*

$$p'(\tau) = X(p(\tau))$$

with the initial condition

$$p(0) = p_o$$

has a unique analytic solution $p(\tau)$ defined in some interval about $\tau = 0$. Denote this solution by $\exp(\tau X)p_o$.

(b) The map $\mathbb{R} \times M \cdots \to M$, $(\tau, p) \to \exp(\tau X)p$ is defined and analytic in a neighbourhood of $\{0\} \times M$ in $\mathbb{R} \times M$. It satisfies

$$\exp((\sigma + \tau)X)p = \exp(\sigma X)\exp(\tau X)p,$$
$$\exp(-\tau X)\exp(\tau X)p = p,$$

where defined.

(c) For every analytic function $\varphi : M \cdots \to \mathbb{R}$, defined in a neighbourhood of p on M,

$$\varphi(\exp(\tau X)p) = \sum_{k=0}^{\infty} \frac{\tau^k}{k!} X^k \varphi(p),$$

the series converges in an interval about $\tau = 0$; this property characterizes $\exp(\tau X)p$ *uniquely.*

Comment. In (c) the vector field X is considered a differential operator. The formula may be taken as justification for the notation $\exp(\tau X)p$. It shows in particular that $\exp(\tau X)p$ really depends only on τX, not on τ and X separately (as may also be verified directly from the definition). The formula (c), conceived of as a formula for $\exp(\tau X)$, may be traced to Lie himself (1888, p. 51).

Proof.[1] (a) Fix a coordinate system $(\xi_1, \xi_2, \ldots, \xi_n)$ around p_o with p_o corresponding to the coordinate point $(0, \ldots, 0)$. Using these coordinates we shall simply write points and tangent vectors in the coordinate domain as n-tuples. In particular, we write out the Taylor series for $X(p)$ at $p = 0$ as well as the (unknown) Taylor series for $p(\tau)$:

$$X(p) = \sum_k c_k p^k, \quad p(\tau) = \sum_j a_j \tau^j$$

where $k = (k_1, \ldots, k_n)$, $k_i \geq 0$, and $p^k = \xi_1^{k_1}, \ldots, \xi_n^{k_n}$. The sum over j starts with $j = 1$ because of the initial condition $p(0) = 0$. The coefficients c_k and a_j are n-tuples. Formally, the differential equation (2) becomes:

$$\sum_j j a_j \tau^{j-1} = \sum_k c_k \left(\sum_i a_i \tau^i \right)^k .$$

[1] Part (a) is the basic existence theorem in the theory of analytic ordinary differential equations. Its proof may well be omitted here.

Comparing the coefficients of powers of τ one finds that

$$a_j = \frac{1}{j} Q_j(c_k, a_i), \tag{3}$$

where $Q_j(c_k, a_i)$ is a polynomial with *positive* integral coefficients in c_k, a_i, for $|k|, i < j$. Here $|k| = k_1 + \cdots + k_n$. In particular, there is a unique formal power series $p(\tau)$ in τ that solves (1) and (2).

To show convergence we use the *method of majorants*. Consider another differential equation of the same sort

$$\tilde{p}'(\tau) = \tilde{X}(\tilde{p}(\tau)), \quad \tilde{p}(0) = p_o \tag{4}$$

with $\tilde{X}(p) = \sum_k \tilde{c}_k x^k$ having non-negative coefficients dominating those of $X(p) : |c_k| \leq \tilde{c}_k$. Write

$$\tilde{p}(\tau) = \sum_j \tilde{a}_j \tau^j.$$

Then

$$|a_j| = \left| \frac{1}{j} Q_j(c_k, a_j) \right| \leq \frac{1}{j} Q_j(|c_k|, |a_j|) \leq \frac{1}{j} Q_j(\tilde{c}_k, \tilde{a}_j) = \tilde{a}_j, \tag{5}$$

the last equality being the analogue of (3). This shows that the series for $p(\tau)$ converges whenever $\tilde{p}(\tau)$ converges.

It remains to construct an appropriate $\tilde{X}(p)$. Since $X(p)$ is convergent, one may find constants $M, R > 0$ so that $\sum |c_k| R^{|k|} < M$. Let $\tilde{c}_k = (M/R^{|k|})1$, where $1 = (1, \ldots, 1)$. Then $|c_k| \leq \tilde{c}_k$ and

$$\tilde{X}(p) = \sum_k M \left(\frac{p}{R} \right)^k 1 = \frac{M}{\prod_i (1 - (\xi_i/R))} 1.$$

A solution of (4) may be obtained by setting $\tilde{p}(t) = \varphi(\tau)1$ where $\varphi(\tau)$ is the solution of

$$\varphi'(\tau) = \frac{M}{(1 - \varphi(\tau)/R)^n}; \tag{6}$$

and by uniqueness this $\tilde{p}(\tau)$, written as power series, is the formal solution of (4).

The solution to (6) may be given explicitly:

$$\varphi(\tau) = R - R \left(1 - (n+1) \frac{M}{R} \tau \right)^{1/(n+1)},$$

as one sees by differentiation.

We now know that eqns (1) and (2) have a unique solution $p(\tau)$ defined by a convergent power series for τ in some interval about 0, and this proves (a). The construction of $p(\tau)$ gives some further information, which we record for reference.

Remark 2. Suppose the vector field X depends analytically on some parameters $s = (\sigma_1, \ldots, \sigma_m)$, say $X(p) = X(p, s)$. Then $p(\tau) = p(\tau, s)$ will also depend analytically on s.

For if one chooses the majorant coefficients so that $|c_k(s)| \le \tilde{c}_k$ for $\|s - s_o\| < \epsilon$, then $p(\tau, \sigma)$ is represented by a series of analytic functions in (τ, s), converging uniformly for $|\tau| < \epsilon$, $\|s - s_o\| < \epsilon$ (some $\epsilon > 0$), hence is analytic in a neighbourhood of $(\tau, s) = (0, s_o)$.

We now use the notation $\exp(\tau X)p_o$ for $p(\tau)$, which is well-defined, since $\exp(\tau(\alpha X))p_o = \exp((\tau\alpha)X)p_o$. The fact that $\exp(\tau X)p$ is analytic in p as well can be verified by the method of the above remark as follows. Dropping the assumption $p_o = 0$, write the coefficients a_k of the power series for $p(\tau) = \exp(\tau X)p_o$ explicitly as $a_k = a_k(p_o)$. They are still given by the recursion formula (3), where the c_k are the coefficients of the series for X in powers of $p - p_o$. They are therefore polynomials in the coordinates of p_o. Thus the series for $\exp(\tau X)p_o$ can be formally rewritten as a power series in (τ, p_o). To justify this procedure it suffices to prove that the estimate

$$|a_k(p_o)| \le \tilde{a}_k,$$

can be arranged to hold uniformly in a neighbourhood of a given p_o. But this is clear from the argument given there.

(b) We now prove the formula

$$\exp(\sigma X)\exp(\tau X)p = \exp((\sigma + \tau)X)p. \tag{7}$$

To this end we fix p and τ and consider both sides of the equation as function of σ only. Assuming τ is within the interval of definition of $\exp(\tau X)p$, both sides of the equation are defined for σ in an interval about 0 and analytic there. We check that the right side satisfies the differential equation and initial condition defining the left side:

$$\frac{d}{d\sigma}\exp((\sigma+\tau)X)p = \frac{d}{d\rho}\exp(\rho X)p\bigg|_{\rho=\sigma+\tau} = X\left(\exp((\sigma+\tau)X)p\right),$$

$$\exp((\sigma+\tau)X)p\,|_{\sigma=0} = \exp(\tau X)p.$$

This proves (7), and $\exp(-\tau X)\exp(\tau X)p = p$ is a consequence thereof.

(c) This is nothing but the Taylor expansion of $\varphi(\exp(\tau X)p)$ as (analytic) function of τ at $\tau = 0$: the equation

$$\frac{d}{d\tau}\exp(\tau X)p = X(\exp(\tau X))p$$

leads to

$$\frac{d}{d\tau}\varphi(\exp(\tau X)p) = X\varphi(\exp(\tau)p).$$

Using this equation repeatedly to calculate the higher-order derivatives one finds that the Taylor series of $\varphi(p(\tau))$ at $\tau = 0$ may be written as

$$\varphi(\exp(\tau X)p) = \sum_k \frac{\tau^k}{k!}X^k\varphi(p),$$

as desired. This property determines $\exp(\tau X)p$, as one sees by taking for φ the coordinate functions $\varphi(p) = \xi_i$ with respect to some coordinate system. QED

For evident reasons $\exp(\tau X)$ is called the *one-parameter group of local transformations of M generated by X*. It is an unfortunate, but unavoidable, complication that these local transformations do not always extend to global transformations, and are not always defined for all τ. For equally evident reasons, vector fields are in this context called *infinitesimal transformations*, although this pleasantly archaic terminology, which may well go back to Lie himself and certainly was favoured by him, seems to have fallen into disgrace in modern times.

The exponential map of a Lie group. We start with some remarks about left and right translations on a Lie group G. The left and right translations of G corresponding to an element $a \in G$ are now denoted by a_l and a_r:

$$a_l(x) = ax \quad a_r(x) = xa^{-1}.$$

By differentiation we get corresponding to each element $X \in \mathrm{g} = T_1G$, two vector fields, called *infinitesimal left and right translation*, denoted X_l and X_r, and defined by

$$X_l(x) = Xx, \quad X_r(x) = -xX.$$

Here Xx and xX are the vectors at $x \in G$ given by

$$Xx = \frac{d}{d\tau}a(\tau)x\bigg|_{\tau=0} \quad \text{and} \quad xX = \frac{d}{d\tau}xa(\tau)\bigg|_{\tau=0},$$

where $a(\tau)$ is any differentiable curve with $a(0) = 1$, $a'(0) = X$. (Thus $X_l(x) = (dx_l)_1X = Xx$ and $X_r(x) = (dx_r)_1X = -Xx$. This is in accordance with the notation for the action of a group on tangent vectors defined in general after Lemma 2, §4.2.)

There is a simple device that can be used to apply methods from linear groups to Lie groups. First, elements of G may be considered as operators on functions on G by left and right translation, again denoted a_l and a_r:

$$a_l\varphi(x) = \varphi(a^{-1}x), \quad a_r\varphi(x) = \varphi(xa).$$

These operators may be applied to analytic functions defined on open subsets of G: if φ has domain U, then $a_l\varphi$ has domain $a_l(U) = aU$ and $a_r\varphi$ has domain $a_r(U) = Ua^{-1}$. Unfortunately, partially defined functions can not be added without restrictions, so the operators a_l and a_r are not linear transformations (and in any case not linear transformations of a finite dimensional space). However, if φ is defined in a neighbourhood of x, so are $a_l\varphi$ and $a_r\varphi$ for a in a neighbourhood of 1 in G, and the sum of two such φ has the same property. For the operators X_l and X_r the situation is even simpler, since $X_l\varphi$ and $X_r\varphi$ will be again analytic with the same domain as φ. These remarks allow one to carry over the methods from linear groups to Lie groups as far as needed, with the operators a_l and X_l taking the place of matrices representing elements of the group and of its Lie algebra. (Naturally, one could equally well use a_r, and X_r.)

As for linear groups, the first thing is to define an exponential map.

Theorem 3. *Let G be a Lie group, g its tangent space at 1. There is a unique analytic map* $\exp\colon \mathsf{g} \to G$ *with* $\exp(0) = 1$ *so that*

$$\frac{d}{d\tau}\exp(\tau X) = \exp(\tau X)X, \tag{8}$$

or equivalently,

$$\frac{d}{d\tau}\exp(\tau X) = X\exp(\tau X), \tag{8'}$$

for all $X \in \mathsf{g}$ and all $\tau \in \mathbb{R}$. This map satisfies

$$\exp((\sigma+\tau)X) = \exp(\sigma X)\exp(\tau X)$$

for all $X \in \mathsf{g}$ and all $\sigma, \tau \in \mathbb{R}$.

Proof. Equation (8) with the initial condition $\exp(0) = 1$ characterizes the solution curve through 1 belonging to the vector field X_l. It follows from Theorem 1 that $\exp(\tau X)$ is the unique analytic curve, defined for τ in an interval about 0, which satisfies (8) and the initial condition. Furthermore, thinking of the components of X with respect to a basis for g as parameters of the vector field X_l, we see from Remark 2 that $\exp(\tau X)$ is an analytic function $\mathbb{R} \times G \cdots \to G$, for the moment defined only on a neighbourhood of $\{0\} \times G$. It follows that $\exp X$ is defined analytic for X in a neighbourhood of 0 in g, and satisfies $\exp((\sigma+\tau)X) = \exp(\sigma X)\exp(\tau X)$ when defined. Define $\exp X$ for *any* $X \in \mathsf{g}$ by the formula $\exp X = (\exp X/p)^p$ with $p \in \mathbb{N}$ large enough so that X/p is sufficiently close to 0. This is well-defined: if $q \in \mathbb{N}$ has the same property, so does pq and $(\exp X/p)^p = (\exp X/pq)^{pq} = (\exp X/q)^q$. Then $\exp : \mathsf{g} \to G$ is analytic on all of g and satisfies $\exp((\sigma+\tau)X) = \exp(\sigma X)\exp(\tau X)$ for all $\sigma, \tau \in \mathbb{R}$. Since the left side is symmetric in σ and τ one gets $\exp((\sigma+\tau)X) = \exp(\sigma X)\exp(\tau X)$ as well. Differentiating this equation with respect to σ at $\sigma = 0$ one finds (8′) and sees that (8′) is equivalent to (8). QED

Remark 4. (a) The formula (8) shows that the differential of $\exp : \mathsf{g} \to G$ at 0 is the identity transformation of g, so that exp is an analytic bijection from a neighbourhood of 0 in g to a neighbourhood of 1 in G, providing *exponential coordinates* around 1 on G.

(b) As in part (c) of Theorem 1 one has formulas

$$\varphi(\exp(-\tau X)a) = \sum_k \frac{\tau^k}{k!} X_l^k \varphi(a),$$

$$\varphi(a\exp(\tau X)) = \sum_k \frac{\tau^k}{k!} X_r^k \varphi(a),$$

valid in an interval about $\tau = 0$ that depends on the function φ defined in a neighbourhood of a. One is tempted to write

$$(\exp X)_l = \sum_k \frac{1}{k!} X_l^k, \quad (\exp X)_r = \sum_k \frac{1}{k!} X_r^k;$$

these relations may be viewed as a first instance of a formula from linear groups adapted to Lie groups.

The Lie algebra of a Lie group. The vector fields $U = X_l$ of infinitesimal left translation are *right invariant* in the sense that

$$U(x)a = U(xa) \tag{9}$$

for all $x, a \in G$. This is seen to be equivalent to

$$a_r U = U a_r \tag{10}$$

as operators on analytic functions. Furthermore, *any* right invariant vector field U is of the form X_l a unique $X \in \mathfrak{g}$, namely $X = U(1)$, as one sees on comparing the equation $U(x) = U(1)x$ with the definition $X_l(x) = Xx$ of X_l. We shall need the following general lemma on vector fields.

Lemma 5. *Let U, V be analytic vector fields defined on an analytic manifold M. There is a unique analytic vector field, denoted $[U, V]$, so that $[U, V] = UV - VU$ as operators on analytic functions defined on open subsets of M.*

Proof. Write locally in coordinates $\xi_1, \xi_2, \ldots, \xi_n$ on M:

$$U = \sum_k u_k \frac{\partial}{\partial \xi_k}, \quad V = \sum_k v_k \frac{\partial}{\partial \xi_k}.$$

By a simple computation using the symmetry of second partials one sees that $W = \sum_k w_k \partial/\partial\xi_k$ satisfies $W\varphi = UV\varphi - VU\varphi$ for all analytic functions φ on the coordinate domain if and only if

$$w_k = \sum_j u_j \frac{\partial v_k}{\partial \xi_j} - v_j \frac{\partial u_k}{\partial \xi_j}.$$

This formula defines $[U, V]$ on the coordinate domain. Because of the uniqueness, the W's defined on the coordinate domains of two coordinate systems agree on the intersection, from which one sees that W is defined globally on the whole manifold (assuming U and V are). QED

Proposition 6. *For any $X, Y \in \mathfrak{g}$ there is a unique element $[X, Y] \in \mathfrak{g}$ so that*

$$[X, Y]_l = X_l Y_l - Y_l X_l.$$

The operation $(X, Y) \to [X, Y]$ makes $\mathfrak{g}$ into a Lie algebra.

Proof. As operators, $X_l = U$, and $Y_l = V$ commute with right translations a_r, hence so does $[U, V] = UV - VU$, and we may define $[X, Y]$ by the requirement that $[X, Y]_l = [X_l, Y_l]$. The Jacobi Identity is a formal consequence of $[U, V] = UV - VU$. QED

The proof of the proposition may be seen as another instance of linear methods adapted to Lie groups: the definition $[X,Y]_l = X_lY_l - Y_lX_l$ for Lie groups takes the place of the definition $[X,Y] = XY - YX$ for linear groups.

One might now go on to rewrite the theory developed for linear groups as a theory for general Lie groups: adjoint representation, the formula for $d\exp$, the Campbell–Baker–Hausdorff Formula, all carry over with minor re-interpretations of the notation. As an illustration we return to the Campbell–Baker–Hausdorff Formula. Consider again

$$\exp(X)\exp(Y) = \exp(Z) \tag{11}$$

as equation for Z in terms of X and Y. Since exp is locally bi-analytic at 0, this equation has a unique solution for Z as a convergent power series in X and Y (i.e. in their components with respect to a basis for $\mathbf{g}$):

$$Z = \sum_{ij} C_{ij}(X,Y) \tag{12}$$

with $C_{ij}(X,Y) \in \mathbf{g}$ homogeneous of degree i in X and j in Y. Equation (11) will then remain valid if X, Y, Z are replaced by the operators X_l, Y_l, and Z_l, in the sense that the operator series on either side of (11) give the same result when applied to an analytic function defined on an open subset of G and evaluated at a point of its domain. X and Y must be restricted to a neighbourhood of 0 in $\mathbf{g}$ (depending on the function and on the point) to insure convergence.

On the other hand, in the operator interpretation exp is the usual exponential series, and one may rewrite the left side of (11) as a bracket series of operators X and Y (with the interpretation just given), as was done for matrices in §1.2. Since the brackets of operators correspond to brackets in the Lie algebra according to the definition $[X,Y]_l = [X_l, Y_l]$ one sees that the solution (12) of (11) is still given by the same bracket series.

The formula for $d\exp$ could be proved by a similar argument, but it is simpler to remark that the proof in §1.2 goes through without change for general Lie groups: one only needs to re-interpret expressions like aX, with $a \in G$ and $X \in \mathbf{g}$, as the action of $a \in G$ on the tangent vector $X \in \mathbf{g}$, rather than as a matrix product. The Lie correspondence also carries over, but must be rephrased: in the setting of Lie groups, $G \to L(G)$ provides a one-to-one correspondence between connected subgroups G of a *fixed* Lie group and subalgebras of its Lie algebra; the inverse is $\mathbf{g} \to \Gamma(\mathbf{g})$ as before. As part of this correspondence it is understood that the subgroups $\Gamma(\mathbf{g})$ are given the manifold structure defined in terms of exp as for linear groups. The basic formula $f(\exp X) = \exp\varphi(X)$ for homomorphisms between linear groups in exponential coordinates holds for Lie groups as well, and the proof remains the same. We shall not go any further into such rather routine extensions of the linear case.

There is one instance, however, where general Lie groups are of genuine help, namely in connection with *quotient groups*. If H is a closed *normal* subgroup of a Lie group, then G/H is a group and carries the structure of an analytic manifold defined in connection with homogeneous spaces for linear groups. The group operations on G/H are analytic, hence G/H is a Lie group. The Lie algebra is the quotient g/h of g by the ideal h.

Example 7 (The universal covering group of $\mathrm{SU}(1,1) \approx \mathrm{SL}(2,\mathbb{R})$**).** Consider the action of $\mathrm{SU}(1,1)$ on $\mathbb{CP}^1$ (Example 3, §4.2). The element

$$\begin{bmatrix} \alpha & \bar{\beta} \\ \beta & \bar{\alpha} \end{bmatrix}, \quad \alpha\bar{\alpha} - \beta\bar{\beta} = 1$$

of SU(1,1) acts as

$$z \longrightarrow \frac{\alpha z + \bar{\beta}}{\beta z + \bar{\alpha}} = \epsilon \frac{z + \delta}{1 + \bar{\delta} z},$$

where $\epsilon = \alpha/\bar{\alpha}$ and $\delta = \bar{\beta}/\alpha$. Note that $|\epsilon| = 1$ and $\delta < 1$.

Consider, in particular, the action of $\mathrm{SU}(1,1)$ on the unit circle $|z| = 1$. There the action may be written as

$$e^{2\pi i \tau} \longrightarrow e^{2\pi i(\tau+\theta)} \frac{1 + \delta e^{-2\pi i \tau}}{1 + \bar{\delta} e^{2\pi i \tau}},$$

or with τ as coordinate:

$$\tau \to \tau + \theta + \frac{1}{2\pi i}(\ln(1 + \delta e^{-2\pi i \tau}) - \ln(1 + \bar{\delta} e^{2\pi i \tau})). \tag{13}$$

$\ln(w)$ is the branch on logarithm that vanishes for $w = 1$, defined off the negative real axis. The expressions involving $\ln(w)$ are then well-defined because $|\delta| < 1$.

We now change point of view: the formula (13) defines a transformation $\tau \to a(\theta, \delta)\tau$ of the real line for all $\theta \in \mathbb{R}$ and $|\delta| < 1$. These transformations commute with the integral translations $\tau \to \tau + k (k \in \mathbb{Z})$, and the induced transformations on $\mathbb{R}/\mathbb{Z}$ give just the action of $\mathrm{SU}(1,1)$ on the circle we started with. As transformation of the circle, the $a = a(\theta, \delta)$ depend only on θ mod 1, but as transformation of $\mathbb{R}$ they depend on $\theta \in \mathbb{R}$ itself.

The transformations $a = a(\theta, \delta)$ $(\tau \in \mathbb{R}, |\delta| < 1)$ of the real line form a group, denoted $\widetilde{SU}(1,1)$. This group is a Lie group with (θ, δ) as (everywhere defined) coordinates. (We omit the verification.) The centre Z of $\widetilde{\mathrm{SU}}(1,1)$ consists of the transformations $a(k, 0)$ with $k \in \mathbb{Z}$; it is thus isomorphic with $\mathbb{Z}$. Passing form the action of $\widetilde{\mathrm{SU}}(1,1)$ on $\mathbb{R}$ to the induced action on the circle $\mathbb{R}/\mathbb{Z}$ one obtains an isomorphism

$$\widetilde{\mathrm{SU}}(1,1)/Z \to \mathrm{SU}(1,1),$$

or equivalently a covering $\widetilde{\mathrm{SU}}(1,1) \to \mathrm{SU}(1,1)$ with kernel Z.

The Lie algebra of $\widetilde{\mathrm{SU}}(1,1)$ is still (isomorphic with) $\mathsf{su}(1,1)$. But $\widetilde{\mathrm{SU}}(1,1)$ *admits no realization as a linear group.* This may be seen as follows. Suppose, to the contrary, that $\widetilde{\mathrm{SU}}(1,1)$ is isomorphic with a linear group and is then thought of as such. Consider the element H in its Lie algebra corresponding to the element

$$\begin{bmatrix} \pi i & 0 \\ 0 & -\pi i \end{bmatrix}$$

in $\mathsf{su}(1,1)$. Then $\exp(\theta H) = a(\theta, 0)$, thus the centre of $\widetilde{\mathrm{SU}}(1,1)$ is $Z = \{\exp(kH) \mid k \in \mathbb{Z}\}$.

Imagine the matrix H reduced to its Jordan form. If x is a generalized eigenvector of H with generalized eigenvalue λ, i.e.

$$(H - \lambda 1)^m x = 0 \quad \text{for some } m \in \mathbb{N},$$

then it is also a generalized eigenvector of $\exp(\tau H)$, with generalized eigenvalue $e^{\tau\lambda}$:

$$(\exp \tau H - e^{\tau\lambda})^m x = \left(\frac{\exp \tau H - e^{\tau\lambda}}{H - \lambda 1} \right)^m (H - \lambda 1)^m x = 0 \quad \text{for some } m \in \mathbb{N}.$$

(The fraction makes sense, because $(e^{\tau z} - e^{\tau\lambda})/(z - \lambda)$ is an analytic function of $z \in \mathbb{C}$ for fixed τ, λ.) The generalized eigenspace of $\exp(\tau H)$ with generalized eigenvalue α is therefore the sum of the generalized eigenspaces of H with eigenvalues λ satisfying $e^{\tau\lambda} = \alpha$. Since the matrices in $\widetilde{\mathrm{SU}}(1,1)$ commute with $\exp(kH)$ for all $k \in \mathbb{Z}$, they leave its generalized eigenspaces invariant for all $k \in \mathbb{Z}$, hence leave the generalized eigenspaces of H itself invariant. Within the generalized eigenspace with eigenvalue λ the linear transformation $\exp(\tau(H - \lambda 1))$ must commute with all matrices of $\widetilde{\mathrm{SU}}(1,1)$ for all $\tau \in \mathbb{Z}$:

$$\exp(\tau(H - \lambda 1))a = a \exp(\tau(H - \lambda 1))$$

for all $a \in \widetilde{\mathrm{SU}}(1,1)$ and all $\tau \in \mathbb{Z}$. For each a, this is a system of *polynomial* equations in τ, because $H - \lambda 1$ is nilpotent in the generalized eigenspace. Since these polynomial equations hold for integral τ, they must hold identically. But then H itself commutes with all elements of $\widetilde{\mathrm{SU}}(1,1)$, hence so does $\exp(\tau H)$ for *all* real τ. This is a contradiction: the centre of $\widetilde{\mathrm{SU}}(1,1)$ consists only of the $\exp(\tau \mathrm{H})$ with *integral* τ.

We have therefore shown that $\mathrm{SU}(1,1)$ admits the covering $\widetilde{\mathrm{SU}}(1,1)$, which is not a linear group. Since $\widetilde{\mathrm{SU}}(1,1)$ is simply connected (being in one-to-one correspondence with $\mathbb{R}\times\{\text{disk}\}$ by the coordinates θ,δ), $\widetilde{\mathrm{SU}}(1,1)$ is in fact the universal covering of $\mathrm{SU}(1,1)$.

The example could naturally also be phrased in terms of $\mathrm{SL}(2,\mathbb{R})$ rather than $\mathrm{SU}(1,1)$, the two groups being isomorphic. As universal covering group of $\mathrm{SL}(2,\mathbb{R})$, $\widetilde{\mathrm{SU}}(1,1)$ is denoted $\widetilde{\mathrm{SL}}(2,\mathbb{R})$.

Example 8 (Nelson's parking problem). This example is intended to give some feeling for the Lie bracket $[X, Y]$. We start with the formula

$$\exp(\tau X)\exp(\tau Y)\exp(-\tau X)\exp(-\tau Y) = \exp(\tau^2[X, Y] + o(\tau^2))$$

proved in the linear case in Proposition 2(c), §1.3). The formula holds of course for elements X, Y in the Lie algebra of any Lie group. In this formula, replace τ by $\sqrt{\tau/k}$, raise both sides to the power $k = 0, 1, 2 \ldots$, and take the limit as $k \to \infty$. There results

$$\lim_{k\to\infty} \left(\exp\left(\sqrt{\tfrac{1}{2}}X\right) \exp\left(\sqrt{\tfrac{1}{2}}Y\right) \exp\left(-\sqrt{\tfrac{1}{2}}X\right) \exp\left(-\sqrt{\tfrac{1}{2}}Y\right) \right)^k$$
$$= \exp(\tau[X, Y]), \tag{14}$$

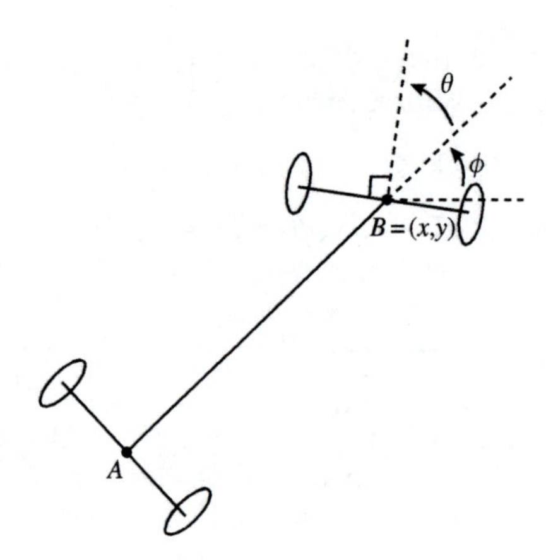

Fig. 1 A car

which we interpret as an iterative formula for the one-parameter group generated by $[X,Y]$. We now join Nelson (1967, p. 33).

"Consider a car. The configuration space of a car is the four-dimensional manifold $M = \mathbb{R}^2 \times \mathbb{T}^2$ parametrized by (x,y,ϕ,θ), where (x,y) are the Cartesian coordinates of the center of the front axle, the angle ϕ measures the direction in which the car is headed, and θ is the angle made by the front wheels with the car. (More realistically, the configuration space is the open submanifold $-\theta_{\max} < \theta < \theta_{\max}$.) See Figure 1.

There are two distinguished vector fields, called Steer and Drive, on M corresponding to the two ways in which we can change the configuration of a car. Clearly

$$\text{Steer} = \frac{\partial}{\partial\theta} \tag{15}$$

since in the corresponding flow θ changes at a uniform rate while x, y and ϕ remain the same. To compute Drive, suppose that the car, starting in the configuration (x,y,ϕ,θ) moves an infinitesimal distance h in the direction in which the front wheels are pointing. In the notation of Figure 2,

$$D = (x + h\cos(\phi+\theta) + o(h), \quad y + h\sin(\phi+\theta) + o(h)).$$

Let $l = \overline{AB}$ be the length of the tie rod (if that is the name of the thing connecting the front and rear axles). Then $\overline{CD} = l$ too since the tie rod does not change length (in non-relativistic mechanics). It is readily seen that $\overline{CE} = l + o(h)$, and since $\overline{DE} = h\sin\theta + o(h)$, the angle BCD (which is the increment in ϕ) is $((h\sin\theta)/l) + o(h)$ while θ remains the same. Let us choose units so that $l = 1$. Then

$$\text{Drive} = \cos(\phi+\theta)\frac{\partial}{\partial x} + \sin(\phi+\theta)\frac{\partial}{\partial y} + \sin\theta\frac{\partial}{\partial\phi}. \tag{16}$$

By (15) and (16)

$$[\text{Steer}, \text{Drive}] = -\sin(\phi+\theta)\frac{\partial}{\partial x} + \cos(\phi+\theta)\frac{\partial}{\partial y} + \cos(\theta)\frac{\partial}{\partial\phi}. \tag{17}$$

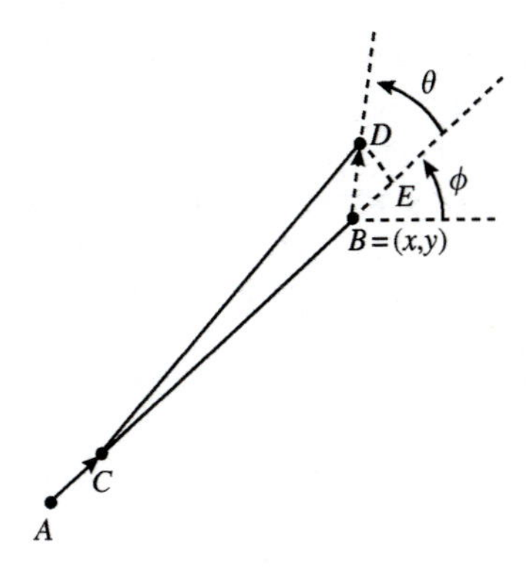

Fig. 2 A car in motion

> Let
>
> $$\begin{aligned}\text{Slide} &= -\sin\phi\frac{\partial}{\partial x} + \cos\phi\frac{\partial}{y},\\ \text{Rotate} &= \frac{\partial}{\partial\phi}.\end{aligned}$$
>
> Then the Lie bracket of Steer and Drive is equal to Slide + Rotate on $\theta = 0$, and generates a flow which is the simultaneous action of sliding and rotating. This motion is just what is needed to get out of a tight parking spot. By formula (14) this motion may be approximated arbitrarily closely, even with the restrictions $-\theta_{\max} < \theta < \theta_{\max}$ with $\theta_{\max}$ arbitrarily small, in the following way: steer, drive reverse steer, reverse drive, steer, drive reverse steer, ... What makes the process so laborious is the square roots in (14).
>
> Let us denote the Lie bracket (17) of Steer and Drive by Wriggle. Then further simple computations show that we have the commutation relations
>
> $$\begin{aligned}[\text{Steer}, \text{Drive}] &= \text{Wriggle},\\ [\text{Steer}, \text{Wriggle}] &= -\text{Drive},\\ [\text{Wriggle}, \text{Drive}] &= \text{Slide},\end{aligned} \tag{18}$$
>
> and the commutator of Slide with Steer, Drive and Wriggle is zero. Thus the four vector fields span a four-dimensional Lie algebra over $\mathbb{R}$.
>
> To get out of an extremely tight parking spot, Wriggle is insufficient because it may produce too much rotation. The last commutation relation shows, however, that one may get out of an arbitrarily tight parking spot in the following way: wriggle, drive, reverse wriggle (this requires a cool head), reverse drive, wriggle, drive, ...".

If we take for granted that the Lie algebra $\mathbf{g}$ spanned by the vector fields in (18) belongs to a Lie group G of transformations of M (as it does), then we may view this example as an illustration of a property of the Lie correspondence: if a Lie algebra $\mathbf{g}$ of a connected group G is generated by certain of its elements (through brackets and linear combinations), then G is generated by the corresponding one-parameter subgroups (through products). Furthermore, if one allows limits of products in the group G, then formula (14) provides an explicit recipe for generating the one parameter group of $[X, Y]$ from those of X and Y themselves.

Problems for §4.3

1. Prove that (9) is equivalent to (10) as asserted.

2. Let G be a Lie group, H a closed, normal subgroup of G. Show that the Lie algebra of G/H may naturally be identified with g/h as asserted.

3. Verify the bracket relations (17) and (18).

4. Prove the assertion: If a Lie algebra g of a connected group G is generated by certain of its elements through (brackets and linear combinations), then G is generated (through products) by the corresponding one-parameter subgroups.

5. Calculate explicitly the one-parameter groups of transformations generated by the vector fields *Drive*, *Wriggle Steer*, *Slide* of Example 8.

6. Find a linear Lie algebra isomorphic with the Lie algebra spanned by the vector fields Drive, Wriggle, Steer, Slide of Example 8.

5

Integration

5.1 Integration on manifolds

Recall the Change of Variables Formula from calculus:

$$\int_D f(x)dx = \int_{\tilde{D}} \tilde{f}(\tilde{x}) \left| \det \frac{\partial x}{\partial \tilde{x}} \right| d\tilde{x}. \tag{1}$$

Explanation. D is the domain of integration in the space of the $x = (\xi_1, \ldots, \xi_n)$. f is an integrable function defined on this domain. Given is a transformation

$$x = \varphi(\tilde{x}), \tag{2}$$

which is assumed to be a C^1 bijection between $\tilde{D}$ and D. $\det(\partial x/\partial \tilde{x}) = \det(\partial \xi_i/\partial \tilde{\xi}_j)$ is its Jacobian. $\tilde{f}$ corresponds to f under φ:

$$\tilde{f}(\tilde{x}) = f(x)$$

if x and $\tilde{x}$ are related by (2).

Now let M be an analytic manifold. The Jacobian factor in the change of variables formula (1) makes it impossible to define the integral of a function on M by simply using a coordinate system, if one wants the integral to come out independent of the coordinate system chosen. What one needs is a *volume element* on M in the following sense. A *volume element* ν_p at a point p on M is a rule that associates to any n-tuple $(v_1, \ldots, v_n)$ of tangent vectors at p a number $\nu_p(v_1, \ldots, v_n) \geq 0$, and not identically $= 0$, subject to the transformation law

$$\nu_p(Av_1, \ldots, Av_n) = |\det A| \nu_p(v_1, \ldots, v_n)$$

for every linear transformation A of T_pM. A *volume element* ν on M associates to each point p on M such a volume element ν_p at p. ν is said to be C^k if for every coordinate system $x = (\xi_1, \ldots, \xi_n)$ on M the function $\nu(\partial p/\partial x)$ is C^k on the coordinate domain. $\partial p/\partial x$ denotes the n-tuple of coordinate basis vector fields:

$$\frac{\partial p}{\partial x} = \left(\frac{\partial p}{\partial \xi_1}, \dots, \frac{\partial p}{\partial \xi_n}\right),$$

The basis vector fields are here denoted $\partial p/\partial \xi_k$ rather than $\partial/\partial \xi_k$ in order to keep the point p in sight, which is considered a function $p = p(x)$ of its coordinates.

One way to construct a volume element at p is to fix a basis for T_pM and to set

$$\nu_p(v_1, \dots, v_n) = |v_1, \dots, v_n|, \tag{3}$$

right side denoting the absolute value of the determinant of the matrix of components of the v_k with respect to this basis. Any volume element at p is a scalar multiple of (3), so that the purpose of the definition is to fix this scalar without explicitly referring to a basis. This normalization becomes important only when one has a volume element at each $p \in M$, because then there is no longer a single normalizing factor.

Assume now M comes equipped with a continuous volume element ν. Let f be a function on M defined on a subset D of M. We assume for the moment that D lies entirely in the domain of a coordinate system x. Define

$$\int_D f(p)\nu(dp) = \int_{D_x} f(p)\nu\left(\frac{\partial p}{\partial x}\right) dx. \tag{4}$$

On the right side $p = p(x)$ is considered a function of x, D_x is the point-set in the space of the coordinates x corresponding to D in M. We assume that the integral exists (as Riemann integral, or, if preferred, as Lebesgue integral); this imposes a condition on D_x and on f.

We have to show that the value of the integral (4) is independent of the coordinate system x. So suppose $\tilde{x}$ is another coordinate system defined on D. Since in matrix notation $\partial p/\partial \tilde{x} = (\partial x/\partial \tilde{x})(\partial p/\partial x)$, (meaning $\partial p/\partial \tilde{\xi}_j = \sum_i (\partial \xi_i/\partial \tilde{\xi}_j)(\partial p/\partial \xi_i)$), one finds

$$\int_{D_{\tilde{x}}} f(p)\nu\left(\frac{\partial p}{\partial \tilde{x}}\right) d\tilde{x} = \int_{D_{\tilde{x}}} f(p)\nu\left(\frac{\partial p}{\partial x}\right) \left|\det \frac{\partial x}{\partial \tilde{x}}\right| d\tilde{x} = \int_{D_x} f(p)\nu\left(\frac{\partial p}{\partial x}\right) dx$$

by the Change of Variables Formula. This proves the desired independence of coordinates.

There remains the question what to do if D does not lie in a single coordinate domain. The most obvious procedure would be to assume that D can be subdivided into a finite number of subsets, each contained in a coordinate domain, and define the integral over D as the sum of the integral over its parts. One then has to show that the result is independent of the subdivision. It is technically easier to use a slightly different approach.

Call a real of complex-valued function f on M *locally integrable* (with respect to the given volume element ν) if around every point of M there is a coordinate system x so that the integral (4) exists when D_x is a sufficiently small coordinate ball around the point in question. (In the present context it suffices to deal with functions defined on all of M: for the purpose of integration, a partially defined function can always be extended by zero to all of M.)

Proposition 1. *Suppose f is a locally integrable function on M vanishing outside a compact set. Then one can write*

$$f = f_1 + \cdots + f_m, \tag{5}$$

where each f_k vanishes outside of some coordinate ball B_k. The number

$$\int_{B_1} f_1(p)\nu(dp) + \cdots + \int_{B_m} f_m(p)\nu(dp) \tag{6}$$

is independent of the representation (5).

Definition. Under the hypothesis of the proposition, the number (6) is denoted $\int_M f(p)\nu(dp)$, or sometimes simply $\int_M f$ if ν is fixed.

The proof is based on the following lemma:

Lemma. Let D be a compact subset of M. There are continuous non-negative functions $g_1, \ldots, g_l$ on M so that

$$g_1 + \cdots + g_l \equiv 1 \text{ on } D, \quad g_k \equiv 0 \text{ outside of some coordinate ball } B_k.$$

Proof. For each point $p \in M$ one can find a continuous function h_p so that $h_p \equiv 0$ outside a coordinate ball B_p around p while $h_p > 0$ on a smaller coordinate ball $C_p \subset B_p$. Since D is compact, it is covered by finitely many of the C_p, say $C_1, \ldots, C_l$. Let $h_1, \ldots, h_l$ be the corresponding functions h_p, $B_1, \ldots, B_l$ the corresponding coordinate balls B_p. The function $\sum h_k$ is > 0 on D, hence one can find a strictly positive continuous function h on M so that $h \equiv \sum h_k$ on D (e.g. $h = \max\{\epsilon, \sum h_k\}$ for sufficiently small $\epsilon > 0$). The functions $g_k = h_k/h$ have all the required properties. QED

Proof of Proposition 1. First we remark that in case f itself vanishes outside of some coordinate domain, the assertion follows from the additivity of the integral on $\mathbb{R}^n$. The general case is reduced to this with the help of the lemma, as follows.

Consider two representations of f of the type (5):

$$f = f_1 + \cdots + f_m, \tag{7}$$
$$f = f'_1 + \cdots + f'_{m'}. \tag{8}$$

Let D be a compact subset of M outside of which all f_j, $f'_{j'}$ vanish and $g_1, \ldots, g_l$ functions as in the lemma. Multiply (7) and (8) by g_k (k fixed) and equate the result:

$$f_1 g_k + \cdots + f_m g_k = f'_1 g_k + \cdots + f'_{m'} g_k.$$

Since g_k vanishes outside a coordinate ball, the special case just mentioned gives

$$\int_M f_1 g_k + \cdots + \int_M f_m g_k = \int_M f'_1 g_k + \cdots + \int_M f'_{m'} g_k.$$

Add these equations for $k = 1, 2, \ldots, l$ and use the fact that each f_j and $f'_{j'}$ also vanishes outside a coordinate ball to write the result as

$$\sum_j \int_M f_j(g_1 + \cdots + g_l) = \sum_{j'} \int_M f'_{j'}(g_1 + \cdots + g_l).$$

Since $g_1 + \cdots + g_l \equiv 1$ on D, this is the desired result:

$$\sum_j \int_M f_j = \sum_{j'} \int_M f'_{j'}.$$

QED

$\int_M f$ is in particular defined if f is a continuous function which vanishes outside of some compact set. A larger class of *integrable* functions may be defined as for $\mathbb{R}^n$ (depending on whether one uses the Riemann integral of the Lebesgue integral).

Example 2 (surface integrals in $\mathbb{R}^3$). Let S be a two-dimensional submanifold of $\mathbb{R}^3$ (*smooth surface*). Define a volume element ν_p at p on S by

$$\nu_p(u, v) = |u, v|_p,$$

where the right side denotes the absolute value of the determinant of the coefficient matrix of the vectors $u, v \in T_pS$ with respect to *any* orthonormal basis for T_pS. Here T_pS is identified with a two-dimensional subspace of $\mathbb{R}^3$; $|u, v|_p$ is independent of the orthonormal basis chosen (and equals $\|u \times v\|$).

Specifically, take $S = S^2$ as in Example 8, §4.1. Since the basis vector fields $\partial/\partial\theta$ and $\partial/\partial\phi$ are already orthogonal, they may be normalized to serve as orthonormal basis for T_pS^2 at any point in the coordinate domain. In this way one finds that

$$\nu\left(\frac{\partial}{\partial\theta}, \frac{\partial}{\partial\phi}\right) = |\sin\phi|.$$

Thus, the integral on S^2 takes the form

$$\int_{S^2} f(p)\nu(dp) = \int_0^\pi \int_0^{2\pi} f(p(\theta, \phi)) \sin\phi \, d\theta \, d\phi,$$

as one knows from calculus.

Remark 3. On any n-dimensional submanifold S of $\mathbb{R}^N$ one may define a volume element by

$$\nu_p(v_1, \ldots, v_n) = |v_1, \ldots, v_n|_p,$$

where the right side denotes the absolute value of the determinant of the coefficient matrix of the vectors $v_1, \ldots, v_n \in T_pS$ with respect to *any* orthonormal basis for T_pS. (QED)

Terminology. If a coordinate system $x = (\xi_1, \ldots, \xi_n)$ is given on a manifold with a volume element ν, then one also refers to the symbol

$$\nu\left(\frac{\partial p}{\partial x}\right) dx = \nu\left(\frac{\partial p}{\partial \xi_1}, \ldots, \frac{\partial p}{\partial \xi_n}\right) d\xi_1 \cdots d\xi_n$$

as *volume element.* The $d\xi_k$ are purely symbolic, not to be interpreted as coordinate differentials. For example, one may say that the volume element on S^2 of Example 2 above is $|\sin\phi|\, d\theta\, d\phi$.

We close this section with some remarks on *orthogonal families of functions.* Assume given a manifold M with a continuous volume element ν. A complex-valued function f on M is said to be *square-integrable*, abbreviated L^2, if $|f|^2$ is integrable. (For our purposes, 'integrable' may be understood in the sense of Riemann, although it is customary to understand it in the sense of Lebesgue in this context.) The *norm* of an L^2-function f on M is defined by

$$\|f\|^2 = \int_M |f|^2.$$

The L^2-functions on M form an infinite-dimensional complex vector space with inner product

$$(f, g) = \int_M f\bar{g}.$$

so that

$$\|f\|^2 = (f, f). \tag{9}$$

We say a sequence f_k of L^2-functions is L^2*-convergent* to an L^2*-function* f if $\|f_k - f\| \to 0$.

Comment. It is customary to identify two L^2-functions f, g for which $\|f - g\| = 0$. Furthermore, if one wants the space of L^2-functions to be *complete* (i.e. all Cauchy sequences converge), then one has to use the Lebesgue integral. This is not essential here.

Suppose $\varphi_k (k = 1, 2, \ldots)$ is an *orthonormal family* of L^2-functions, meaning

$$(\varphi_j, \varphi_k) = \delta_{jk}. \tag{10}$$

Consider the problem of approximating an arbitrary L^2-function f by linear combinations $\sum_k c_k\varphi_k$ so as to minimize $\|f-\sum_k c_k\varphi_k\|$. A simple manipulation gives

$$\left\|f-\sum_k c_k\varphi_k\right\|^2 = (f,f)-\sum_{k=1}^{N}|(f,\varphi_k)|^2+\sum_{k=1}^{N}|c_k-(f,\varphi_k)|^2.$$

It follows from this equation that the minimum is taken on for $c_k=(f,\varphi_k)$. In particular, if the orthonormal family is *complete*, in the sense that $\|f-\sum_k c_k\varphi_k\|$ can be made arbitrarily small for every L^2-function f, then

$$\lim_{N\to\infty}\left\|f-\sum_{k=1}^{N}(f,\varphi_k)\varphi_k\right\|=0.$$

We write this last relation as

$$f\equiv\sum_{k=1}^{\infty}(f,\varphi_k)\varphi_k, \tag{11}$$

even though the series need not converge pointwise to f. The expansion (11) together with the relations (10) imply further that

$$\|f\|^2=\sum_{k=1}^{\infty}|(f,\varphi_k)|^2,$$

known as *Parseval's Identity.*

Finally, we remark that an orthonormal family φ_k is complete if on compact sets every continuous function can be uniformly approximated by finite linear combinations of the φ_k. (This condition is sufficient, but not necessary.)

Reference. A more detailed, but elementary, discussion of orthogonal families can be found in Section 10.2 of Marsden (1974), for example. The discussion there applies to functions of a real variable, but the extension to the present case is immediate.

Problems for §5.1

1. Show that every volume element ν_p at p on M is of the form

   $$\nu_p(v_1,\ldots,v_n)=\text{const. }|v_1,\ldots,v_n|,$$

   where $|v_1,\ldots,v_n|$ is defined as in eqn (3).

2. Verify that the definition of Remark 3 does define a volume element ν on S.

3. Consider SO(3) as a submanifold of $M_3(\mathbb{R})=\mathbb{R}^{3\times3}$. As such, SO(3) has a volume element ν defined as in Remark 3. Show that in terms of the Euler angles θ,ϕ,ψ this volume element is

   $$|\sin\phi|\,d\theta\,d\phi\,d\psi.$$

   (See Example 4(b), §2.3.)

4. Show that

$$\int_{\mathrm{SO}(3)} f(a)\,da = \frac{1}{8\pi^2}\int_{S^2}\left\{\int_0^{2\pi} f(ab(\psi))\,d\psi\right\}d(ae_3),$$

where $b(\psi) = \exp(\psi E_3)$ is the rotation about e_3 with angle ψ. The integral over SO(3) is the one defined in problem 3, normalized so that SO(3) has unit volume; the integral over S^2 is the one defined in Example 2. Explain the meaning of the outer integral on the right.

5. Denote by ν^S the volume element on a submanifold S of $\mathbb{R}^N$ defined in Remark 3. (a) Show that the corresponding integral has the following property. Suppose T is a Euclidean isometry of $\mathbb{R}^N$, i.e. a transformation of the form

$$Tx = ax + b$$

where a is an orthogonal linear transformation and $b \in \mathbb{R}^N$. Then

$$\int_{T(S)} f(T^{-1}y)\nu^{T(S)}(dy) = \int_S f(x)\nu^S(dx),$$

for any submanifold S and any integrable function f on S.

(b) Use part (a) to construct a volume element ν on any closed subgroup of $\mathrm{O}(n)$ so that

$$\int_G f(bac)\nu(da) = \int_G f(a)\nu(da)$$

for all integrable functions f on G and all $b, c \in G$. [*Any* compact linear group can be realized as a closed subgroup of $O(n)$ for some n.]

5.2 Integration on linear groups and their homogeneous spaces

Let G be a linear group, M a homogeneous space for G. Fix a *base point* o for M and let $H = \{a \in G \mid a \cdot o = o\}$ be its stabilizer. Fix also a basis for the tangent space at o. For an n-tuple $(v_1, \ldots, v_n)$ of tangent vectors at o ($n = \dim M$), let

$$\nu_o(v_1, \ldots, v_n) = |v_1, \ldots, v_n|;$$

the right side denotes the absolute value of the determinant of the component matrix of v_k with respect to the fixed basis. As noted before, ν_o is a volume element at o, i.e.

$$\nu_o(Av_1, \ldots, Av_n) = |\det A|\nu_o(v_1, \ldots, v_n), \tag{1}$$

for any linear transformation A of T_oM.

The stabilizer H of o acts on T_oM by linear transformations. Equation (1) gives in particular that for $a \in H$,

$$\nu_o(a\dot{v}_1, \ldots, a\dot{v}_n) = |\det_o a| \nu_o(v_1, \ldots, v_n),$$

where $\det_o a$ is the determinant of the linear transformation of T_oM corresponding to $a \in H$.

Under the identification $T_oM \approx \mathfrak{g}/\mathfrak{h}$ the action of H on T_oM corresponds to the action of H on $\mathfrak{g}/\mathfrak{h}$ induced by the adjoint representation on $\mathfrak{g}$. Consequently,

$$\det_o a = \det \mathrm{Ad}_{\mathfrak{g}/\mathfrak{h}}\, a;$$

$\mathrm{Ad}_{\mathfrak{g}/\mathfrak{h}}$ is the action of H on $\mathfrak{g}/\mathfrak{h}$ just mentioned.

From now on we shall assume that

$$|\det_o a| = 1, \text{ i.e. } |\det \mathrm{Ad}_{\mathfrak{g}/\mathfrak{h}}\, a| = 1 \quad \text{for all } a \in H. \tag{2}$$

Define a volume element ν_p at all points p of M by the condition

$$\nu_p(v_1, \ldots, v_n) = \nu_o(a \cdot v_1, \ldots, a \cdot v_n),$$

where $a \in G$ is any element with

$$a \cdot p = o; \tag{3}$$

$a \cdot v_i \in T_oM$ is the image of $v_i \in T_pM$ under the differential of the transformation $p \to a \cdot p$ of M. Note that ν_p is well defined: any other element a' of G with $a' \cdot p = o$ is of the form $a' = ba$ for some $b \in H$ and then

$$\nu_o(a' \cdot v_1, \ldots, a' \cdot v_n) = \nu_o(a \cdot v_1, \ldots, a \cdot v_n)$$

in view of the condition (2). To check that ν_p is a volume element, observe that for any linear transformation A of T_pM

$$\begin{aligned}\nu_p(Av_1, \ldots, Av_n) &= \nu_o(a \cdot Av_1, \ldots, a \cdot Av_n)\\ &= \nu_o(aAa^{-1}a \cdot v_1, \ldots, aAa^{-1}a \cdot Av_n)\\ &= |\det aAa^{-1}| \nu_o(a \cdot v_1, \ldots, a \cdot v_n)\\ &= |\det A| \nu_p(v_1, \ldots, v_n).\end{aligned}$$

Thus, we have a volume element ν on M. One may check that this volume element is analytic in the sense explained in connection with the definition of 'volume element' (assuming the action of G on M is analytic, as usual).

The main point about the integral with respect to the volume element ν is that it is invariant under the action of the group G in the following sense.

Theorem 1. *For any integrable function f on M and any $c \in G$,*

$$\int_M f(c \cdot p)\nu(dp) = \int_M f(p)\nu(dp). \tag{4}$$

Comment. There is no need to consider separately integrals over domains D in M, as an integrable function on a domain D can always be extended to an integrable function on all of M by making it zero outside of D.

Proof. We may assume f vanishes outside the domain D of some coordinate system $x = (\xi_1, \ldots, \xi_n)$ on M. Then formula (4) becomes

$$\int\limits_{c^{-1}\cdot D} f(c \cdot p)\nu(dp) = \int\limits_{D} f(p)\nu(dp). \tag{5}$$

Using the letter q rather than p to denote a general point in $c^{-1}D$, write eqn (5) as

$$\int\limits_{D} f(p)\nu(dp) = \int\limits_{c^{-1}\cdot D} f(c \cdot q)\nu(dq). \tag{6}$$

Introduce coordinates y on $c^{-1}D$ by taking as coordinates y of a point q in $c^{-1}D$ the coordinates x of the point $p = c \cdot q$ in D. With the coordinates x in D and the coordinates y in $c^{-1}D$, eqn (6) becomes

$$\int\limits_{R} f(p)\nu\left(\frac{\partial p}{\partial x}\right) dx = \int\limits_{R} f(c \cdot q)\nu\left(\frac{\partial q}{\partial y}\right) dy.$$

The integrals are over the same region R of $\mathbb{R}^n$, which corresponds both to the coordinates x of points p in D and to coordinates y of points q in $c^{-1}D$. Since $c \cdot q = p$ when $y = x$, it suffices to show that $\nu(\partial q/\partial y) = \nu(\partial p/\partial x)$ when $y = x$. This follows directly from the definitions, as follows.

From the identity $q = c^{-1} \cdot p$ we get $(\partial q/\partial y) = c^{-1} \cdot (\partial p/\partial x)$ when $y = x$ by differentiation. If we fix p and choose $a \in G$ so that $a \cdot p = o$, then $(ac) \cdot q = (ac) \cdot (c^{-1} \cdot p) = o$, and at these points

$$\nu_q\left(\frac{\partial q}{\partial y}\right) = \left|(ac)\frac{\partial q}{\partial y}\right| = \left|(ac)c^{-1}\frac{\partial p}{\partial x}\right| = \left|a\frac{\partial p}{\partial x}\right| = \nu_p\left(\frac{\partial p}{\partial x}\right).$$

QED

Remark 2. The formula for the invariant integral on a homogeneous space M becomes particularly simple if one sets $M = G/H$ and introduces coordinates x on G/H of the form $aH = a(x)H$, where $a(x) \in G$ depends analytically on x. (For example, one can choose a supplementary subspace s for h in g and set $a = a_o \exp(x)$, using x in a neighborhood of 0 in s as coordinate point.) For a function f vanishing outside the coordinate domain one then has

$$\int\limits_{G/H} f(aH)\nu(daH) = \int f(aH)\left|a^{-1}\frac{\partial a}{\partial x}\right| dx. \tag{7}$$

On the right side, we set $a = a(x)$. $a^{-1}(\partial a)/(\partial x)$ is an n-tuple of vectors in g, and $|a^{-1}(\partial a/\partial x)|$ is the absolute value of the determinant of the matrix of components of the images of these vectors in g/h, components with respect to a fixed basis of g/h.

We may apply the above construction, in particular, when $M = G$, considered as homogeneous space by either left or right translations. In this case, the stabilizer H of $o = 1$ is $\{1\}$, so the assumption '$|\det_o a| = 1$' is trivially satisfied. Accordingly, we get a left- and right-invariant volume element ν^l and ν^r on G giving a left-invariant integral and a right-invariant integral, respectively:

$$\int_G f(ca)\, d_l(a) = \int_G f(a)\, d_l a$$
$$\int_G f(ac)\, d_r(a) = \int_G f(a)\, d_r a$$

for all integrable functions f on G. (We have written $d_l a$ for $\nu^l(da)$.) To see the relation between these two integrals, choose coordinates x on G and assume f vanishes outside the coordinate domain. Then from (7):

$$\int_G f(a)\, d_l(a) = \int f(a) \left| a^{-1} \frac{\partial a}{\partial x} \right| dx$$

and

$$\int_G f(a)\, d_r(a) = \int f(a) \left| \frac{\partial a}{\partial x} a^{-1} \right| dx.$$

Since

$$\left(\frac{\partial a}{\partial x}\right) a^{-1} = a \left(a^{-1} \left(\frac{\partial a}{\partial x}\right)\right) a^{-1} = \mathrm{Ad}(a) \left(a^{-1} \left(\frac{\partial a}{\partial x}\right)\right),$$

one finds that

$$\left| \frac{\partial a}{\partial x} a^{-1} \right| = |\det \mathrm{Ad}(a)| \left| a^{-1} \frac{\partial a}{\partial x} \right|.$$

Thus,

$$\begin{aligned} \int_G f(a)\, d_r a &= \int_G f(a) |\det \mathrm{Ad}(a)| d_l a, \\ \int_G f(a)\, d_l a &= \int_G f(a) |\det \mathrm{Ad}(a)|^{-1} d_r a. \end{aligned} \tag{8}$$

In particular, if $|\mathrm{Ad}(a)| = 1$ for all $a \in G$, then the integrals are the same and therefore bi-invariant; this bi-invariant integral is simply denoted

$$\int_G f(a)\, da.$$

Such a group is said to be *unimodular*. Compact groups are always unimodular.

Lemma 3. *If G is compact, then $|\det \mathrm{Ad}(a)| = 1$ for all $a \in G$.*

Proof. Suppose to the contrary that G is compact, but $|\det \mathrm{Ad}(a)| \neq 1$, say $|\det \mathrm{Ad}(a)| > 1$, for some $a \in G$. Since G is compact, the sequence a^k, $k = 1, 2, \ldots$, has a limit point, say b. Since

$$|\det \mathrm{Ad}(a^k)| = |\det \mathrm{Ad}(a)|^k \to \infty,$$

we get $|\det \mathrm{Ad}(b)| = \infty$, which is nonsense. And $|\det \mathrm{Ad}(a)| < 1$ is equally impossible for similar reasons. QED

We also note that the integral of a continuous function on a compact group always exists. In particular, we set

$$\int_G 1\, da = |G|.$$

The integral may be *normalized* so that $|G| = 1$ (by adjusting the arbitrary positive factor which comes from the choice of the volume element at 1).

Example 4 (The invariant integral on S^2 and on SO(3)). We consider S^2 as homogeneous space for $\mathrm{SO}(3) : S^2 = \mathrm{SO}(3)/\mathrm{SO}(2)$, SO(2) realized as subgroup of rotations about e_3, which we take as base-point. The point $p = p(\theta, \phi)$ on S^2 with coordinates θ, ϕ (Example 8, §4.1) may be written as

$$p(\theta, \phi) = a(\theta, \phi) e_3,$$

where $a(\theta, \phi) = a_3(\theta) a_2(\phi) = \exp(\theta E_3) \exp(\phi E_2)$. According to formula (7) the volume element at $p = p(\theta, \phi)$ is

$$\begin{aligned}\left| a^{-1}\frac{\partial a}{\partial \theta}, a^{-1}\frac{\partial a}{\partial \phi} \right| &= |\mathrm{Ad}(a_2(\phi)^{-1}) E_3, E_2| \\ &= |-\sin\phi E_1 + \cos\phi E_3, E_2| \\ &= |\sin\phi|\, |E_1, E_2|.\end{aligned}$$

The term $\cos\phi E_3$ has been dropped because in accordance with the remark after (7) the vectors are considered mod $\mathsf{so}(2)$ and $E_3 \equiv 0 \bmod \mathsf{so}(2)$. The constant $|E_2, E_1|$ depends on the choice of the basis at the base-point and may be assumed to be equal to 1. Then the invariant volume element on S^2 is $|\sin\phi|\, d\theta\, d\phi$, which agrees with Example 2, §5.1, as it must.

Consider now the invariant integral on SO(3). Use the Euler angles θ, ϕ, ψ as coordinates (Example 4(b), §2.3) and formula (7) for the invariant integral. The formulas for the partials of $a = a(\theta, \phi, \psi)$ in Example 4(b), §2.3 show that

$$\left| a^{-1}\frac{\partial a}{\partial \theta}, a^{-1}\frac{\partial a}{\partial \phi}, a^{-1}\frac{\partial a}{\partial \psi} \right| = |\sin\phi|.$$

Compare the formulas for the invariant integrals on S^2 and on SO(3):

$$\int_{S^2} f(p)dp = \int_0^{2\pi}\int_0^{\pi} f(p(\theta,\phi)) \sin\phi\, d\phi\, d\theta,$$

$$\int_{\mathrm{SO}(3)} f(a)da = \int_0^{2\pi}\int_0^{\pi}\int_0^{2\pi} f(a(\theta,\phi,\psi)) \sin\phi\, d\psi\, d\phi\, d\theta.$$

These formulas express a relation between the invariant integral of SO(3) and on its homogeneous space $S^2 \approx \mathrm{SO}(3)/\mathrm{SO}(2)$, which illustrates the following general proposition.

Proposition 5. *Let G be a linear group, H a closed subgroup of G satisfying*

$$|\det \mathrm{Ad}_{\mathsf{g}/\mathsf{h}}\, b| = 1$$

for all $b \in H$. Then the left-invariant integrals on G, H, and G/H satisfy

$$\int_G f(a)\, d_l a = \int_{G/H} \left\{ \int_H f(cb)\, d_l b \right\} dcH.$$

Explanation. More precisely, the formula should be understood as follows. The function of $c \in G$ in parentheses on the right depends only on the coset $cH \in G/H$, because of the left-invariance of the integral over H. The assertion is that the integral on the left side exists if and only if the inner integral on the right exists as an integrable function on G/H. Suitable normalizations of the invariant integrals are presupposed throughout, as will become clear from the proof.

Proof. In a neighborhood of any point on G one can introduce coordinates (x, y) of the type $a = c(x)b(y)$, where $x = (\xi_1, \ldots, \xi_p)$ is the coordinate point of $c(x)H$ on G/H and $y = (\eta_1, \ldots, \eta_q)$ the coordinate point of $b(y)$ on H. For example, let s be a subspace of g supplementary to h; then in a neighborhood of a_o on G, write (uniquely)

$$a = a_o \exp X \exp Y$$

with $X \in \mathsf{s}$, $Y \in \mathsf{h}$, and set $c(X) = a_o \exp X$, $b(Y) = \exp Y$. Use formula (7) for the invariant integral: the volume element on G is given by

$$\left| a^{-1}\frac{\partial a}{\partial x}, a^{-1}\frac{\partial a}{\partial y} \right| = \left| b^{-1}\left(c^{-1}\frac{\partial c}{\partial x}\right) b, \left(b^{-1}\frac{\partial b}{\partial y}\right) \right|.$$

Write $\mathsf{g} = \mathsf{s}+\mathsf{h}$ (direct sum on vector spaces). With respect to a basis compatible with this decomposition, the right side is the absolute value of the determinant of the matrix

$$\begin{bmatrix} [\mathrm{Ad}\, b^{-1}(c^{-1}\frac{\partial c}{\partial x})]_{\mathsf{s}} & 0 \\ * & b^{-1}\frac{\partial b}{\partial y} \end{bmatrix},$$

where $[\operatorname{Ad} b^{-1}(c^{-1}(\partial c)/(\partial c))]_{\mathsf{s}}$ denotes the coefficient matrix of the component in s of the element of g in brackets. Under the identification $\mathsf{s} \approx \mathsf{g}/\mathsf{h}$ it becomes $[\operatorname{Ad} b^{-1}(c^{-1}(\partial c)/(\partial c))]_{\mathsf{g}/\mathsf{h}}$, where $\operatorname{Ad} b^{-1}$ denotes the action of $b^{-1} \in H$ on g/h coming from the adjoint representation on g. Thus,

$$\begin{aligned}
\left| a^{-1}\frac{\partial a}{\partial x}, a^{-1}\frac{\partial a}{\partial y} \right|_{\mathsf{g}} &= \left| \operatorname{Ad} b^{-1}\left(c^{-1}\frac{\partial c}{\partial x} \right) \right|_{\mathsf{g}/\mathsf{h}} \left| b^{-1}\frac{\partial b}{\partial y} \right|_{\mathsf{h}} \\
&= \left| \det \operatorname{Ad}_{\mathsf{g}/\mathsf{h}} b^{-1} \right| \left| c^{-1}\frac{\partial c}{\partial x} \right|_{\mathsf{g}/\mathsf{h}} \left| b^{-1}\frac{\partial b}{\partial y} \right|_{\mathsf{h}} \\
&= \left| c^{-1}\frac{\partial c}{\partial x} \right|_{\mathsf{g}/\mathsf{h}} \left| b^{-1}\frac{\partial b}{\partial y} \right|_{\mathsf{h}} .
\end{aligned}$$

(The subscripts indicate to which space the vectors in question belong.) Thus, for a function f on G that vanishes outside of the coordinate domain

$$\int_G f(a)\, d_l a = \iint f(c(x)b(y)) \left| c^{-1}\frac{\partial c}{\partial x} \right| \left| b^{-1}\frac{\partial b}{\partial y} \right| dx\, dy = \int_{G/H} \left\{ \int_H f(cb) d_l b \right\} dcH.$$

This proves the formula under the assumption that all integrals in question exist and that f vanishes outside of the coordinate domain. The first assumption is taken care of by Fubini's Theorem, which guarantees that the left side exists if and only if the right side exists in the sense explained above. (See Marsden (1974), for example.) The second assumption is taken care of by the definition of the integral on a manifold: one is reduced to integrals over coordinate domains for coordinate systems that may be taken to be of the above type. QED

Problems for §5.2

1. Verify that the volume element defined after (2) is analytic, as stated.

2. Show that the volume elements invariant integrals on the multiplicative groups $\mathbb{R}^\times, \mathbb{C}^\times$, and $\mathbb{H}^\times$ are given as follows: (a) $\mathbb{R}^\times : d\tau/|\tau|$, (b) $\mathbb{C}^\times : d\tau/|\tau|^2$, (c) $\mathbb{H}^\times : d\tau/|\tau|^4$. In each case, $d\tau$ is the usual volume element on $\mathbb{R}$, $\mathbb{C} = \mathbb{R}^2$, $\mathbb{H} = \mathbb{R}^4$, and $|\tau| = (\tau\bar{\tau})^{1/2}$.

3. (a) Let $G = \mathrm{GL}(n, \mathbb{R})$. Show that

$$\int_G f(a)\, d_l(a) = \int_G f(a)\, d_r(a) = \int_G f(a) |\det a|^{-n}\, da,$$

where the last integral is the usual integral on the matrix space $M_n(\mathbb{R})$.

(b) Let G be the group of real upper triangular matrices with non-zero diagonal entries. Show that:

$$\int_G f(a)d_l(a) = \int_G f(a)|j_l(a)|^{-1}da,$$

$$\int_G f(a)d_r(a) = \int_G f(a)|j_r(a)|^{-1}da.$$

The integrals on the right are the usual integrals on the space of all upper triangular matrices and the functions j_l, j_r are given by

$$j_l(a) = (\alpha_1)^{-1}(\alpha_2)^{-2}\cdots(\alpha_n)^{-n},$$
$$j_r(a) = (\alpha_1)^{-n}(\alpha_2)^{-(n-1)}\cdots(\alpha_n)^{-1},$$

where $(\alpha_1, \alpha_2, \ldots, \alpha_n)$ are the diagonal entries of $a \in G$.

In both (a) and (b) calculate the left- and right-invariant integral independently and verify that the result agrees with formula (7).

4. Let G be a linear group. Let $\mathfrak{g}_o$ be an open subset of $\mathfrak{g}$ which $\exp : \mathfrak{g} \to G$ maps bijectively onto an open subset G_o of G. Show that for every integrable function f on G_o

(a)

$$\int_{G_o} f(a)\, d_l a = \int_{\mathfrak{g}_o} f(\exp X)|j_l(X)|\, dX$$

where $j_l(X) = (1 - \exp - \operatorname{ad} X)/\operatorname{ad} X$.

(b)

$$\int_{G_o} f(a)\, d_r a = \int_{\mathfrak{g}_o} f(\exp X)|j_r(X)| dX$$

where $j_r(X) = \det(\exp \operatorname{ad} X - 1)/\operatorname{ad} X$.

Explain the normalization of the volume elements on G for which these equations hold.

5. Use the definition of the invariant integral on a group to show that the *normalized* invariant integral on SO(3) is

$$\int_G f(a)\, da = \frac{1}{(2\pi)^2} \int_{\{\|X\|<\pi\}} f(\exp X) \frac{1 - \cos\|X\|}{\|X\|^2},$$

where

$$j(X) = \frac{1 - \cos\|X\|}{\|X\|^2}$$

[*Normalized* integral means $\int_G da = 1$. The norm is $\|X\| = (1/2)\operatorname{tr}(X^*X)$, as in Example 4, §2.3.]

6. Write $b \in \mathrm{SU}(2)$ as $b = b(\theta, \phi, \psi) = b_1(\theta) b_2(\phi) b_3(\psi)$ as in problem 5, §2.3. Show that the normalized invariant integral on SU(2) is given by

$$\int_{\mathrm{SU}(2)} f(b)\, db = \frac{1}{8\pi^2} \int_0^{2\pi}\int_0^{2\pi}\int_0^{\pi} f(b(\theta, \phi, \psi)) \sin\phi \, d\phi \, d\theta \, d\psi.$$

7. Use the covering $\mathrm{SU}(2) \to \mathrm{SO}(3)$ to show that the formula of problem 6 is equivalent to the formula for the invariant integral on SO(3) of Example 4.

8. Write $a \in \mathrm{SL}(2, \mathbb{R})$ as $a = k(\theta) n(\sigma) a(\tau)$ as in problem 6, §2.3. Show that the invariant integral on $\mathrm{SL}(2, \mathbb{R})$ is given by

$$\text{(a)} \quad \int_{\mathrm{SL}(2,\mathrm{R})} f(a)\, da = \int_0^{2\pi}\int_{-\infty}^{\infty}\int_0^{\infty} f\left(k(\theta) n(\sigma) a(\tau)\right) d\tau \, d\sigma \, d\theta.$$

$$\text{(b)} \quad \int_{\mathrm{SL}(2,\mathrm{R})} f(a)\, da = \int_0^{2\pi}\int_{-\infty}^{\infty}\int_0^{\infty} f\left((k(\theta) a(\tau) n(\sigma)\right) e^{2\tau} d\tau \, d\sigma \, d\theta.$$

9. Write $a \in \mathrm{SL}(2, \mathbb{R})$ as $a = k(\theta) a(\tau) k(\phi)$ with $\tau > 0$ as in problem 7, §2.3) Show that the invariant integral on $\mathrm{SL}(2, \mathbb{R})$ is

$$\int_{\mathrm{SL}(2,\mathrm{R})} f(a)\, da = \int_0^{2\pi}\int_0^{\infty}\int_0^{2\pi} f(k(\theta) a(\tau) k(\phi)) (e^{\tau} - e^{-\tau}) d\theta \, d\tau \, d\phi.$$

10. *The real classical groups are unimodular.* Prove this (a) for $\mathrm{SL}(n, \mathbb{R})$, $\mathrm{SL}(n, \mathbb{C})$, and $\mathrm{SL}(n, \mathbb{H})$; (b) for $\mathrm{Aut}(\varphi)$, φ a non-degenerate bilinear or sesquilinear form on a vector space over $\mathbb{R}$, $\mathbb{C}$, or $\mathbb{H}$.

    [Suggestion: for (b) $\mathrm{Aut}(\varphi)$ leaves invariant a non-degenerate $\mathbb{R}$–bilinear form on its Lie algebra $\mathrm{aut}(\varphi)$.]

5.3 Weyl's integration formula for U(n)

Let $K = \mathrm{U}(n) = \{a \in M_n(\mathbb{C}) \mid a^* a = 1\}$. We start with some observations about the stucture of U(n). We know that every element $a \in K$ is conjugate to a diagonal matrix $t = (\epsilon_1, \ldots, \epsilon_n)$, $\epsilon_j = e^{2\pi i \theta_j}$, $\theta_j \in \mathbb{R}$. t is uniquely determined by a up to a permutation of the eigenvalues ϵ_j of a. Thus every element $a \in K$ is conjugate to an element of the *Cartan subgroup*[1] of diagonal matrices

$$T = \mathbb{T}^n = \{t = (\epsilon_1, \ldots, \epsilon_n \mid \epsilon_j = e^{2\pi i \theta_j}, \ \theta_j \in \mathbb{R}\}.$$

[1]The results of §3.2 are not required.

The *normalizer* $N_K(T) = \{n \in G \mid nTn^{-1}\} = T$ of T in K acts on T by conjugation. Since T is Abelian, we actually have a well-defined action of the *Weyl group*

$$W = N_K(T)/T$$

on T, with $nT \in N_K(T)/T$ acting as $t \to ntn^{-1}$.

Lemma 1. *$W \approx S_n$ (symmetric group) and acts on T by permuting the diagonal entries.*

Proof. Let $n \in N_k(T)$. We show first that n can be uniquely written as $n = st$ where s is a permutation matrix and $t \in T$ is diagonal. By assumption,

$$n(\epsilon_1, \ldots, \epsilon_n)n^{-1} = (\epsilon_{\sigma^{-1}(1)}, \ldots, \epsilon_{\sigma^{-1}(n)}),$$

where $\sigma \in S_n$ is a permutation of $\{1, 2, \ldots, n\}$. Let s be the corresponding permutation matrix:

$$se_j = e_{\sigma(j)}, \quad j = 1, \ldots, n.$$

Then $n(\epsilon_1, \ldots, \epsilon_n)n^{-1} = s(\epsilon_1, \ldots, \epsilon_n)s^{-1}$ for all $(\epsilon_1, \ldots, \epsilon_n) \in T$. So $s^{-1}n$ commutes with all diagonal matrices hence is itself diagonal, i.e. $s^{-1}n = t$ for some $t \in T$. It follows that $n = st$ as required.

To prove uniqueness, suppose $a = s_1t_1 = s_2t_2$. Then $s_2^{-1}s_1 = t_2t_1^{-1}$ is a diagonal permutation matrix, hence$= 1$. Thus $s_1 = s_2$ and $t_1 = t_2$. If one identifies permutations and permutation matrices, then $sT \to s$ gives the required isomorphism $W \approx S_n$. QED

Remark 2.

(a) We record explicitly how W is identified with S_n: a permutation matrix s is identified with a permutation of $\{1, \ldots, n\}$ so that

$$s(\epsilon_1, \ldots, \epsilon_n)s^{-1} = (\epsilon_{s^{-1}(1)} \cdots \epsilon_{s^{-1}(n)}).$$

The inverse (already met in the proof of the lemma) is needed so that multiplication of permutation matrices corresponds to composition of permutations in the proper order.

(b) One may wonder why W is defined as the quotient group $N_K(T)/T$ rather than directly as S_n, or as the group of permutation matrices. The reason is that the definition $W = N_K(T)/T$ has an analogue for all semisimple Lie groups.

We return to the representation of an element $a \in K$ as $a = ctc^{-1}$. We observed that t is unique up to conjugation by an element from W. But even for fixed t, c may be replaced by cb where b commutes with t, and this is exactly the extent of the ambiguity. When t has distinct eigenvalues, then the only such b commuting with t are the elements of T, because such ab must preserve the (by assumption one-dimensional) eigenspaces of t. Generally, an element of K with distinct eigenvalues is called *regular*. The regular elements are denoted K_{reg}, those in T are denoted T_{reg}. The discussion so far proves the following proposition.

Proposition 3. *The map $p: K/T \times T \to K$, $(cT, t) \to ctc^{-1}$ is surjective. When $a = ctc^{-1}$ is regular, its inverse image $p^{-1}(a)$ consists of the $|W|$ points $(cs^{-1}T, sts^{-1})$, $s \in W$.*

For example, consider the case when $n = 2$. In order to visualize the map $K/T \times T \to K$ we further replace U(2) by SU(2) and consider the map $K/T \times T \to K$ in exponential cooordinates. It then becomes the usual spherical coordinate map $S^2 \times \mathbb{R} \to \mathbb{R}^3, (u, \rho) \to \rho u$, restricted to $|\rho| \leq 2\pi$, since SU(2) corresponds to a ball of radius 2π (see Example 2, §2.1). The regular elements in SU(2) correspond to the interior of the ball with the origin excluded. There are evidently exactly two points $\pm(u, \rho)$ above each regular point, but the whole sphere $S^2 \times \{0\}$ gets collapsed into the singular point $0 \in \mathbb{R}^3$, corresponding to the identity in SU(2).

This example illustrates that the map $K/T \times T \to K$ may generally be considered as a kind of 'spherical coordinate system' on $K = \mathrm{U}(n)$, with the regular conjugacy classes playing the rôle of spheres and T the rôle of a radial coordinate.

Define a complex-valued function $a \to J(a)$ on $\mathrm{U}(n)$ by the condition that

$$\det(\mathrm{Ad}_{\mathsf{k}}\, a - 1 + \lambda) = J(a)\lambda^r + \text{ higher terms in } \lambda, \quad J(a) \equiv 0.$$

We shall see in the proof of the lemma below that $r = n$ and that $J(a)$ is explicitly given by the formula

$$J(a) = \prod_{j \neq k} (\epsilon_j \epsilon_k^{-1} - 1),$$

where $\epsilon_1, \ldots, \epsilon_n$ are the eigenvalues of a. But from this expression it is not immediately clear that $J(a)$ is analytic, while it follows from its definition that $J(a)$ is in fact a polynomial in the matrix coefficients of a.

Lemma 4. *$a \in K$ is regular if and only if it satisfies the following equivalent conditions:*

(a) The centralizer $Z_K(a) = \{b \in K \mid ab = ba\}$ is conjugate to T.

(b) $J(a) \neq 0$.

Proof. The linear transformations commuting with $a \in K$ are precisely those that leave the eigenspaces of a invariant. From this observation one sees that $a \in K$ is regular (i.e. has one-dimensional eigenspaces) if and only if the $b \in K$ commuting with a are all simultaneously diagonal with a, i.e. (a) holds.

To prove the equivalence with (b), suppose $a \in K$ has eigenvalues $\epsilon_1, \ldots, \epsilon_n$. We may as well assume $a = t = (\epsilon_1, \ldots, \epsilon_n)$ is itself diagonal. The eigenvalues of $\mathrm{Ad}(t)$ as linear transformation of $\mathsf{k} = \mathsf{u}(n)$ are those of $\mathrm{Ad}(t)$ as linear transformation of $\mathsf{gl}(n, \mathbb{C}) : \mathsf{gl}(n, \mathbb{C}) = \mathsf{u}(n) + i\mathsf{u}(n)$ may be identified with the complexification $\mathsf{u}(n) \oplus i\mathsf{u}(n)$ of $\mathsf{u}(n)$ in the sense of the Appendix to §1.1.

$\mathsf{gl}(n, \mathbb{C})$ has a basis consisting of the matrices E_{jk} with jk-entry equal to 1 and all other entries equal to 0. Since

$$tE_{jk}t^{-1} = \epsilon_j \epsilon_k^{-1} E_{jk},$$

this basis consists of eigenvectors for $\mathrm{Ad}(t)$ with eigenvalues $\epsilon_j \epsilon_k^{-1}$. Therefore

$$\det(\mathrm{Ad}\, t - 1 + \lambda) = \lambda^n \prod_{j \neq k} (\epsilon_j \epsilon_k^{-1} - 1 + \lambda).$$

and

$$J(t) = \prod_{j \neq k} (\epsilon_j \epsilon_k^{-1} - 1).$$

Thus, $J(t) \neq 0$ if and only if the ϵ_i are distinct. QED

Remark 5.

(a) We record from the proof that for $t \in T$, $J(t)$ is the determinant of $\mathrm{Ad}\, t - 1$ as linear transformation of k/t.

(b) One sees from Lemma 4 that the regular points fill out K except for the 'analytic hypersurface' with equation $J(a) = 0$, the *singular elements* of K. Such an analytic hypersurface has *measure zero*, in the sense that it may be neglected for the purpose of integration on K. (We shall take this for granted, as one does in elementary calculus in similar situations in $\mathbb{R}^n$.)

Addendum to Proposition 3. The map $p : K/T \times T_{\mathrm{reg}} \to K_{\mathrm{reg}}$ is locally bi-analytic.

Proof. It suffices to show that p has an invertible differential at points of $K/T \times T_{\mathrm{reg}}$ (Inverse Function Theorem). We now compute this differential.

Recall that we introduced coordinates in a neighborhood of cT on the homogeneous space K/T by writing $\mathsf{k} = \mathsf{s} \oplus \mathsf{t}$ and then using $X \in \mathsf{s}$ as coordinates of $c\exp(X)T$ on K/T. Choose for s the matrices in k with diagonal entries all equal to 0. (Observe that then $\mathrm{Ad}(t)\mathsf{s} \subset \mathsf{s}$ for all $t \in T$.) Introduce coordinates in a neighborhood of (cT, t) on $K/T \times T$ by using $(X, Y) \in \mathsf{s} \times \mathsf{t}$ as coordinates of the point $(c\exp(X)T, t\exp(Y))$ on $K/T \times T$. Consider the curve

$$(c\exp(\tau X)T, t\exp(\tau Y))$$

on $K/T \times T_{\mathrm{reg}} (X \in s, Y \in t, t \in T_{\mathrm{reg}})$ and compute:

$$\begin{aligned}
\frac{d}{d\tau} p(c\exp(\tau X)T, t\exp(\tau Y)) \,|_{\tau=0} &= \frac{d}{d\tau} c\exp(\tau X) t \exp(\tau Y) \exp(-\tau X) c^{-1} \Big|_{\tau=0} \\
&= cXtc^{-1} + ctYc^{-1} - ctXc^{-1} \\
&= ct[(t^{-1}Xt - X) + Y]c^{-1}. \qquad (1)
\end{aligned}$$

The term inside the bracket may be thought of as a linear transformation

$$\mathsf{s} \times \mathsf{t} \to \mathsf{k}, \quad (X, Y) \to (\mathrm{Ad}(t^{-1}) - 1)X + Y.$$

Here $\mathsf{s} \times \mathsf{t}$ represents the tangent space to $K/T \times T$ at the point (cT, t) in the coordinate system introduced above. The formula (1) says that the differential of p at (cT, t) is the composite of this linear transformation with the left translation by ct and the right translation by c^{-1}. So the differential of p is invertible at (cT, t) if and only if $\mathrm{Ad}\, t^{-1} - 1$ is invertible as a linear transformation of s, or equivalently of k/t. This happens precisely when a is regular, as we shall see in the proof of the next theorem. QED

We now turn to integrals. We normalize the invariant integrals by requiring that

$$\int_K da = 1, \quad \int_T dt = 1, \quad \int_{K/T} dcT = 1.$$

If we write a general element of T as $t = (\epsilon_1, \ldots, \epsilon_n)$ with $\epsilon_j = e^{2\pi i \theta_j}$, then the integral on T is given by the formula

$$\int_T f(t)\, dt = \int_0^1 \cdots \int_0^1 f(t)\, d\theta_1, \ldots, d\theta_n \tag{2}$$

(problem 1).

Theorem 6 (Weyl's Integration Formula for $\mathrm{U}(n)$**).** *For every function* f *on* K,

$$\int_K f(a)\, da = \frac{1}{|W|} \int_{K/T} \left\{ \int_T f(ctc^{-1}) \Delta(t) \bar{\Delta}(t)\, dt \right\} dcT,$$

where

$$\Delta(t) = \prod_{j<k} (\epsilon_j - \epsilon_k).$$

Remark 7.

(a) The formula is understood in the sense that the integral on the left exists if and only if the integral on the right exists, and if so the integrals are equal.

(b) Note that $\Delta(t)\bar{\Delta}(t) = |\prod_{j \neq k} (\epsilon_j - \epsilon_k)| = |J(t)|$.

Unlike J, Δ is not defined on all of K; it is an antisymmetric function of the ϵ_j:

$$\Delta(sts^{-1}) = \mathrm{sgn}(s) \Delta(t)$$

for all $s \in W$ and all $t \in T$. $\mathrm{sgn}(s)$ is the sign of s considered as a permutation. It will become apparent in §6.4 why one writes $|J(t)|$ in the form $\Delta(t)\bar{\Delta}(t)$.

Proof. In terms of a coordinate system x on K, the integral of a function f that vanishes outside the coordinate domain is

$$\int_K f(a)\, da = \int f(a) \left| a^{-1} \frac{\partial a}{\partial x} \right| dx$$

where $a = a(x)$ is considered as a function of the coordinates and $\partial a/\partial x$ is the $(\dim K)$-tuple of basis vector fields of the coordinates, obtained by taking the partial derivatives of $a(x)$ with respect to the coordinates (eqn (7), §5.2).

We make a particular choice of coordinates as follows. Choose coordinates y on K/T and z on T and write $cT = c(y)T$ and $t = t(z)$. ($c(y) \in K$ is assumed to depend analytically on y, as may by arranged in view of the definition of coordinates on K/T.) Set

$$a = c(y)t(z)c(y)^{-1}.$$

We may take the pair (y, z) as coordinates in a neighborhood of any regular element $a \in G$ of this form (Proposition 3). We restrict y and z to sufficiently small domains, denoted $(K/T)'$ and T', so that p maps $(K/T)' \times T'$ bi-analytically onto an open subset of K. We may then use (y, z) as coordinates on the image in K. Calculate (using the differentiation rule $(c^{-1})' = -c^{-1}c'c^{-1}$ from the Appendix to §1.1):

$$\frac{\partial a}{\partial y} = \frac{\partial c}{\partial y}tc^{-1} - ctc^{-1}\frac{\partial c}{\partial y}c^{-1}, \quad \frac{\partial a}{\partial z} = c\frac{\partial t}{\partial z}c^{-1}.$$

One finds that

$$a^{-1}\frac{\partial a}{\partial y} = \mathrm{Ad}(c)\left(\mathrm{Ad}\,t^{-1} - 1\right)c^{-1}\left(\frac{\partial c}{\partial y}\right), \quad a^{-1}\frac{\partial a}{\partial z} = \mathrm{Ad}(c)\left(t^{-1}\frac{\partial t}{\partial z}\right).$$

Choose a subspace s of k complementary to t and a basis for k composed of a basis for s and for t. These bases are used to define the volume elements at 1 on K, T, and at $1T$ on K/T (the tangent space of K/T at $1T$ being identified with $\mathsf{s} \approx \mathsf{k}/\mathsf{t}$ for this purpose). The matrix of components of the $\dim\mathsf{s} + \dim\mathsf{t}$ vectors $a^{-1}\partial a/\partial y$, $a^{-1}\partial a/\partial z$ in k is then of the form

$$(\mathrm{Ad}\,c)\begin{bmatrix} (\mathrm{Ad}\,t^{-1} - 1)\left[c^{-1}\dfrac{\partial c}{\partial y}\right]_{\mathsf{s}} & 0 \\ * & \left[t^{-1}\dfrac{\partial t}{\partial z}\right], \end{bmatrix}$$

the subscript s indicating the component in s according to $\mathsf{k} = \mathsf{s} + \mathsf{t}$. The absolute value of the determinant of this matrix is

$$\begin{aligned} \left|a^{-1}\frac{\partial a}{\partial y}, a^{-1}\frac{\partial a}{\partial z}\right| &= |\det \mathrm{Ad}(c)|\,|\det \mathrm{Ad}_{\mathsf{k}/\mathsf{t}}(t^{-1} - 1)|\left|c^{-1}\frac{\partial c}{\partial y}\right|\left|t^{-1}\frac{\partial t}{\partial z}\right| \\ &= |J(t)|\left|c^{-1}\frac{\partial c}{\partial y}\right|\left|t^{-1}\frac{\partial t}{\partial z}\right|. \end{aligned}$$

Here $|c^{-1}(\partial c/\partial y)|$ is calculated in k/t relative to the basis consisting of the vectors corresponding to the chosen basis for $\mathsf{s} \approx \mathsf{k}/\mathsf{t}$. (Since K is compact, $|\det \mathrm{Ad}(c)| = 1$, by Lemma 3 of §5.2.) The last expression involves the volume elements on K/T at $1T$ and on T at 1. Assuming f vanishes outside of the coordinate domain, we get

$$\begin{aligned}\int_K f(a)\,da &= \iint f(ctc^{-1})\left|a^{-1}\frac{\partial a}{\partial y}, a^{-1}\frac{\partial a}{\partial z}\right| dz\,dy \\ &= \iint f(ctc^{-1})|J(t)|\left|c^{-1}\frac{\partial c}{\partial y}\right|\left|t^{-1}\frac{\partial t}{\partial z}\right| dz\,dy \\ &= \int\left\{\int f(ctc^{-1})|J(t)|\left|t^{-1}\frac{\partial t}{\partial z}\right| dz\right\}\left|c^{-1}\frac{\partial c}{\partial y}\right| dy \\ &= \int_{(K/T)'}\left\{\int_{T'} f(ctc^{-1})|J(t)|\,dt\right\}dcT.\end{aligned}$$

i.e.

$$\int_K f(a)da = \int_{(K/T)'}\left\{\int_{T'} f(ctc^{-1})|J(t)|dt\right\}dcT. \tag{3}$$

As a function of (cT, t), the integrand $f(ctc^{-1})|J(t)|$ does not vanish outside of $(K/T)' \times T'$, as $(K/T)' \times T'$ is only one of the $|W|$ regions of $K/T \times T$ which map bijectively onto the coordinate domain in K, the other ones being obtained from $(K/T)' \times T'$ by the transformations $(cT, t) \to (cs^{-1}T, sts^{-1})$, $s \in W$. But the integrals over all of these $|W|$ regions are the same, since there is nothing special about $(K/T)' \times T'$ among them (and as one can see explicitly by using y as coordinate of $c(y)s^{-1}T$ and z as coordinate of $st(z)s^{-1}$ in the region corresponding to $s \in W$). Adding the $|W|$ equations analogous to (3) gives

$$|W|\int_K f(a)\,da = \int_{K/T}\left\{\int_T f(ctc^{-1})|J(t)|dt\right\}dcT. \tag{4}$$

This proves Weyl's Formula when f vanishes outside a sufficiently small region in K_{reg}. It then holds for every integrable f on K_{reg}, as one sees by subdividing the region of integration into sufficiently small pieces. It therefore holds for every integrable function on K, in accordance with part (b) of Remark 5.

There remains to check the normalization of the integrals. Formula (4) was proved under the assumption that the basis for k used to define the volume element on K was put together from the bases for s and t used to define the volume elements on K/T and on T. Without this assumption the formula would hold up to a positive factor, say

$$\int_K f(a)\,da = \gamma\int_{K/T}\int_T f(ctc^{-1})|J(t)|\,dt\,dcT.$$

Using the normalizations

$$\int_K da = 1, \quad \int_T dt = 1, \quad \int_{K/T} dcT = 1,$$

we get

$$1 = \gamma \int_T |J(t)| dt = \gamma \int_T \Delta(t)\bar{\Delta}(t) dt. \tag{5}$$

We have

$$\begin{aligned}\Delta(t) &= \prod_{j<k} (\epsilon_j - \epsilon_k) \\ &= \det(\epsilon_j^{k-1}) \quad \text{[Vandermonde]} \\ &= \sum_{s \in S_n} \operatorname{sgn}(s) \epsilon_{s(1)}^{n-1} \cdots \epsilon_{s(n-1)}^{1} \epsilon_{s(n)}^{0} \quad \text{[expansion of `det']} \\ &= \sum_{s \in S_n} \operatorname{sgn}(s) e^{2\pi i((n-1)\theta_{s(1)} + 1\theta_{s(n-1)} + 0\theta_{s(n)})}. \end{aligned} \tag{6}$$

The terms in the above sum form an orthonormal family of functions on

$$0 \le \theta_1 \le 1, \quad \ldots, \quad 0 \le \theta_n \le 1.$$

So if one expands the product $\Delta(t)\bar{\Delta}(t)$ in (5) using (6) and integrates over T using Remark 8, one finds that

$$1 = \gamma |W|,$$

which is the desired result.

Implicit so far was the assumption that the integral on the left of Weyl's formula exists. It is a consequence of Fubini's Theorem (see Marsden (1974), for example) applied to the integrals (3) that this is the case if and only if the integral on the right exists. QED

Comment. One will notice that during the proof we have again calculated the differential of $K/T \times T \to K$, calculated previously in the proof of the Addendum to Proposition 3. The argument can easily be re-arranged so as to eliminate this redundancy, but it seemed clearer to proceed as we did.

We record separately the special case of Weyl's formula when f is a *class function* on K, i.e. $f(a)$ depends only on the conjugacy class of $a \in K$:

$$f(cac^{-1}) = f(a)$$

for all $c \in K$.

Theorem 8 (Weyl's Integration Formula for Class Functions). *For every integrable class function f on K,*

$$\int_K f(a)\, da = \frac{1}{|W|} \int_T f(t)\Delta(t)\bar{\Delta}(t)\, dt,$$

where

$$\Delta(t) = \prod_{j<k} (\epsilon_j - \epsilon_k).$$

One may think of the situation as follows. Intersection with T provides a one-to-one correspondence between K-conjugacy classes in K and W-orbits in T. The above formula gives the expression for the integral of a function of the K-conjugacy classes when considered a function of the W-orbits.

Problems for §5.3

1. Prove formula (2), using the definition of the invariant integral on a matrix group in terms of volume elements.

2. Let $K = \mathrm{SO}(3)$, T the subgroup of matrices of the form

$$t = \begin{bmatrix} \cos 2\pi\theta & -\sin 2\pi\theta & 0 \\ \sin 2\pi\theta & \cos 2\pi\theta & 0 \\ 0 & 0 & 1 \end{bmatrix}.$$

Let $W = N_K(T)/T$. Show that $|W| = 2$ and prove that for every integrable function f on K

$$\int_K f(a)\,da = \frac{1}{|W|} \int_{K/T} \Big\{ \int_T f(ctc^{-1}) \Delta(t) \bar{\Delta}(t)\,dt \Big\} dcT$$

where $\Delta(t) = (e^{\pi i\theta} - e^{-\pi i\theta})$.

3. (a) Let $K = \mathrm{SO}(2n)$, T the subgroup of block-diagonal matrices $t = (e_1, \ldots, e_n)$, where

$$e_j = \begin{bmatrix} \cos 2\pi\theta_j & -\sin 2\pi\theta_j \\ \sin 2\pi\theta_j & \cos 2\pi\theta_j \end{bmatrix}.$$

Show that for every integrable function f on K

$$\int_K f(a)\,da = \text{const.} \int_{K/T} \Big\{ \int_T f(ctc^{-1}) |J(t)|\,dt \Big\} dcT,$$

where $J(t) = \det\big(\mathrm{Ad}_{\mathsf{k}/\mathsf{t}}(t^{-1}) - 1\big)$.

(b) Verify that $|J| = \Delta\bar{\Delta}$, where

$$\Delta(t) = \prod_{j<k} \left(e^{\pi i(\theta_j+\theta_k)} - e^{-\pi i(\theta_j+\theta_k)}\right) \left(e^{\pi i(\theta_j-\theta_k)} - e^{-\pi i(\theta_j-\theta_k)}\right).$$

4. Let $K = \mathrm{SO}(2n+1)$, T the subgroup of block-diagonal matrices $t = (e_1, \ldots, e_n, 1)$, where the e_j are as above and the 1 is a 1×1 block with entry 1. Prove the same formulas as in (a) and (b) of problem 3 with

$$\Delta(t) = \prod_{j<k} \left(e^{\pi i(\theta_j+\theta_k)} - e^{-\pi i(\theta_j+\theta_k)}\right) \left(e^{\pi i(\theta_j-\theta_k)} - e^{-\pi i(\theta_j-\theta_k)}\right) \times \prod_j \left(e^{\pi i\theta_j} - e^{-\pi i\theta_j}\right).$$

(It suffices to indicate how the proof of problem 3 has to be modified.)

5. Let $K = \mathrm{Sp}(n)$, the group of unitary linear transformations of $\mathbb{C}^{2n}$ preserving the skew-symmetric form

$$x_1 y_{n+1} - x_{n+1} y_1 + \cdots + x_n y_{2n} - x_{2n} y_n.$$

Let T be the subgroup of diagonal matrices $t = (\epsilon_1, \ldots, \epsilon_n, \epsilon_1^{-1}, \ldots, \epsilon_n^{-1})$ with $\epsilon_j = e^{2\pi i \theta_j}, \theta_j \in \mathbb{R}$. Prove the same formulas as in (a) and (b) of problem 3 with

$$\Delta(t) = \prod_{j<k} \left(e^{\pi i(\theta_j+\theta_k)} - e^{-\pi i(\theta_j+\theta_k)}\right)\left(e^{\pi i(\theta_j-\theta_k)} - e^{-\pi i(\theta_j-\theta_k)}\right)$$
$$\times \prod_j \left(e^{2\pi i\theta_j} - e^{-2\pi i\theta_j}\right).$$

(It suffices to indicate how the proof of problem 3 has to be modified.)

Suggested reading. In Section VII.4 of his book on the classical groups (1939), Weyl gives an intriguing geometric argument for his integration formula.

6

Representations

6.1 Representations: definitions

A *representation* of a group G on a vector space V associates to each $a \in G$ an invertible linear transformation $\pi(a)$ of V so that

$$\pi(ab) = \pi(a)\pi(b),$$

i.e. π is a group homomorphism from G into the group $\mathrm{GL}(V)$ of invertible linear transformations of V. When G is a linear group we also assume that the homomorphism $\pi : G \to \mathrm{GL}(n, \mathbb{C})$ is *differentiable*.

Unless explicitly stated otherwise we shall assume that V is a *finite dimensional, complex* vector space. The dimension of V is called the *degree* of the representation, denoted $\deg \pi$. After choice of a basis for V we may think of V as $\mathbb{C}^n$ and of $\pi(a) \in \mathrm{GL}(n, \mathbb{C})$ as a matrix. We often refer to the pair V, π as a representation. V is also referred to as the *representation space* for π. If π is fixed, $\pi(a)x$ is sometimes written as $a \cdot x$. Along with this notation goes the terminology *G-module* for the representation space V.

Example 1. Any linear group comes automatically with a representation, as it consists already of matrices. (We may think of real matrices as special complex matrices to satisfy the requirement that the representation space V should be complex. This amounts to passing to the complexification, explained in the appendix to §1.1.)

For example, $G = \mathrm{SO}(3)$ comes with its *rotation representation* on $\mathbb{R}^3$, which is extended to $\mathbb{C}^3$ in accordance with the last remark. One obtains an infinite-dimensional representation τ of $\mathrm{SO}(3)$ on the space of all complex valued functions f on $\mathbb{R}^3$ via $\tau(a)f(x) = f(a^{-1}x)$. To cut down to finite dimensions we take for f only polynomial functions of degree $\leq d$ (some d), or even only homogeneous polynomial functions of degree d. We can also let $\mathrm{SO}(3)$ act on complex-valued functions on the unit sphere S^2 by the same recipe. To cut down to finite dimensions we may consider only function on S^2 that are restrictions of (homogeneous) polynomial functions on $\mathbb{R}^3$ of degree $\leq d$.

In general, when a group G acts on a set X , we get a representation τ of G on the space $\mathbb{C}^X$ of all complex-valued functions on X by setting $\tau(a)f(x) = f(a^{-1}x)$. $\mathbb{C}^X$ is of course infinite dimensional if X is infinite. If X is a subset of some $\mathbb{R}^n$ and the action on X comes from a linear action (= representation) of G on $\mathbb{R}^n$, the same device with polynomials produces finite-dimensional representations. The procedure works also with $\mathbb{R}^n$ replaced by $\mathbb{C}^n$. It applies in particular when G is a linear group, when we take $X = G$ with either the left or right translation action. The corresponding representations are called the *left regular* or *right regular* representation of G on $\mathbb{C}^G$, denoted L and R, respectively:

$$L(a)f(x) = f(a^{-1}x), \quad R(a)f(x) = f(xa).$$

Again one may restrict attention to restrictions to G of polynomial functions on the surrounding matrix space. In this context one should recall:

Theorem 2 (Weierstrass's Approximation Theorem). *Any continuous function on a compact subset of $\mathbb{R}^n$ may be uniformly approximated by polynomials.*

Comments.

(a) f must be continuous in the relative topology from $\mathbb{R}^n(|f(x) - f(x_o)| < \epsilon$ if $\|x - x_o\| < \delta)$ and the uniform approximation on X means that for every $\epsilon > 0$ there is a polynomial function p on $\mathbb{R}^n$ so that $|f(x) - p(x)| < \epsilon$ for all $x \in X$.

(b) The theorem may be applied to a *compact* subset X of $\mathbb{C}^n \approx \mathbb{R}^{2n}$, with $z = x + iy$. One should note, however, that the polynomials in question will then be polynomials in (the components of) x and y, or in $z = x + iy$ and in $\bar{z} = x - iy$, not in z alone.

(c) The theorem applies in particular to a *compact* linear group G. (We know that in this case the group topology on G, defined using the exponential map, is the same as the relative topology as a subset of the matrix space.)

A representation V, π of G is *unitary* if V has a positive-definite, Hermitian inner product (x, y) with respect to which the $\pi(a)$ are unitary, i.e.

$$(\pi(a)x, \pi(a)y) = (x, y)$$

for all $a \in G$, $x, y \in V$. Such an inner product is said to be *invariant* under G.

A subspace U of V is *invariant*, or *stable* (under π or G), if $\pi(a)U \subset U$ (hence $= U$) for all $a \in G$. In that case the restriction of the $\pi(a)$ to U gives a representation of G on U, called a *subrepresentation* of V, π.

If V admits such an invariant subspace U other than $\{0\}$ and V itself, then the representation V, π is said to be *reducible*. Otherwise, if V has no invariant subspaces other than $\{0\}$ and V itself, the representation V, π is said to be *irreducible*, except that $V = \{0\}$ is not considered irreducible, by convention (but the *trivial* representation $V = \mathbb{C}$, $\pi(a) \equiv 1$ is considered irreducible).

If $V = V_1 \oplus V_2 \oplus \cdots \oplus V_k$ is the direct sum of invariant subspaces, we say the representation V, π is the *direct sum of the subrepresentations of G on the V_j*. We say V, π *decomposes* into these subrepresentations.

Remark 3.

(a) If V, π is unitary, then the $\pi(a)$ are represented by unitary matrices with respect to an orthonormal basis for V.

(b) If $U \subset V$ is invariant, then the $\pi(a)$ are represented by matrices of the form

$$\begin{bmatrix} * & * \\ 0 & * \end{bmatrix}$$

with respect to a basis for V whose first $\dim U$ basis vectors form a basis for U.

(c) If the representation V, π is a direct sum $V = V_1 \oplus V_2$, then the $\pi(a)$ are represented by matrices of the form

$$\begin{bmatrix} * & 0 \\ 0 & * \end{bmatrix}$$

with respect to a basis for V composed of a basis for V_1 and a basis for V_2. For $V = V_1 \oplus V_2 \oplus \cdots \oplus V_k$ there are k blocks along the diagonal.

(d) In a unitary representation V, π the orthogonal $U^\perp$ of an invariant subspace is again invariant. In that case we may decompose V into an orthogonal direct sum of irreducibles by successively splitting off invariant subspaces. (QED)

The last remark is especially important in view of the following proposition.

Proposition 4. *Any representation of a compact group admits an invariant inner product.*

Proof. Start with any inner product (x, y) on V. Define a new inner product $(x, y)^o$ by the formula

$$(x, y)^o = \int_G (\pi(a)x, \pi(a)y)\, da.$$

$(x, y)^o$ is again positive-definite, Hermitian. But most importantly, it is also invariant:

$$\begin{aligned}(\pi(b)x, \pi(b)y)^o &= \int_G (\pi(ab)x, \pi(ab)y)\, da \\ &= \int_G (\pi(a)x, \pi(a)y)\, da \quad \text{[invariance of the integral]} \\ &= (x, y)^o.\end{aligned}$$

QED

In view of the Remark 3(d) above we see that

Corollary 5. *Any representation of a compact group decomposes as a direct sum of irreducibles.*

Remark 6.

(a) The corollary is also pharased as: any representation of a compact group is *completely reducible*. Such is generally not the case for non-compact groups: it suffices to consider the representation on $\mathbb{C}^2$ of the subgroup of $\mathrm{GL}(2,\mathbb{C})$ consisting of the matrices

$$\begin{bmatrix} 1 & \alpha \\ 0 & 1 \end{bmatrix}, \quad \alpha \in \mathbb{C},$$

for example.

(b) The decomposition into irreducibles is generally not unique. (One may think of $\pi(a) \equiv 1$ on $\mathbb{C}^n$ as an extreme case.)

A representation V, π of G has a *contragredient* representation $V^*, \check{\pi}$ on the dual space V^* of linear functions on V. It is defined by

$$\langle \check{\pi}(a)x, y\rangle = \langle x, \pi(a^{-1})y\rangle.$$

Here $\langle x, y\rangle$ is the value of $x \in V^*$ on $y \in V$. We use the same letters x, y for elements of V and V^* and also write $\langle y, x\rangle$ for $\langle x, y\rangle$.

Remark 7. With respect to dual bases in V and V^*, the matrix of $\check{\pi}(a)$ is the inverse-transpose of the matrix of $\pi(a)$. (QED)

Lemma 8. *A representation V, π is irreducible if and only if its contragredient $V^*, \check{\pi}$ is irreducible.*

Proof. Associating with each subspace U of V its orthogonal $U^\perp = \{x \in V^* \mid \langle x, y\rangle = 0 \text{ for all } y \in U\}$ in V^*, one gets a one-to-one correspondence between invariant subspaces of V and V^*. QED

The *tensor product* $V \otimes W$ of two finite-dimensional vector spaces may be defined as follows. Take a basis $\{e_j | j = 1, \ldots, m\}$ for V and a basis $\{f_k \mid k = 1, \ldots, n\}$ for W. Create a vector space, denoted $V \otimes W$, with a basis consisting of mn symbols $e_j \otimes f_k$. Generally, for $x = \sum_j \xi_j e_j \in V$ and $y = \sum_k \eta_k f_k \in W$, let

$$x \otimes y = \sum_{ik} \xi_j \eta_k e_j \otimes f_k \in V \otimes W.$$

The space $V \otimes W$ is independent of the bases $\{e_j\}$ and $\{f_k\}$ in the following sense. Any other bases $\{\tilde{e}_j\}$ and $\{\tilde{f}_k\}$ lead to symbols $\tilde{e}_j \tilde{\otimes} \tilde{f}_k$ forming a basis of $V \tilde{\otimes} W$. We may then identify $V \tilde{\otimes} W$ with $V \otimes W$ by sending the basis $\tilde{e}_j \tilde{\otimes} \tilde{f}_k$ of $V \tilde{\otimes} W$ to the vectors $\tilde{e}_j \otimes \tilde{f}_k$ in $V \otimes W$, and vice versa.

The space $V \otimes W$ is called the *tensor product* of V and W. The vector $x \otimes y \in V \otimes W$ is called the *tensor product* of $x \in V$ and $w \in W$. (One should keep in mind that an element of $V \otimes W$ looks like $\sum \zeta_{ij} e_j \otimes f_k$ and may not be expressible as a single tensor product $x \otimes y = \sum \xi_j \eta_k e_j \otimes f_k$.)

Given a linear transformation A of V and a linear transformation B of W we define their *tensor product* $A \otimes B$ as the linear transformation $A \otimes B$ of $V \otimes W$ given by the formula

$$(A \otimes B)(x \otimes y) = (Ax) \otimes (By)$$

on simple tensor products $x \otimes y$ and extended to all of $V \otimes W$ by linearity. To see that $A \otimes B$ is well-defined one may first use this definition for the basis elements

$e_j \otimes f_k$ only, and check that it then holds for all $x \otimes y$. The same definition works when A or B is a linear transformation between two different spaces.

When V carries a representation π of a group G and W a representation ρ of another group H, then $V \otimes W$ carries a representation $\pi \otimes \rho$ of the product group $G \times H$ defined by

$$(\pi \otimes \rho)(a, b) = \pi(a) \otimes \rho(b).$$

When $G = H$ we also get a representation of G itself, also denoted $\pi \otimes \rho$, by setting $(\pi \otimes \rho)(a) = \pi(a) \otimes \rho(a)$. This is just the restriction of the representation $\pi \otimes \rho$ of $G \times G$ to the diagonal subgroup, identified with G. The representation $\pi \otimes \rho$ of $G \times H$ or of G is naturally called the *tensor product* of π and ρ.

The definitions extend to *multiple tensor products*: If U, V, W are three vector spaces we may identify $(U \otimes V) \otimes W$ with $U \otimes (V \otimes W)$ by sending the basis elements $(d_i \otimes e_j) \otimes f_k$ to the basis elements $d_i \otimes (e_j \otimes f_k)$. The resulting triple tensor product may therefore be denoted $U \otimes V \otimes W$ without ambiguity. In the same way one may form the tensor product, in a definite order, of any finite number of vector spaces. In particular, the n-fold tensor product of V with itself is denoted $T^k(V) = V \otimes \cdots \otimes V$. It has a basis consisting of the n^k tensor products $e_{i_1} \otimes \cdots \otimes e_{i_k}$, where $\{e_j | j = 1, \ldots, n\}$ is a basis for V.

Remark 9. If the linear transformation A of V has matrix (α_{ij}) with respect to the basis $\{e_j\}$ of V and the linear transformation B of W has matrix (β_{kl}) with respect to the basis $\{f_k\}$ for W, then the tensor product transformation $A \otimes B$ has matrix $(\alpha_{ij}\beta_{kl})$ with respect to the basis $e_j \otimes f_k$ of $V \otimes W$. (QED)

Problems for §6.1

1. Prove Remark 3.

2. Prove Remark 7.

3. Prove Remark 9.

4. Define a linear map C from $V^* \otimes W$ to the space $L(V, W)$ linear transformations $V \to W$ so that

$$C(x \otimes y)z = \langle x, z\rangle y.$$

(a) Show that C is well-defined and bijective.

(b) Suppose V and W carry representations π and ρ of a group G. Find a formula for the representation of G on $L(V, W)$ corresponding to the representation $\check{\pi} \otimes \rho$.

5. Define a linear map B from $V^* \otimes W^*$ to the space $\mathrm{BIL}(V, W)$ of bilinear forms on $V \times W$ so that

$$B(x \otimes y)(u, v) = \langle x, u\rangle\langle y, v\rangle.$$

(a) Show that B is well-defined and bijective.

(b) Suppose V and W carry representations π and ρ of a group G. Find a formula for the representation of G on $\mathrm{BIL}(V, W)$ corresponding to the representation $\check{\pi} \otimes \rho$ of G on $V^* \otimes W^*$ under the bijection B.

6. *Representations of* $\mathrm{GL}(n, \mathbb{C})$ *and* S_m *on* $T^m(\mathbb{C}^n)$. $T^m(\mathbb{C}^n)$ denotes the m-fold tensor product of $\mathbb{C}^n$ with itself. $\mathrm{GL}(n, \mathbb{C})$ acts on $T^m(\mathbb{C}^n)$ through its action on the tensor factors:

$$a(x_1 \otimes \cdots \otimes x_m) = (ax_1) \otimes \cdots \otimes (ax_m).$$

S_m acts on $T^m(\mathbb{C}^n)$ by permuting the tensor factors

$$s(x_1 \otimes \cdots \otimes x_m) = x_s^{-1}(1) \otimes \cdots \otimes x_s^{-1}(n).$$

(The formulas are extended to linear combinations of tensor products $x_1 \otimes \cdots \otimes x_m$ by linearity.) These actions are representations of $\mathrm{GL}(n, \mathbb{C})$ and S_m on $T^m(\mathbb{C}^n)$.

(a) Show that the actions of $\mathrm{GL}(n, \mathbb{C})$ and S_m on $T^m(\mathbb{C}^n)$ commute, i.e. $asx = sax$ for all $a \in \mathrm{GL}(n, \mathbb{C})$, $s \in S_m$, $x \in T^m(\mathbb{C}^n)$. Deduce that $cT^m(\mathbb{C}^n)$ is $\mathrm{GL}(n, \mathbb{C})$ invariant for any operator c on $T^m(\mathbb{C}^n)$ of the form $c = \sum_{s \in S_n} \gamma_s s$, with $\gamma_s \in \mathbb{C}$.

(b) Define an inner product on $T^m(\mathbb{C}^n)$ by setting

$$(x_1 \otimes \cdots \otimes x_m, x_1 \otimes \cdots \otimes x_m) = (x_1, y_1) \cdots (x_m, y_m)$$

for all $x_j, y_k \in \mathbb{C}^n$. Show that this inner product on $T^m(\mathbb{C}^n)$ is well defined and that

$$(ax, y) = (x, a^* y)$$

for all $x, y \in T^m(\mathbb{C}^n)$ and all $a \in \mathrm{GL}(n, \mathbb{C})$. Deduce (i) that the representation $\mathrm{GL}(n, \mathbb{C})$ on $T^m(\mathbb{C}^n)$ is unitary when restricted to $\mathrm{U}(n)$, and (ii) that the representation $\mathrm{GL}(n, \mathbb{C})$ on $T^m(\mathbb{C}^n)$ is *completely reducible*, i.e. decomposes as a direct sum of irreducible subrepresentations.

7. *Symmetric tensors.* Let

$$S^m(\mathbb{C}^n) = \{x \in T^m(\mathbb{C}^n) | sx = x\} \quad \text{for all } s \in S_m,$$

the space of *symmetric tensors* of degree m on $\mathbb{C}^n$. Let

$$\mathrm{sym} = \frac{1}{m!} \sum_{s \in S_m} s$$

as operator on $T^m(\mathbb{C}^n)$.

(a) Show that sym is a projection of $T^m(\mathbb{C}^n)$ onto $S^m(\mathbb{C}^n)$, i.e. $S^m(\mathbb{C}^n) = \mathrm{sym}\, T^m(\mathbb{C}^n)$, $\mathrm{sym}\, x = x$ for $x \in S^m(\mathbb{C}^n)$. Notation: $x_1 x_2 \cdots x_m = \mathrm{sym}\, x_1 \otimes x_2 \otimes \cdots \otimes x_m$.

(b) Show that the sym-monomials $e_1^{m_1} e_2^{m_2} \cdots e_n^{m_n}$, $m_1+m_2+\cdots+m_n = m$, in the basis vectors $e_1, e_2, \ldots, e_n$ of $\mathbb{C}^n$ form a basis for $S^m(\mathbb{C}^n)$.

[The symbol $e_1^{m_1}$ means the tensor product of m_1 factors e_1. Recall that the $e_{i_1} \otimes e_{i_2} \otimes \cdots \otimes e_{i_m}$ form a basis for $T^m(\mathbb{C}^n)$.]

(c) Show that

$$\dim S^m(\mathbb{C}^n) = \binom{m+n-1}{m} = \frac{(m+n-1)!}{n!m!},$$

is the number of m-tuples with distinct entries of $m+n-1$ elements. [In part (b), order the indices on $\operatorname{sym}(e_{i_1} \otimes e_{i_2} \otimes \cdots \otimes e_{i_m})$ so that $i_1 \geq i_2 \geq \cdots \geq i_m$. Then

$$i_1 + (m-1), \quad i_2 + (m-2), \quad \ldots, \quad i_m + 0$$

are m distinct numbers from $\{n+m-1, n+m-2, \quad \ldots, \quad 1\}$.]

(d) Show that for any diagonalizable $a \in \mathrm{GL}(n, \mathbb{C})$ the trace $\psi_m(a)$ of a as linear transformation on $S^m(\mathbb{C}^n)$ is

$$\psi_m(a) = \sum \epsilon_1^{m_1} \epsilon_2^{m_2} \cdots \epsilon_n^{m_1}.$$

The sum is over all n-tuples $(m_1, m_2, \ldots, m_n)$ of integers ≥ 0 with $m_1 + m_2 + \cdots + m_n = m$; $\epsilon_1, \epsilon_2, \ldots, \epsilon_n$ are the eigenvalues of $a \in \mathrm{GL}(n, \mathbb{C})$.

(e) Verify that for diagonalizable $a \in \mathrm{GL}(n, \mathbb{C})$ $\psi_m(a)$ can be calculated from the generating function

$$\frac{1}{\det(1 - za)} = \sum_{m=0}^{\infty} \psi_m(a) z^m.$$

Deduce that the formula for $\psi_m(a)$ holds for all $a \in \mathrm{GL}(n, \mathbb{C})$. [Suggestion: use Remark 5(b), §5.3.]

8. *Alternating tensors.* (a) Let

$$\wedge^m(\mathbb{C}^n) = \{x \in T^m(\mathbb{C}^n) \mid sx = \operatorname{sgn}(s)x\} \quad \text{for all } s \in S_m,$$

the space of *alternating tensors* of degree m on $\mathbb{C}^n$. Let

$$\operatorname{alt} = \frac{1}{m!} \sum_{s \in S_m} \operatorname{sgn}(s) s$$

(b) Show that alt is a projection of $T^m(\mathbb{C}^n)$ onto $\wedge^m(\mathbb{C}^n)$, i.e. $\wedge^m(\mathbb{C}^n) = \operatorname{alt} T^m(\mathbb{C}^n)$, $\operatorname{alt} x = x$ for $x \in \wedge^m(\mathbb{C}^n)$. Notation: $x_1 \wedge x_2 \wedge \cdots \wedge x_m = \operatorname{alt} x_1 \otimes x_2 \otimes \cdots \otimes x_m$.

(c) Show that $x_1 \wedge x_2 \wedge \cdots \wedge x_m = 0$ if and only if $x_1, x_2, \ldots, x_m$ are linearly independent. [Suggestion: Show first that $x_1 \wedge x_2 \wedge \cdots \wedge x_m = 0$ if two x_k

coincide. On the other hand, if $x_1, x_2, \ldots, x_m$ are linearly dependent, one may as well assume $x_1 = e_1, \ldots, x_m = e_m$.]

(d) Show that the ordered alt-monomials $e_{i_1} \wedge e_{i_2} \wedge \cdots \wedge e_{i_m}, i_1 < i_2 < \cdots < i_m$, in the basis vectors $e_1, e_2, \ldots, e_n$ of $\mathbb{C}^n$ form a basis for $\wedge^m(\mathbb{C}^n)$.

(e) Show that

$$\dim \Lambda^m(\mathbb{C}^n) = \binom{n}{m} = \frac{n!}{m!(n-m)!},$$

the number of m-element subsets from a set of n elements.

(f) Show that for diagonalizable $a \in \mathrm{GL}(n, \mathbb{C})$ the trace $\varphi_m(a)$ of a as linear transformation on $\wedge^m(\mathbb{C}^n)$ is the m-th elementary symmetric function in the eigenvalue $\epsilon_1, \epsilon_2, \ldots, \epsilon_n$ of a:

$$\varphi_m(a) = \sum \epsilon_{i_1} \epsilon_{i_2} \cdots \epsilon_{i_3}.$$

The sum is over all ordered m-tuples $(i_1, i_2, \ldots, i_m), i_1 < i_2 < \cdots < i_m$ of indices.

(g) Verify that $\varphi_m(a)$ for diagonalizable $a \in \mathrm{GL}(n, \mathbb{C})$ can be calculated from the *generating function*

$$\det(1 + za) = \sum_{m=0}^{n} \varphi_m(a) z^m.$$

Deduce that the formula holds for *all* $a \in \mathrm{GL}(n, \mathbb{C})$. [Suggestion: remark 6(b), §5.3, again.]

9. Show that $T^m(V^*)$ is isomorphic with the space of *m-linear forms* on V, i.e. the space of scalar-valued functions on $V \times V \times \cdots \times V$ (m copies that are linear in each variable separately). Describe the images of $S^m(V^*)$ and of $\wedge^m(V^*)$ under this isomorphism. Write out the formula for the representation of G on the space of m-linear forms on V corresponding to the representation on $T^m(V^*)$ constructed from a representation of G on V.

10. Let G be a group, V, π be a representation of G. Let $R(V)$ be the space of all holomorphic polynomials on V (i.e. all functions on V which are expressible as polynomials in the coordinates with respect to a fixed basis), $R^m(V)$ the subspace of homogeneous polynomials of degree m. G acts on V by the rule $af(x) = f(a^{-1}x)$. Show that there is a G-isomorphism $S^m(V^*) \to R^m(V)$ (homogeneous of degree m) sending the symmetric tensors $\varphi_1 \varphi_2 \cdots \varphi_m$ in $S^m(V^*)(\varphi_j \in V^*)$ to $\varphi_1 \varphi_2 \cdots \varphi_m$ considered as function on V.

11. *Representations of* SO(n) *on polynomials.* SO(n) acts on functions on $\mathbb{R}^n$ by

$$af(x) = f(a^{-1}x)$$

(a) Show that on real- or complex-valued C^2 functions the action of $\mathrm{SO}(n)$ commutes with the *Laplace operator*

$$\Delta = \frac{\partial^2}{\partial \xi_1^2} + \cdots + \frac{\partial^2}{\partial \xi_n^2}.$$

(b) Let $R^m(\mathbb{R}^n)$ be the space of homogeneous, complex-valued polynomial functions of degree m on $\mathbb{R}^n$. Show that the formula

$$(f, g) = \int_{S^{n-1}} f(x)\overline{g(x)}\nu(dx)$$

defines a positive-definite, $\mathrm{SO}(n)$-invariant inner product on $R^m(\mathbb{R}^n)$. [The integral over S^{n-1} is with respect to the invariant volume element ν on S^{n-1} considered as homogeneous space for $\mathrm{SO}(n)$.]

(c) Show that the representation of $\mathrm{SO}(n)$ on $R^m(\mathbb{R}^n)$ is reducible for $n \geq 2$ and $m \geq 2$. [Suggestion: Calculate $\Delta(\xi_1 + i\xi_2)^m$.]

6.2 Schur's lemma, Peter–Weyl theorem

Let V, π and W, ρ be two representations of a group G. A *G-map*, or *intertwining operator* is a linear map $A : V \to W$ satisfying

$$A \circ \pi(a) = \rho(a) \circ A$$

for all $a \in G$. An invertible G-map is called a *G-isomorphism*. If there is a G-isomorphism $V \to W$, then the representations π and ρ are said to be *equivalent* and we write $\pi \approx \rho$ or $V \approx W$. If $\pi \approx \rho$, then $\pi(a)$ and $\rho(a)$ are represented by the same matrix with respect to bases related by a G-map.

We start with a simple but crucial observation.

Theorem 1 (Schur's Lemma). *Let V, π and W, ρ be two irreducible representations of $G, A : V \to W$ a G-map.*

(a) If $\pi \not\approx \rho$, then $A = 0$.

(b) If $\pi = \rho$, then $A = \alpha 1$ is a scalar multiple of the identity.

Comment. Assertion (b) requires $\pi = \rho$ not only $\pi \approx \rho$ to make sense. But one may rephrase (b) as follows: If $\pi \approx \rho$, then any two non-zero G-maps are scalar multiples of each other.

Proof.

(a) Both the kernel and the image of A are invariant subspaces, hence $\{0\}$ or the whole space, by irreducibility. This gives (a).

(b) Let α be an eigenvalue of the G-map $A : V \to V$. The corresponding eigenspace is invariant, hence all of V, by irreducibility. This gives (b). QED

Corollary 2. *An irreducible representation of an abelian group is one-dimensional.*

Example 3. Let $T = \mathbb{T}^n$ be n-torus, the group of $n \times n$ diagonal matrices with complex entries of modulus 1. As previously, we use the notation

$$t = (\epsilon_1, \ldots, \epsilon_n), \quad \epsilon_j = e^{2\pi i \theta_j},$$

for elements of T. The n-tuple (diagonal matrix)

$$H = (2\pi i \theta_1, \ldots, \pi i \theta_n)$$

represents a general element of t and $t = \exp H$ means $\epsilon_j = e^{2\pi i \theta_j}$.

By the corollary, an irreducible representation of T is a homomorphism $\omega : T \to \mathrm{GL}(1, \mathbb{C}) = \mathbb{C}^\times$. These were determined in Example 4, §2.6:

$$\omega(\exp H) = e^{\lambda(H)} \tag{1}$$

where $\lambda : t \to \mathbb{C}$ is an $\mathbb{R}$-linear function of the form

$$\lambda(H) = 2\pi i (l_1 \theta_1 + \cdots + l_n \theta_n) \tag{2}$$

with

$$l_j \in \mathbb{Z} \quad \text{for all } j. \tag{3}$$

Conclusion. The irreducible characters of T are *exactly* the ω defined by (1) with λ satisfying (3).

Notation. For $\lambda \in it^*$ of the form (2) satisfying (3), the corresponding homomorphism $T \to \mathbb{C}^\times$ is denoted e^λ:

$$e^\lambda(\exp H) = e^{\lambda(H)}.$$

This notation may be justified as follows. T may be identified with the quotient of $t = i\mathbb{R}^n$ by $\{H \in t \exp H = 1\} \approx 2\pi i \mathbb{Z}^n$. In this way functions on T become functions on t *periodic* with respect to $2\pi i \mathbb{Z}^n$. The notation e^λ is in agreement with this identification. We record the explicit formula for e^λ:

$$e^\lambda(t) = \epsilon_1^{l_1} \cdots \epsilon_n^{l_n}$$

if $t = (\epsilon_1, \ldots, \epsilon_n)$.

For the remainder of this section G *denotes a compact linear group.*

Corollary 4. *Let V, π and W, ρ be two irreducible representations of G, and $A : V \to W$ a linear map. Let*

$$A^o = \frac{1}{|G|} \int\limits_G \rho(a) A \pi(a^{-1})\, da.$$

(a) If $\pi \not\approx \rho$, then $A^o = 0$.

(b) *If* $\pi = \rho$, *then* $A^o = \alpha 1$ *is a scalar multiple of the identity and* $\alpha = \mathrm{tr}(A)/\deg(\pi)$.

Proof. A^o is a G-map:

$$\rho(b)A^o\pi(b^{-1}) = \frac{1}{|G|}\int_G \rho(ba)A\pi(a^{-1}b^{-1})\,da = A^o$$

by left-invariance of the integral. Schur's Lemma gives $A^o = 0$ if $\pi \not\approx \rho$ and $A^o = \alpha 1$ if $\pi = \rho$. In the latter case:

$$\mathrm{tr}(A^o) = \frac{1}{|G|}\int_G \mathrm{tr}\,\rho(a)A\pi(a^{-1})\,da = \frac{1}{|G|}\int_G \mathrm{tr}(A)\,da[\rho = \pi] = \mathrm{tr}(A).$$

But also

$$\mathrm{tr}(A^o) = \mathrm{tr}(\alpha 1) = \alpha \deg(\pi).$$

Comparing the two expressions for $\mathrm{tr}(A^o)$ one finds $\alpha = \mathrm{tr}(A)/\deg(\pi)$. QED

Corollary 5. *Let* V, π *and* W, ρ *be two irreducible representations of* G. *Let* $x \in V$, $x' \in V^*$, $y \in W$, $y' \in W^*$. *Then*

$$\frac{1}{|G|}\int_G \langle \pi(a)x, x'\rangle\langle \rho(a^{-1})y, y'\rangle\,da = \begin{cases} 0 & \text{if } \pi \not\approx \rho, \\ \dfrac{1}{\deg(\pi)}\langle x, y'\rangle\langle y, x'\rangle & \text{if } \pi = \rho. \end{cases}$$

Proof. Define $A : V \to W$ by $Av = \langle v, x'\rangle y$. Then

$$\rho(a)A\pi(a^{-1})x = \langle \pi(a^{-1})x, x'\rangle \rho(a)y.$$

Evaluate y' on both sides of this equation:

$$\langle \rho(a)A\pi(a^{-1})x, y'\rangle = \langle \pi(a^{-1})x, x'\rangle\langle \rho(a)y, y'\rangle.$$

Integrate over G:

$$\frac{1}{|G|}\int_G \langle \pi(a^{-1})x, x'\rangle\langle \rho(a)y, y'\rangle\,da = \frac{1}{|G|}\int_G \langle \rho(a)A\pi(a^{-1})x, y'\rangle\,da = \langle A^o x, y'\rangle.$$

By Corollary 4, this equals 0 if $\pi \not\approx \rho$ and equals $\alpha\langle x, y'\rangle$ if $\pi = \rho$ with $\alpha = \mathrm{tr}(A)/\deg(\pi)$. Finally, $\mathrm{tr}(A) = \langle y, x'\rangle$. QED

A function of the form $a \to \langle \pi(a)x, x'\rangle$ is called a *matrix coefficient* of the representation π. It is evidently a linear combination of the matrix coefficients $a \to \pi_{jk}(a) = \langle \pi(a)e_j, e^k\rangle$ of $\pi(a)$ with respect to any basis $\{e_j\}$ of V, with dual basis $\{e^k\}$ for V^*.

Corollary 6. *Let V, π and W, ρ be two irreducible unitary representations of G. Then for all $x_1, x_2 \in V$ and $y_1, y_2 \in W$,*

$$\frac{1}{|G|}\int\limits_G (\pi(a)x_1, x_2)\overline{(\rho(a)y_1, y_2)}\, da = \begin{cases} 0 & \text{if } \pi \not\approx \rho, \\ \dfrac{1}{\deg(\pi)} & (x_1, y_1)\overline{(x_2, y_2)} \quad \text{if } \pi = \rho. \end{cases}$$

Proof. In Corollary 5, take $x = x_1$, $x' = (\ , x_2)$, $y = y_2$, and $y' = (\ , y_1)$. QED

Theorem 7 (Schur's Orthogonality Relations for Matrix Coefficients). *Let V, π and W, ρ be two irreducible unitary representations of G, $[\pi_{ij}(a)]$ and $[\rho_{kl}(a)]$ the matrices of $\pi(a)$ and $\rho(a)$ with respect to orthonormal bases for V and W. Then*

$$\frac{1}{|G|}\int\limits_G \pi_{ij}(a)\overline{\rho_{kl}(a)}\, da = \begin{cases} 0 & \text{if } \pi \not\approx \rho \text{ or if } \pi = \rho \text{ but } (ij) \neq (kl), \\ \dfrac{1}{\deg(\pi)} & \text{if } \pi = \rho \text{ and } (ij) = (kl). \end{cases}$$

Proof. Apply Corollary 6 to $\pi_{ij}(a) = (\pi(a)e_j, e_i)$ and $\rho_{kl}(a) = (\rho(a)f_l, f_k)$. QED

For any L^2-function f on G, set

$$\|f\|^2 = \frac{1}{|G|}\int\limits_G |f|^2\, da.$$

The inner product of two L^2-functions f, g on G is

$$(f, g) = \frac{1}{|G|}\int\limits_G f(a)\overline{g(a)}\, da.$$

Theorem 8 (Peter–Weyl Theorem for Linear Groups). *Let G be a compact linear group, $\{\pi^\lambda\}_\lambda$ a complete set of pairwise inequivalent, irreducible, unitary matrix representations of G. The matrix entries $\{\pi^\lambda_{ij}\}_{\lambda,ij}$ form a countable complete set of orthogonal functions on G. More precisely,*

$$(\pi^\lambda_{ij}, \pi^\mu_{kl}) = \begin{cases} 0 & \text{if } \pi^\lambda_{ij} \neq \pi^\mu_{kl}, \\ \dfrac{1}{\deg(\pi^\lambda)} & \text{if } \pi^\lambda_{ij} = \pi^\mu_{kl} \end{cases} \tag{4}$$

Every L^2-function f on G admits an L^2-convergent Fourier expansion

$$f \equiv \sum_{\lambda,ij} \deg(\pi^\lambda)(f, \pi^\lambda_{ij})\pi^\lambda_{ij} \tag{5}$$

Proof. (4) is a re-statement of Schur's orthogonality relations for matrix coefficients. Assuming countability, the relation (5) will follow from the discussion of orthonormal families of functions at the end of §5.1, provided the orthonormal family $(\deg \pi^\lambda)^{1/2}\pi^\lambda_{ij}$ of irreducible matrix coefficients is complete. For this it suffices to show that every continuous function on G can be uniformly approximated by linear combinations of the π^λ_{ij}, or by linear combinations of matrix coefficients $a \to \langle \pi(a)x, y\rangle$ of arbitrary finite-dimensional representations π of G; and by Weierstrass's Approximation Theorem it suffices to show that every polynomial function f on G is a linear combination of this form. But this is clear: $f(a) = \langle \pi(a)f, e\rangle$ where π is the representation on restrictions to G of polynomials of degree d (any $d \geq$ degree of f) by right translation $[\pi(a)g(b) = g(ba)]$ and e is the linear functional given by $\langle g, e\rangle = g(1)$. That the orthonormal family of irreducible matrix coefficients is countable is also clear now from the argument with polynomials: for each d, the space of restrictions to G of homogeneous polynomials of degree d is finite dimensional. QED

Comment. The Peter-Weyl theorem as stated above holds for arbitrary compact groups (not necessarily linear groups, nor even necessarily Lie groups), but is much harder to prove in the general case. For linear groups as here, the completeness assertion (which is the heart of the theorem) is a corollary of Weierstrass's Approximation Theorem as we have seen. The remaining assertion is a corollary of Schur's Orthogonality Relations. In its general form the theorem may be used to prove that *every compact Lie group is isomorphic to a linear group.*

One may re-formulate the above theorem without reference to an explicit unitary matrix realization of the irreducible representations π^λ using the following lemma.

Lemma 9. *Let V, π be a representation of G. Let*

(a) $M(V) = \{$the space of functions f on G spanned by the matrix coefficients $a \to \langle \pi(a)x, y\rangle, x \in V, y \in V^*\}$ with $(G \times G)$-action

$$(s,t)\cdot f(a) = f(t^{-1}as)$$

(b) $L(V) = \{$the space of all linear transformations A of $V\}$ with $(G \times G)$-action

$$(s,t)A = \pi(s)A\pi(t^{-1})$$

(c) $V \otimes V^* = \{$the tensor product of V and $V^*\}$, as usual, with $(G\times G)$-action

$$(s,t)\cdot(x \otimes y) = (\pi(s)x) \otimes (\check{\pi}(t)y).$$

These three representations of $G \times G$ are equivalent.

Proof. The isomorphisms are:

(c) $\to$ (a): $(x \otimes y) \to$ (matrix coefficient $c_{x,y}(a) = \langle \pi(a)x, y\rangle$).

(c) $\to$ (b): $(x \otimes y) \to$ (linear transformation $C_{x,y}(z) = \langle z, y\rangle x$). QED

Theorem 8′ (Peter–Weyl Theorem, Re-formulated) *$L^2(G)$ is an orthogonal, L^2-sum of the subspaces $M(V^\lambda)$ spanned by the matrix coefficients of the irreducible representations π^λ of G:*

$$L^2(G) = \oplus_\lambda M(V^\lambda).$$

These subspaces $M(V^\lambda)$ are invariant under the left and right regular representation of G, and the resulting representation of $G \times G$ on $M(V^\lambda)$ is irreducible and isomorphic with the representation $\pi^\lambda \otimes (\pi^\lambda)^\vee$ of $G \times G$ on $V^\lambda \otimes (V^\lambda)^*$.

Explanation and proof. The term *orthogonal* L^2-sum means that the subspaces $M(V^\lambda)$ of $L^2(G)$ are mutually orthogonal and that their span is L^2-dense in $L^2(G)$ in the sense that every $f \in L^2(G)$ admits an L^2-convergent expansion $f = \sum_\lambda f_\lambda$ with $f_\lambda \in M(V^\lambda)$. This is a reformulation of Theorem 8. The irreducibility of the $M(V^\lambda) \approx V^\lambda \otimes (V^\lambda)$ will be proved later. (Proposition 10, §6.3.). QED

Example 10. If one specializes the Fourier expansion (5) to the n-torus $\mathbb{T}$ of Example 3 one gets the classical expansion of a periodic L^2-function $f(t) = f(\theta_1, \ldots, \theta_n)$ of n variables into an L^2-convergent multiple Fourier series:

$$f(\theta_1, \ldots, \theta_n) = \sum_{(l_1,\ldots,l_n)} c_{l_1,\ldots,l_n} e^{2\pi i(l_1\theta_1+\cdots+l_n\theta_n)}$$

with

$$c_{l_1,\ldots,l_n} = \int_0^1 \cdots \int_0^1 f(\theta_1, \ldots, \theta_n) e^{-2\pi i(l_1\theta_1+\cdots l_n\theta_n)} d\theta_1 \ldots d\theta_n.$$

Problems for §6.2

1. *Matrix coefficients.* Let G be a compact matrix group, f a complex-valued function on G. Show that the following statements are equivalent.

   (a) The left translates of f, $L(a)f$, $a \in G$, span a finite-dimensional space. (Example 1, §6.1.)

   (b) The right translates of f, $R(a)f$, $a \in G$, span a finite-dimensional space.

   (c) f is a matrix-coefficient of a finite-dimensional representation V, π of G (i.e. $f(a) = \langle \pi(a)x, y \rangle$, $x \in V$, $y \in V^*$).

   (d) f is a polynomial function on G.

   The statement (d) means $f(a)$ may be written as a polynomial in the real and imaginary parts of the matrix coefficients of a. Show further that $f(a)$ may be written as a polynomial in the matrix coefficients of a and of a^{-1}, hence also as a polynomial in the matrix coefficients of a and in $\det(a)$.

   Notation and terminology. Such functions f will be called *matrix coefficients* of G and their totality denoted $M(G)$.

2. *Convolution.* (a) Let G be a *finite* group. The *group algebra* of G, denoted $\mathbb{C}[G]$, consists of all formal sums $\sum_{a\in G} f(a)a$ of elements of G with coefficients $f(a) \in \mathbb{C}$. Such sums may be added and multiplied in an obvious way. Show that

$$\left(\sum_{a\in G} f(a)a\right)\left(\sum_{a\in G} g(a)a\right) = \sum_{a\in G} f * g(a)a,$$

where

$$f * g(a) = \sum_{b\in G} f(ab^{-1})g(b).$$

[Thus $\mathbb{C}[G]$ may be identified with the algebra of all functions on G under the *convolution product* $f * g$. Finite G should be kept in mind as a special case of compact G, discussed below.]

(b) Let G be a *compact* matrix group. Define the *convolution product* of two complex-valued functions f, g on G by

$$f * g(a) = \int_G f(ab^{-1})g(b)\, db$$

whenever the integral exists (as it does, for example, when f and g are continuous). The invariant integral on G need not be normalized here.

Show that

$$(f * g) * h = f * (g * h)$$

for all $f, g, h \in C(G) = \{\text{continuous functions } G \to \mathbb{C}\}$. [Thus $C(G)$ forms an algebra over $\mathbb{C}$: the remaining algebra axioms are obviously satisfied. This algebra is *not* commutative, unless G is, and does *not* have a unit element, unless G is finite.]

(c) For any finite-dimensional representation V, π of G and any $f \in C(G)$ define an operator $\pi(f)$ on V by the formula

$$\pi(f) = \int_G f(a)\pi(a)\, da.$$

The integral may be interpreted coefficientwise with respect to a matrix representation of the $\pi(a)$.

Show that

$$\pi(f * g) = \pi(f)\pi(g).$$

[One says: $f \to \pi(f)$ is a *representation* of the algebra $C(G)$.]

(d) Show that the space $M(G)$ of matrix coefficients on G (see problem 1) forms a subalgebra of $C(G), *$, i.e if $f, g \in M(G)$ then $f * g \in M(G)$.

3. *Structure of the group algebra.* Let A be an associative algebra. An element $e \in A$ is an *idempotent* if $e^2 = e$. Two idempotents e, e' are *orthogonal idempotents* if $ee' = e'e = 0$. An idempotent is *primitive idempotent* if it is not the sum of two non-zero orthogonal idempotents. A (possibly infinite) set $\{e^\lambda\}_\lambda$ of orthogonal idempotents is *complete* if for any $a \in A$ one has $a = \sum_\lambda ae^\lambda = \sum_\lambda e^\lambda a$ with the understanding that only finitely many terms is these sums are non-zero. (When A has a unit element 1 this is equivalent to $1 = \sum_\lambda e^\lambda$.) An idempotent e is *central* if $ae = ea$ for all $a \in A$.

Notation. G: a compact matrix group;

$|G| = \int_G dg$;

$M(G)$: the algebra of matrix coefficients of G (with the convolution product);

V^λ, π^λ: a complete set of inequivalent, irreducible, unitary representations of G;

$\chi^\lambda(a) = \operatorname{tr} \pi^\lambda(a)$;

$\{e_j^\lambda \mid j = 1, 2, \dots, d_\lambda = \deg \pi^\lambda\}$: an orthonormal basis for V^λ;

π_{jk}^λ the matrix coefficient defined by $\pi^\lambda(a)x_k^\lambda = \sum_j \pi_{jk}^\lambda(a)e_j^\lambda$; $M_d(\mathbb{C})$: the algebra of all complex $d \times d$ matrices;

$\sum_\lambda M_{d_\lambda}(\mathbb{C})$: a direct sum of algebras; its elements are finite direct sums $\sum_\lambda m_\lambda$ of matrices $m_\lambda \in M_{d_\lambda}$. (One may think of $\sum_\lambda m_\lambda$ either as a formal sum, or as a block-diagonal matrix with diagonal blocks m_λ; there may be infinitely many blocks, but only finitely m_λ can be non-zero.)

(a) Show that the map $M(G) \to \sum_\lambda M_{d_\lambda}(\mathbb{C})$, $f \to \sum_\lambda \pi^\lambda(f)$, is an isomorphism of algebras. [Suggestion: Fourier expansion.]

(b) Under the isomorphism in (a), the element $e_{jk}^\lambda \in M(G)$ defined by

$$e_{jk}^\lambda(a) = \frac{d_\lambda}{|G|}\pi_{kj}^\lambda(a^{-1})$$

corresponds to the jk-matrix unit in $M_{d_\lambda}(\mathbb{C})$, i.e.

$$\pi^\lambda(e_{jk}^\lambda)e_l^\lambda = \delta_{kl}e_j^\lambda.$$

[Suggestion: Schur Orthogonality.]

(c) The elements e_{jj}^λ form a complete set of orthogonal primitive idempotents. [Suggestion: $M(G) \approx \sum_\lambda M_{d_\lambda}(\mathbb{C})$.]

(d) Every primitive idempotent of $M(G)$ is of the form $e = e_{jj}^\lambda$ for some λ and j and some basis $\{e_j^\lambda\}$ of V^λ, which may depend on e. [Suggestion: Pick a basis $\{e_j^\lambda\}$ for each V^λ which diagonalizes $\pi^\lambda(e)$. Why is this possible?]

(e) Let $L_j^\lambda = M(G) * e_{jj}^\lambda$ be the left ideal generated by e_{jj}^λ. Show that the map

$$L_j^\lambda \to V^\lambda, \quad f \to \pi^\lambda(f)e_j^\lambda$$

is a G-isomorphism (with the left regular representation of G on L_j^λ). Conclude that the L_j^λ are minimal left ideals.

(f) Under the isomorphism $M(G) \approx \sum_\lambda M_{d_\lambda}(\mathbb{C})$ of (a), the element e^λ of $M(G)$ defined by

$$e^\lambda = \frac{d_\lambda}{|G|}\bar{\chi}^\lambda$$

corresponds to the identity matrix in the λ-block (and the zero matrix elsewhere). [Suggestion: use part (a).]

(g) The elements e^λ form a complete set of orthogonal, central idempotents of $M(G)$. [Suggestion: $M(G) \approx \sum_\lambda M_{d_\lambda}(\mathbb{C})$.]

(h) Let $M^\lambda = M(G)^* e^\lambda = e^{\lambda *} M(G)$, the ideal generated by e^λ. Show that the map

$$M^\lambda \to M_{d_\lambda}(\mathbb{C}) f \to \pi^\lambda(f)$$

is a $G \times G$ isomorphism. ($G \times G$ acts on M^λ by the *biregular* representation $\beta(a,b)f(c) = f(a^{-1}cb)$ and on $M_{d_\lambda}(\mathbb{C})$ by $\theta^\lambda(a,b)m = \pi^\lambda(a)m\pi^\lambda(b^{-1})$.)

6.3 Characters

Let π be a representation of a group G. The *character* of π is the function $\chi = \chi_\pi$ on G defined by

$$\chi(a) = \operatorname{tr} \pi(a).$$

χ is said to be *irreducible* if π is.

The first thing to notice is that χ is invariant under conjugation:

$$\chi(cac^{-1}) = \chi(a).$$

Any function f on G with this property is called a *class function* since $f(a)$ is only a function of the conjugacy class of a. Next, observe that equivalent representations have the same character (they have the same matrices with respect to suitable bases). Equally evident are the following properties of characters.

Lemma 1.

(a) $\chi_{\pi\oplus\rho}(a) = \chi_\pi(a) + \chi_\rho(a)$,
(b) $\chi_{\pi\otimes\rho}(a) = \chi_\pi(a)\chi_\rho(a)$,
(c) $\chi_{\check{\pi}}(a) = \chi_\pi(a^{-1})$, *and if* π *is unitary, then also* $\chi_{\check{\pi}}(a) = \overline{\chi_\pi(a)}$.

Proof.

(a) Remark 3(c), §6.1,
(b) Remark 9, §6.1,
(c) Remark 7, §6.1. QED

Remark 2. If π and ρ are representations of two different groups G and H and if $\pi \otimes \rho$ is considered a representation of the product group $G \times H$, then (b) must be replaced by

$$\chi_{\pi\otimes\rho}(a, b) = \chi_{\pi}(a)\chi_{\rho}(b).$$

For the remainder of this section G will denote a *compact linear group*. As previously, the inner product of two L^2-functions f and g on G is

$$(f, g) = \frac{1}{|G|} \int_G f(a)\bar{g}(a)\, da.$$

Theorem 3 (Schur's Orthogonality Relations for Characters). *Let π and ρ be two irreducible representations of G. Then*

$$(\chi_{\pi}, \chi_{\rho}) = \begin{cases} 0 & \text{if } \pi \not\approx \rho, \\ 1 & \text{if } \pi \approx \rho. \end{cases}$$

Proof. Using matrices:

$$\begin{aligned}(\chi_{\pi}, \chi_{\rho}) &= \sum_{ij} (\pi_{ii}(a), \rho_{jj}(a)) \\ &= \begin{cases} 0 & \text{if } \pi \not\approx \rho \\ \sum_i \dfrac{1}{\deg(\pi)} = 1 & \text{if } \pi = \rho, \end{cases}\end{aligned}$$

by the orthogonality relations for matrix coefficients. (When $\pi \approx \rho$ we may as well assume $\pi = \rho$ here, since equivalent representations have the same character.) QED

Corollary 4. *Let $V \approx m_1V_1 \oplus \cdots \oplus m_kV_k$ be a decomposition of V into multiples of inequivalent irreducibles V_i. ($m_iV_i = V_i \oplus \cdots \oplus V_i, m_i$ copies). Then the characters satisfy*

$$\chi = m_1\chi_1 + \cdots + m_k\chi_k$$

and the multiplicities m_i *are given by*

$$m_i = (\chi, \chi_i).$$

Proof. This is clear from the orthogonality relations for characters. QED

Corollary 5. *Two representations are equivalent if and only if they have the same character.*

Proof. 'Only if' we know, 'if' by Corollary 4. QED

Corollary 6 (Irreducibility Criterion). *A character χ is irreducible if and only if $(\chi, \chi) = 1$.*

Proof. Immediate from Corollary 4. QED

Let V, π be a representation of G. For any integrable function f on G, let

$$\pi(f) = \frac{1}{|G|}\int_G f(a)\pi(a)\,da.$$

Lemma 7. *Suppose f is a class-function and π is irreducible. Then $\pi(f) = \alpha 1$ where $\alpha = 1/\deg(\pi)(f, \bar{\chi})$.*

Proof. $\pi(f)$ is a G-map:

$$\begin{aligned}\pi(b)\pi(f)\pi(b^{-1}) &= \frac{1}{|G|}\int_G f(a)\pi(bab^{-1})\,da \\ &= \frac{1}{|G|}\int_G f(a)\pi(a)\,da \\ &\quad [\text{replace } a \text{ by } bab^{-1} \text{ and use } f(bab^{-1}) = f(a)] \\ &= \pi(f).\end{aligned}$$

Therefore, $\pi(f) = \alpha 1$, by Schur's Lemma. Then $\operatorname{tr}\pi(f) = \alpha \deg(\pi)$, which gives

$$\begin{aligned}\alpha &= \frac{1}{\deg(\pi)}\frac{1}{|G|}\int_G f(a)\chi(a)\,da \\ &= \frac{1}{\deg(\pi)}(f, \bar{\chi}).\end{aligned}$$

QED

Theorem 8 (Fourier Expansion of Class Functions). *Let f be an L^2-class function on G. Then*

$$f \equiv \sum_\lambda (f, \chi^\lambda)\chi^\lambda$$

(L^2-convergent series). The irreducible characters χ^λ form an orthonormal L^2-basis for the space of L^2-class-functions on G.

Proof. Start with the Fourier expansion

$$f \equiv \sum_{\lambda ij} \deg(\pi^\lambda)\left(\frac{1}{|G|}\int_G f(a)\overline{\pi_{ij}^\lambda(a)}\,da\right)\pi_{ij}^\lambda.$$

Since $\overline{\pi_{ij}(a)} = \pi_{ji}(a^{-1})$, by unitarity, this may be written as

$$f \equiv \sum_\lambda \deg(\pi^\lambda)\operatorname{tr}\left\{\left(\frac{1}{|G|}\int_G f(a)\pi^\lambda(a^{-1})da\right)\pi^\lambda\right\}.$$

When f is a class function, the last lemma gives (after replacing a by a^{-1}):

$$\frac{1}{|G|}\int_G f(a)\pi^\lambda(a^{-1})\,da = \frac{1}{\deg(\pi^\lambda)|G|}\int_G f(a^{-1})\chi^\lambda(a)\,da\cdot 1$$
$$= \frac{1}{\deg(\pi^\lambda)}(f,\chi^\lambda)\cdot 1,$$

from which the given expansion for f follows. The second assertion is a restatement of the first. QED

Consider once more the decomposition

$$V \approx m_1V_1 \oplus \cdots \oplus m_kV_k$$

into multiples of irreducibles. Here $\approx$ means 'G-equivalent'. If one wants to replace $\approx$ by $=$, one must write

$$V = \left[V_1^{(1)} + V_1^{(2)} + \cdots + V_1^{(m_1)}\right] + \cdots + \left[V_k^{(1)} + V_k^{(2)} + \cdots + V_k^{(m_k)}\right]$$

where each $V_1^{(1)}, V_1^{(2)}, \ldots$, etc., is a subspace of V G-isomorphic with V_1. The individual subspaces $V_1^{(1)}$, $V_1^{(2)}, \ldots$, etc are *not* unique. But their sum $[V_1^{(1)} + V_1^{(2)} + \cdots + V_1^{(m_1)}]$ is unique. This will follow from the following proposition.

Proposition 9. *Let $V = \bigoplus_\rho V(\rho)$ be a decomposition of V into G-invariant subspaces $V(\rho)$ so that the subrepresentation of G on $V(\rho)$ is equivalent with a multiple of irreducible representation ρ, the ρ's being inequivalent. Then the projection of V onto the summand $V(\rho)$ is given by*

$$E_\rho = \frac{\deg(\rho)}{|G|}\int_G \bar\chi_\rho(a)\pi(a)\,da = \deg(\rho)\pi(\bar\chi_\rho).$$

Proof. By Lemma 7 above and the orthogonality of irreducible characters, $\pi(\bar\chi_\rho)$ acts by the scalar $1/\deg(\rho)$ in the irreducible representation ρ and by 0 in any other irreducible representation. Hence $\pi(\bar\chi_\rho)$ acts equally by the scalar $1/\deg(\rho)$ in the multiple $V(\rho)$ of ρ and by 0 in a multiple of any other irreducible representation. QED

The subspace $V(\rho)$ of V is called the *ρ-isotypic* (or *ρ-primary*) component of V. $V(\rho)$ may also be characterized as the smallest subspace that contains all the irreducible subspaces of type ρ. Each $V(\rho)$ decomposes further into a direct sum of irreducible subspaces of type ρ, but these are not unique, unless ρ occurs with multiplicity one. The *multiplicity* of ρ in π, denoted $[\pi:\rho]$, is of course the number of irreducible subspaces in such a decomposition. This is also the multiplicity of the irreducible character χ_ρ in the character χ of π, given by $[\pi:\rho] = (\chi_\pi, \chi_\rho)$.

We now discuss the representations of the direct product $G \times H$ of two compact linear groups. Recall that for a representation V, π of G and a representation W, ρ of H we have a representation $V \otimes W$, $\pi \otimes \rho$ of $G \times H$ defined by $\pi \otimes \rho(a,b) = \pi(a) \otimes \rho(b)$.

Proposition 10. *The irreducible representations of $G \times H$ are precisely the representations $\pi \otimes \rho$ where π is an irreducible representation of G and ρ is an irreducible representation of H.*

Proof. Let π be an irreducible representation of G, ρ that of H. Then

$$\frac{1}{|G|}\int_G |\chi_\pi|^2 = 1 \quad \text{and} \quad \frac{1}{|H|}\int_H |\chi_\rho|^2 = 1.$$

Therefore, also

$$\frac{1}{|G\times H|}\int_{G\times H} |\chi_{\pi\otimes\rho}|^2 = \frac{1}{|G||H|}\int_G\int_H |\chi_\pi(a)|^2|\chi_\rho(b)|^2\, da\, db = 1.$$

This shows that $\pi \otimes \rho$ is irreducible as well.

To show that every irreducible character of $G \times H$ is of this form, it suffices to show that there is no continuous class-function f on $G \times H$ which is L^2-orthogonal to all these $\chi_\pi \otimes \chi_\rho$, except $f \equiv 0$ (because any other irreducible character would have this property). To see this, note that

$$\frac{1}{|G\times H|}\int_{G\times H} f(a,b)\bar{\chi}_{\pi\otimes\rho}(a,b)\, da\, db = \frac{1}{|G|}\int_G \left\{\frac{1}{|H|}\int_H f(a,b)\bar{\chi}_\rho(b) db\right\}\bar{\chi}_\pi(a)\, da.$$

If this equals 0 for all irreducible π, ρ, then the inner integral must be equal to 0 for all $a \in G$ (because of the completeness of the χ_π), and then $f(a,b)$ must be 0 for all $a \in G$, $b \in H$ (because of the completeness of the χ_ρ). QED

Corollary 11 (Burnside's Theorem). *Let V, π be an irreducible representation of G. Then every linear transformation of V is of the form*

(a) $\sum_i \alpha_i \pi(a_i)$ *(finite sum) for some* $a_i \in G$, $\alpha_i \in \mathbb{C}$,

(b) $\pi(f) = \frac{1}{G}\int_G f(a)\pi(a)\, da$ *for some continuous function f on G (which may even be taken a linear combination of matrix coefficients of π).*

Proof. By the proposition, the representation of $G \times G$ on the space $L(V) \approx V \otimes V^*$ of linear transformations of V is irreducible. The subspaces described in (a) and (b) are $G \times G$-invariant, hence all of $L(V)$. QED

Problems for §6.3

1. Let G be a *finite* group.

   (a) Show that the number of irreducible characters of G equals the number of conjugacy classes of G.

(b) The characters of the left and right regular representation of G (Example 1, §6.1) are both given by

$$\chi(a) = \begin{cases} \frac{1}{|G|} & \text{if } a = 1, \\ 0 & \text{otherwise.} \end{cases}$$

Prove and explain.

(c) Suppose G acts on the finite set X. Let π be the associated permutation representation on the space $\mathbb{C}^X$ of all complex-valued function on X given by $\pi(a)f(x) = f(a^{-1}x)$. Show that the character of π is given by

$$\chi(a) = |X^a|,$$

where $X^a = \{x \in X \mid a \cdot x = x\}$ is the fixed-point set of a in X.

2. Let G be a compact matrix group, π a representation of G, χ its character. Show that $|\chi(a)| \leq \chi(1)$ with equality only if $\pi(a)$ is a scalar.

3. Let G be a compact matrix group, V, π a representation of G, χ its character. Let $\mathrm{Sym}^2 V$ (resp. $\mathrm{Alt}^2 V$) be the space of *symmetric* (resp. *alternating*) tensors of degree 2 on V, i.e. the subspace of $V \otimes V$ spanned by tensors $x \otimes y + y \otimes x$ (resp. $x \otimes y - y \otimes x$). Let $\mathrm{Sym}^2\chi$ (resp. $\mathrm{Alt}^2\chi$) be the character of the representation of G on $\mathrm{Sym}^2 V$ (resp. $\mathrm{Alt}^2 V$).

(a) Show that

$$\mathrm{Sym}^2\chi(a) = \tfrac{1}{2}(\chi(a)^2 + \chi(a^2)),$$
$$\mathrm{Alt}^2\chi(a) = \tfrac{1}{2}(\chi(a)^2 - \chi(a^2)).$$

(b) Deduce from (a) that $V \otimes V = \mathrm{Sym}^2 V \oplus \mathrm{Alt}^2 V$; prove this directly.

4. Consider the representation ρ of $\mathrm{GL}(n, \mathbb{C}) \times \mathrm{GL}(n, \mathbb{C})$ on the space $R(M_n(\mathbb{C}))$ of holomorphic polynomial functions on $M_n(\mathbb{C})$ given by $\rho(a, b)f(z) = f(a^{-1}zb)$. ('Holomorphic polynomial' means $f(z)$ is a polynomial in the matrix entries of $z \in M_n(\mathbb{C})$.) Show that the character of ρ is given by the formula

$$\mathrm{ch}\,\rho(a^{-1}, b) = \frac{1}{\det(1 - a \otimes b)}$$

in the following sense. The character of the representation on the finite-dimensional subspace $R^m(M_n(\mathbb{C}))$ of homogeneous polynomials of degree m is a homogeneous polynomial of degree m in the matrix entries of (a, b). The formula is understood in the sense that this homogeneous part of degree m of $\mathrm{ch}\,\rho(a, b)$ equals the homogeneous part of degree m on the right side, when the right side is written as a power series in the matrix entries of (a, b), convergent for $\|a\|, \|b\|$ sufficiently small. $a \otimes b$ is the tensor product of matrices, as in Remark 9, §6.1. [Suggestion: it suffices to consider diagonal a, b.]

5. (a) Show that the irreducible characters of $\mathbb{Z}_n = \mathbb{Z}/n\mathbb{Z}$ are given by

$$\chi_b(a) = e^{2\pi i ab/n}$$

with $b \in \mathbb{Z}_n$.

(b) Let p be a prime. For $m \in \mathbb{Z}_p$ set

$$\left(\frac{m}{p}\right) = \begin{cases} 1 & \text{if } m \neq 0 \text{ is a square in } \mathbb{Z}_p, \\ -1 & \text{if } m \neq 0 \text{ is not a square in } \mathbb{Z}_p, \\ 0 & \text{if } m = 0 \text{ in } \mathbb{Z}_p. \end{cases}$$

Show that for all $x \in \mathbb{Z}_p$,

$$\sum_{m \in \mathbb{Z}_p} \left(\frac{m}{p}\right) e^{2\pi i m x/p} = \sum_{m \in \mathbb{Z}_p} e^{2\pi i m^2 x/p}$$

(These sums are called *Gauss sums.*)

6. *Early history of character theory.* Characters of finite commutative groups (defined as systems of roots of unity depending multiplicatively on the group elements) were used in an essential way in number theory already by Gauss, and later especially by Dirichlet and Dedekind. The theory of characters of non-commutative groups began with a letter from Dedekind to Frobenius, mentioned by Frobenius in his paper '*Über Gruppencharactere*' of 1896:

> 'In April of this year Dedekind communicated to me a problem, which he came upon in the year 1880, and which in his opinion should interest me, belonging both to the theory of groups and of determinants, while he himself would be taken too far from his arithmetical investigations by its further consideration. Its solution, which I hope to report on shortly, lead me to a generalization of the notion of character to arbitrary finite groups.'

Namely, according to Frobenius, a character of a finite group H (Frobenius's notation) is an assignment of a complex number $\chi(R)$ to each element of R of h satisfying the following conditions.

(1) $\chi(E) = f$, a positive integer;

(2) $\chi(AB) = \chi(BA)$ for all elements A, B of H;

(3) $\mathsf{H}\chi(A)\chi(B) = f \sum_R \chi(AR^{-1}BR)$, h the order of H;

(4) $h = \sum_R \chi(R)\chi(R^{-1})$.

Dedekind's problem concerns '*Die Primfactoren der Gruppendeterminante*' (Frobenius, 1896). The determinant is

$$\Theta = \det(x_{PQ^{-1}}),$$

a polynomial in independent variables x_R indexed by the elements R of the finite group H. Frobenius shows that Θ factors as

$$\Theta = \prod \Phi^f,$$

the number of *prime factors* (irreducible polynomials) Φ being the number of conjugacy classes. The exponent f is at the same time the degree of the polynomial Φ (whence the expression *degree of a character*), and

$$\sum f^2 = h,$$

the order of H. If one sets equal the variables x_R for which R belongs to same conjugacy class (or, what amounts to the same thing, restricts the variables by $x_{AB} = x_{BA}$), then the prime factors Φ become powers of linear forms ξ,

$$\Phi = \xi^f.$$

The linear forms $f\Phi$ have Frobenius's characters χ as coefficients:

$$\xi = \frac{1}{f}\sum_R \chi(R)x_R,$$

always under the restriction $x_{AB} = x_{BA}$.

So much for the genesis of character theory. One will appreciate that these results must have appeared wonderful to Frobenius's contemporaries, especially to Dedekind, who had worked them out for the symmetric group of order six and the quaternion group of order eight.

Problems: (a) Show that Frobenius's characters χ of a finite group H are exactly the irreducible characters of H.

(b) Prove Frobenius's results on the prime factors of the group determinant. [Suggestion: Think of the x_R as coordinates on the group algebra $\mathbb{C}[\mathsf{H}]$, so that $x = \sum_R x_R R$ represents the 'general' element of $\mathbb{C}[\mathsf{H}]$. Consider left multiplication by x as linear transformation of $\mathbb{C}[\mathsf{H}]$; use problem 3, §6.2. It will be necessary to know that $\det(X_{ij})$ is an irreducible polynomial in the variables X_{ij}. To prove this, suppose $\sum \operatorname{sgn}(\sigma)X_{1\sigma(1)}\cdots X_{n\sigma(n)} = f(X)g(X)$, argue that each X_{ij} must occur in exacly one of the factors on the right, and deduce that one factor must contain all X_{ij}'s.]

6.4 Weyl's character formula for $\mathrm{U}(n)$

In three famous papers of 1925–1926, entitled *Theorie der Darstellung kontinuielicher halb-einfacher Gruppen durch lineare Transformationen*, Weyl determined the irreducible characters of the compact semisimple groups, describing along the way the structure of these groups and of their representations. The representations of semisimple groups had been studied previously by E. Cartan

(1913) from an algebraic view point, where the Lie algebra plays the main rôle. Cartan gave a complete classification of the irreducible representations (even though his arguments were incomplete at some points, as Weyl remarked), without producing a formula for their characters or their degrees. Weyl's point of view is global: he operates with the representations of the groups themselves, rather than with those of their Lie algebras. An essential tool in Weyl's approach is integration over the group, a method which had long been familiar for finite groups from the work of Frobenius and Schur, extended to compact groups by Hurwitz (1897) to treat a problem in invariant theory, and applied by Schur (1924, one year before Weyl) to the orthogonal groups to prove the character formula for that case. We shall illustrate the method for the group U(n), following Weyl. We proceed by some simple observations.

Observation 1. A character χ of U(n) is uniquely determined by its restriction to the diagonal subgroup

$$T = \{t = (\epsilon_1, \ldots, \epsilon_n) \mid \epsilon_j = e^{2\pi i \theta_j}\}.$$

Explanation. Every element of U(n) is conjugate to an element of T. Characters are class functions.

Observation 2. On T, χ is of the form $\chi = \sum_\lambda m_\lambda e^\lambda$, finite sum over certain $\lambda = (l_1, \ldots, l_n)$ with $l_j \in \mathbb{Z}$, $e^\lambda(t) = \epsilon_1^{l_1} \ldots \epsilon_n^{l_n} = e^{2\pi i(l_1\theta_1 + \cdots + l_n\theta_n)}$, m_λ nonnegative integral.

Explanation. Since T is abelian, its irreducible characters are one-dimensional and therefore of the form e^λ. (Notation as in example 3, §6.2.) A representation V, π of U(n) may be decomposed under T into multiples of one-dimensional subspaces:

$$V \approx \oplus_\lambda V_\lambda;$$

$V_\lambda \approx m_\lambda \mathbb{C}_\lambda$ consists of $m_\lambda = \dim V_\lambda$ copies of $\mathbb{C}_\lambda = \{\mathbb{C} \text{ with } T \text{ acting by } e^\lambda\}$. m_λ is called the multiplicity of the weight λ in the representation (or in its character χ). Thus, on T, $\chi = \sum_\lambda m_\lambda e^\lambda$.

Observation 3. On T, χ is *symmetric* under the Weyl group $W = S_n$, i.e. $\chi(sts^{-1}) = \chi(t)$ for all $s \in W$, $t \in T$. Hence, its coefficients m_λ may be assumed to satisfy $m_{s\cdot\lambda} = m_\lambda$ for all $s \in W$. Here $s \in W \approx S_n$ operates on the n-tuples λ by permuting the components:

$$s \cdot (l_1, \ldots, l_n) = (l_{s^{-1}(1)}, \ldots, l_{s^{-1}(n)}).$$

Explanation. χ is a class function.

Observation 4. If χ_1 and χ_2 are two irreducible characters, then

$$\frac{1}{|G|}\int_G \chi_1\bar{\chi}_2 = \frac{1}{|W|}\int_T \chi_1\bar{\chi}_2\Delta\bar{\Delta}$$
$$= \begin{cases} 1 & \text{if } \chi_1 = \chi_2, \\ 0 & \text{if } \chi_1 \neq \chi_2, \end{cases}$$

where $\Delta = \prod_{j<k}(\epsilon_j - \epsilon_k)$.

Explanation. From Schur's orthogonality relations for irreducible characters:

$$\frac{1}{|G|}\int\limits_G \chi_1\bar{\chi}_2 = \begin{cases} 1 & \text{if } \chi_1 = \chi_2, \\ 0 & \text{if } \chi_1 \neq \chi_2. \end{cases}$$

From Weyl's integration formula

$$\int\limits_G \chi_1\bar{\chi}_2 = \frac{1}{|W|}\int\limits_T \chi_1\bar{\chi}_2\Delta\bar{\Delta}.$$

In view of Observation 1 we may identify characters χ of $\mathrm{U}(n)$ with its restriction to T.

With any character χ associate the function $\xi = \chi\Delta$ on T. Such a ξ is *antisymmetric* under W, i.e. $\xi(sts^{-1}) = \mathrm{sgn}(s)\xi(t)$, because Δ is. In view of Observation 4, the ξ-functions ξ_1, ξ_2 belonging to two irreducible characters χ_1, χ_2 satisfy.

Observation 5.

$$\frac{1}{|W|}\int\limits_T \xi_1\bar{\xi}_2 = \begin{cases} 1 & \text{if } \xi_1 = \xi_2, \\ 0 & \text{if } \xi_1 \neq \xi_2. \end{cases}$$

Any $\lambda = (l_1, \ldots, l_n)$, $l_j \in \mathbb{Z}$, is called a *weight.* We order them *lexicographically*:

> λ is *lexicographically higher* than μ if the first component l_j of λ that differs from the corresponding component m_j of μ satisfies $l_j > m_j$.

Observation 6. Any weight is W-conjugate to a unique weight λ satisfying $l_1 \geq l_2 \geq \cdots \geq l_n$. Such a weight is called *dominant, strictly dominant* if the inequalities are all strict.

Explanation. This just says that the n-tuple λ may be rearranged decreasingly.

For any weight $\kappa = (k_1, \ldots, k_n)$, define an *elementary antisymmetric sum* ξ_κ:

$$\xi_\kappa = \sum_{s\in W} \mathrm{sgn}(s)\, e^{s\cdot\kappa} \tag{1}$$

One may write ξ_κ as a determinant

$$\xi_\kappa = \det(\epsilon_i^{k_j}) \tag{2}$$

since (1) is nothing but the definition of the determinant (2) in terms of an alternating sum. But it is usually preferable to use the more explicit form (1) for ξ_κ.

Recall that Δ is such an elementary antisymmetric sum (Vandermonde determinant; eqn (6), §5.3):

$$\Delta = \sum_{s\in W} \mathrm{sgn}(s)\, e^{s\cdot\rho}, \tag{3}$$

where ρ is the weight: $\rho = (n-1, \ldots, 1, 0)$.

Observation 7. Every elementary antisymmetric sum $\xi_\kappa(t)$ is W-antisymmetric both in t and in κ:

$$\xi_\kappa(sts^{-1}) = \mathrm{sgn}(s)\,\xi_\kappa(t)$$
$$\xi_{s\cdot\kappa}(t) = \mathrm{sgn}(s)\,\xi_\kappa(t).$$

Explanation. Clear.

So ξ_κ changes sign when two components of κ are interchanged, hence $= 0$ if two components of κ are equal. For these reasons we shall from now on consider only ξ_κ's corresponding to strictly dominant weights κ.

Observation 8.

$$\frac{1}{|W|}\int_T \xi_{\kappa_1}\bar{\xi}_{\kappa_2} = \begin{cases} 1 & \text{if } \kappa_1 = \kappa_2 \\ 0 & \text{if } \kappa_1 \neq \kappa_2. \end{cases}$$

Explanation. Expand the product

$$\xi_{\kappa_1}\bar{\xi}_{\kappa_2} = \sum_{s_1, s_2} \mathrm{sgn}(s_1)\,\mathrm{sgn}(s_2)\,e^{s_1\cdot\kappa_1}\,e^{-s_2\cdot\kappa_2}$$

and integrate over T using

$$\int_T e^{\kappa}e^{-\kappa'} = \int_0^1 \cdots \int_0^1 e^{2\pi i[(\kappa_1-\kappa_1')\theta_1+\cdots+(\kappa_n-\kappa_n')\theta_n]}d\theta_1 \ldots d\theta_n$$
$$= \begin{cases} 1 & \text{if } \kappa = \kappa, \\ 0 & \text{if } \kappa \neq \kappa'. \end{cases}$$

Comparison of Observations 5 and 8 suggests that the ξ_κ are precisely the ξ-functions of the irreducible characters χ. This is

Theorem 1 (Weyl's Character Formula for U(n)). *The irreducible characters of* U(n) *are precisely*

$$\chi_\lambda = \frac{1}{\Delta}\sum_{s\in W} \mathrm{sgn}(s)\,e^{s\cdot(\lambda+\rho)},$$

λ a dominant weight.

Proof. Let χ be an irreducible character of U(n), $\xi = \chi\Delta$ its ξ-function. Let $\lambda = (l_1, \ldots, l_n)$ be the lexicographically highest weight in χ, i.e.

$$\chi = m_\lambda e^\lambda + \text{lower terms}$$

with $m_\lambda \neq 0$ and 'lower terms' indicating terms $m_\mu e^\mu$ with μ lexicographically lower than λ. If one multiplies out the product $\xi = \chi\Delta$, using $\Delta = \sum_s \mathrm{sgn}(s)e^{s\cdot\rho}$ one sees that

$$\xi = m_\lambda e^{\lambda+\rho} + \text{lower terms}.$$

Since χ is symmetric, it contains all weights $\mu = s \cdot \lambda$ with the same multiplicity $m_\mu = m_\lambda$ as λ. Since λ is the highest among these, it must be dominant:

$$l_1 \geq l_2 \geq \cdots \geq l_n.$$

Adding to $\lambda = (l_1, \ldots, l_n)$ the weight $\rho = (n-1, \ldots, 0)$ we get a weight $\kappa = \lambda + \rho$ which is strictly dominant:

$$k_1 = l_1 + (n-1) > \cdots > k_n = l_n + 0.$$

Since ξ is antisymmetric it contains all terms $\mathrm{sgn}(s)e^{s \cdot \kappa}$ along with e^κ with the same multiplicity m_λ, which we now denote by m^κ when $\kappa = \lambda + \rho$. Grouping these terms together one gets

$$\xi = m^\kappa \xi_\kappa + \cdots$$

where the dots indicate terms lower than κ. Combining these terms again into elementary antisymmetric sums we get an expansion

$$\xi = m^\kappa \xi_\kappa + m^{\kappa'} \xi_{\kappa'} + \cdots .$$

The $m^\kappa, m^{\kappa'}, \ldots,$ are all integers (they are coefficients in $\chi\Delta$) and $m^\kappa > 0$, as we saw. From the orthogonality relations in Observations 5 and 8 we find that

$$1 = (m^\kappa)^2 + (m^{\kappa'})^2 + \cdots ,$$

which leaves us with $m^\kappa = 1$ and all other $m^{\kappa'} = 0$. Thus $\xi = \xi_\kappa$ and therefore $\chi = (1/\Delta)\xi_\kappa$. This proves that every irreducible character χ is of the form $\chi = \chi_\lambda$ where $\chi_\lambda = (1/\Delta)\xi_\kappa$ with $\kappa = \lambda + \rho$.

It remains to show that each $\chi_\lambda = (1/\Delta)\xi_\kappa$ $(\kappa = \lambda + \rho)$ is an irreducible character. This is a consequence of the completeness of the irreducible characters χ in the following way. A class function $f(a)$ on $\mathrm{U}(n)$ depends only on the eigenvalues of a and may therefore be considered as a symmetric function of the eigenvalues of a, i.e. as a W-invariant function on T. Considered as such, the L^2-inner product of class functions on $\mathrm{U}(n)$ is given by

$$\frac{1}{|W|} \int_T f_1 \bar{f}_2 \Delta \bar{\Delta}. \tag{4}$$

It follows that the ξ-functions $\xi = \chi\Delta$ corresponding to the irreducible characters of $\mathrm{U}(n)$ form an orthonormal basis with respect to the usual inner product

$$\frac{1}{|W|} \int_T g_1 \bar{g}_2 \tag{5}$$

on L^2-functions g on T of the form $g = f\Delta$ where f is W-invariant and L^2 with respect to the inner product (4). These g are just the W-anti-invariant L^2 functions for the usual inner product (5). As the ξ_κ are orthogonal among themselves, any ξ_κ which is not a ξ would have to be orthogonal to all the ξ. But this is impossible in view of the completeness of the ξ just mentioned. Thus every ξ_κ is a ξ. QED

In the course of the proof we have also shown that

$$\chi_\lambda = e^\lambda + \text{lower terms}.$$

λ is the lexicographically highest weight in χ_λ and occurs with multiplicity one. This gives:

Theorem 2 (Theorem of the Highest Weight for U(n)**).** *The highest weight of an irreducible representations of* U(n) *occurs with multiplicity one and determines the representation uniquely up to equivalence. These highest weights are exactly the dominant weights* $\lambda = (l_1, \ldots, l_n)$, $l_1 \geq l_2 \geq \cdots \geq l_n$.

Remark 3.

(a) There are two natural parameters for the irreducible characters, λ and κ, related by $\kappa = \lambda + \rho$. λ is called the *highest weight*, κ the *infinitesimal character*. λ is dominant, κ strictly dominant. Every strictly dominant weight κ is of the form $\kappa = \lambda + \rho$ for an (obviously unique) dominant weight λ.

(b) The proof of Weyl's character formula depends only on (1) Weyl's integration formula, (2) Schur's Orthogonality Relations for Characters, (3) completeness of characters. Nothing about the representations themselves is required.

(c) We remarked that the elementary antisymmetric sum $\xi_\kappa = \sum_s \operatorname{sgn}(s) e^{s\cdot\kappa}$ may be written as a determinant: $\xi_\kappa = \det(\epsilon_j^{k_k})$. For these determinants Weyl uses the notation $\xi_\kappa = |\epsilon^{k_1} \cdots \epsilon^{k_n}|$. With this notation Weyl's character formula reads

$$\chi_\lambda = \frac{|\epsilon^{k_1} \ldots \epsilon^{k_n}|}{|\epsilon^{n-1} \ldots \epsilon^0|}$$

with $\kappa = \lambda + \rho$.

A corollary of Weyl's character formula is:

Theorem 4 (Weyl's Dimension Formula). *The degree of the irreducible character* χ_λ *is*

$$\frac{\prod_{i<j}(k_i - k_j)}{\prod_{i<j}(j - i)}.$$

where $\kappa = \lambda + \rho = (k_1, \ldots, k_n)$.

Proof. The degree of χ is simply $\chi(1)$. However, one cannot immediately set $\epsilon_j = 1$ in Weyl's formula, as that would give $0/0$. Rather, one must take the limit as all $\epsilon_j \to 1$. To do this, choose special $\bar{\epsilon}_j s$, namely set $\epsilon_j = \epsilon^j = e^{2\pi i j\theta}$ and let $\theta \to 0$. Then the numerator of Weyl's formula also becomes a Vandermonde determinant:

$$|\epsilon^{k_1}\dots\epsilon^{k_n}| = \begin{vmatrix} \epsilon^{k_1} & \epsilon^{k_2} & \dots & \epsilon^{k_n} \\ \epsilon^{2k_1} & \epsilon^{2k_2} & \dots & \epsilon^{2k_n} \\ \vdots & \vdots & \dots & \vdots \\ \epsilon^{nk_1} & \epsilon^{2k_2} & \dots & \epsilon^{nk_n} \end{vmatrix}$$

$$= \prod_{p<q}(\epsilon^{k_p} - \epsilon^{k_q})$$

$$= \prod_{p<q}(2\pi i k_p\theta - 2\pi i k_q\theta) + \text{higher terms in } \theta.$$

Divide by

$$|\epsilon^{n-1}\dots\epsilon^{0}| = \prod_{p<q}(2\pi i q\theta - 2\pi i p\theta) + \text{higher terms in } \theta,$$

cancel a factor θ, and take the limit $\theta \to 0$ to obtain the desired formula. QED

Problems for §6.4

1. Representations of SU(n). (a) Show that the restriction to SU(n) of an irreducible representation of U(n) remains irreducible.

   (b) Show that an irreducible representation of SU(n) extends to a (necessarily irreducible) representation of U(n).

   (c) Show that two irreducible characters χ_λ, χ_μ of U(n) coincide on SU(n) if and only if λ and μ differ by a weight of the form $(l, l, \dots, l)$. Deduce that the irreducible representations of SU(n) are in one-to-one correspondence with dominant n-tuples $\lambda = (l_1, l_2, \dots, l_n)$, $l_1 \geq l_2 \geq \dots \geq l_n$, provided n-tuples differing by weights of the form $(l, l, \dots, l)$ are identified. (One may normalize the weights by $l_1 + l_2 + \dots + l_n = 0$, for example.)

2. (a) Show that the irreducible characters of SU(2) are of the form

$$\chi_l = \epsilon^{-l} + \epsilon^{-l+2} + \dots + \epsilon^{l-2} + \epsilon^{l}$$

   with $l = 1, 2, \dots$ (the formula gives the value of χ_l on a diagonal matrix

$$\begin{bmatrix} \epsilon & 0 \\ 0 & \epsilon^{-1} \end{bmatrix}$$

   in SU(2)).

   (b) Let V_l, π_l be the irreducible representation of SU(2) with character χ_l. Show that

$$V_l \otimes V_m \approx \sum_k V_k,$$

where k runs from $l+m$ to $|l-m|$ in steps of 2:

$$k = l+m, l+m-2, \dots, |l-m|.$$

(This decomposition is called the Clebsch–Gordan series, a term sometimes also applied to the decomposition of tensor products for other groups.)

3. *Characters on symmetric tensors.* (a) Prove the determinant identity

$$\left|\frac{1}{1-z\epsilon}\epsilon^{n-1}, \epsilon^{n-2}, \dots, \epsilon, 1\right| = \left(\frac{1}{\prod_j 1-z\epsilon_j}\right)|\epsilon^{n-1}, \epsilon^{n-2}, \dots, \epsilon, 1|.$$

[Suggestion: use the relation

$$\frac{1}{1-z\epsilon} = 1 + \frac{z\epsilon}{1-z\epsilon}$$

to sucessively replace ϵ^{n-2} by $(1/(1-z\epsilon))\epsilon^{n-2}$, ϵ^{n-3} by $(1/(1-z\epsilon))\epsilon^{n-3}$, etc.]

(b) Show that the character ψ_m of the representation of $\mathrm{U}(n)$ on the space $S^m(\mathbb{C}^n)$ of symmetric m-tensors is the irreducible character with highest weight

$$m\lambda_1 = (m, 0, \dots, 0).$$

($S^m(\mathbb{C}^n)$ is therefore irreducible under $\mathrm{U}(n)$, hence under $\mathrm{GL}(n, \mathbb{C})$.) [See problem 7, §6.1.]

4. *Characters on alternating tensors.* (a) Prove the determinant identity

$$|\epsilon^{n-1} + z\epsilon^n, \epsilon^{n-2} + z\epsilon^{n-2}, \dots, \epsilon + z\epsilon^2, 1 + z\epsilon|$$
$$= \sum_{k=0}^{n} |z\epsilon^n, z\epsilon^{n-1}, \dots, z^{n-k+1}, \epsilon^{n-k-1}, \dots, \epsilon, 1|$$

(The j-th rows of the determinants are obtained for the ones written down by replacing ϵ by ϵ_j.)

(b) Prove the determinant identity

$$\prod_j (1 + z\epsilon_j) = \sum_{k=0}^{n} \frac{|z\epsilon^n, z\epsilon^{n-1}, \dots, z\epsilon^{n-k+1}, \epsilon^{n-k-1}, \dots, \epsilon, 1|}{|\epsilon^{n-1}, \epsilon^{n-2}, \dots, \epsilon, 1|}.$$

[Suggestion: write the determinant on the left in (a) as a product of determinants.]

(c) Show that the character φ_m of the representation of $\mathrm{U}(n)$ on the space $\wedge^m(\mathbb{C}^n)$ of alternating m-tensors is the irreducible character with highest weight

$$\lambda_1 + \lambda_2 + \cdots + \lambda_m = (1, \dots, 1, 0, \dots, 0).$$

λ_k is the weight whose k-th component is 1, the others 0. ($\wedge^m(\mathbb{C}^n)$ is therefore irreducible under $\mathrm{U}(n)$, hence under $\mathrm{GL}(n, \mathbb{C})$.) [See problem 8, §6.1].

5. *Cauchy's determinant identity reads*:

$$\det\left(\frac{1}{1-\alpha_i\beta_j}\right) = \frac{\prod_{i<j}(\alpha_i-\alpha_j)\prod_{i<j}(\beta_i-\beta_j)}{\prod_{ij}(1-\alpha_i\beta_j)}. \tag{5.1}$$

(a) Show that this identity is equivalent to

$$\sum_{k_1>\dots k_n>0} \frac{|\alpha^{k_1}\cdots\alpha^{k_n}|\,|\beta^{k_1}\cdots\beta^{k_n}|}{|\alpha^{n-1}\cdots\alpha^0|\,|\beta^{n-1}\cdots\beta^0|} = \frac{1}{\prod_{ij}(1-\alpha_i\beta_j)}. \tag{5.2}$$

The formula may be understood as an identity of functions on an appropriate domain, or as an identity of formal power series in the α_i, β_j, when both sides of the equation are written as such series. [Suggestion: Expand the coefficients $(1/1-\alpha_i\beta_j)$ of the determinant on the left of (5.1) using

$$\frac{1}{1-x} = \sum_{k=0}^{\infty} x^k.$$

Write the determinant as a power series in the β_j using the multilinearity of det; use the alternating property of det to express the terms obtained from an *ordered* monomial $\beta_1^{k_1}\beta_2^{k_2}\cdots\beta_n^{k_n}$, $k_1 > k_2 > \cdots > k_n > 0$; by a permutation of the exponents again as a determinant. For a simple proof of Cauchy's identity itself, see Weyl (1939, p. 202).]

(b) Explain in what sense Cauchy's identity is equivalent to Weyl's character formula for $\mathrm{U}(n)$. [Suggestion: Compare the left side of (5.2) with Theorem 8′, §6.2, the right side with problem 4, §6.3. The homogeneous part on the left of (5.2) of degree l in the α and β (as formal power series) is the character of the biregular representation of $\mathrm{U}(n) \times \mathrm{U}(n)$ at (a^{-1}, b) on the subspace of $M(\mathrm{U}(n))$ on which scalar matrices (α^{-1}, β) act by $\alpha^l\beta^l$. The same homogeneous part on the right is the character of the representation of $\mathrm{U}(n) \times \mathrm{U}(n)$ at (a^{-1}, b) on the subspace of homogeneous holomorphic polynomials of degree l on the matrix space $M_n(\mathbb{C})$. Not all irreducible characters occur in (5.2). Account for the remaining ones using det.]

6. *Jacobi's determinant identity reads*

$$\frac{|\epsilon^{k_1}\dots\epsilon^{k_n}|}{|\epsilon^{n-1}\dots\epsilon^0|} = |\,\psi_{k-(n-1)},\dots,\psi_{k-1},\psi_k|. \tag{6.1}$$

The right side is the determinant of the matrix whose j-th row is obtained from the one written down by replacing k by k_j, as usual. The ψ_m are given by

$$\psi_m = \sum \epsilon_1^{m_1}\epsilon_2^{m_2}\dots\epsilon_n^{m_n}.$$

The sum is over all n-tuples $(m_1, m_2, \dots, m_n)$ of integers ≥ 0 with $m_1 + m_2 + \cdots + m_n = m$. We agree that $\psi_m = 0$ if $m < 0$. The ψ_m can be calculated from the *generating function*

$$\frac{1}{\prod_i (1-\epsilon_i z)} = \sum_{m=0}^{\infty} \psi_m z^m. \tag{6.2}$$

(a) Deduce Jacobi's identity (6.1) from Cauchy's identity (5.2). [Suggestion: deduce from (5.2) that

$$\frac{|\epsilon^{k_1} \dots \epsilon^{k_n}|}{|\epsilon^{n-1} \dots \epsilon^0|}$$

is the coefficient of $z_1^{k_1} z_2^{k_2} \cdots z_n^{k_n}$ in the expansion of

$$\frac{|z^{n-1} \dots z^0|}{\prod_{ij} (1-\epsilon_i z_j)}$$

as a power series in the z_j. Then calculate this expansion directly and compare terms. It may help to write a monomial in the z_j as above as z^κ and to write determinants as alternating sums over S_n.]

Historical note. Jacobi's determinant identity goes back to 1841, long before the advent of character theory (Jacobi, 1841). As a character identity it may be written as

$$\chi_\lambda = \sum_{s \in W} \operatorname{sgn}(s) \psi_{s(\lambda+\rho)-s\rho}, \tag{6.3}$$

where generally $\psi_\mu = \psi_{m_1}, \psi_{m_2}, \dots, \psi_{m_n}$ if $\mu = (m_1, m_2, \dots, m_n)$. This formula for the irreducible characters of $\mathrm{U}(n)$ goes back to Schur's dissertation of 1901.

(b) Show that the ψ_κ on the right side of (6.3) are characters of $\mathrm{U}(n)$. Deduce that these ψ_κ form a $\mathbb{Z}$-basis for the characters of $\mathrm{U}(n)$ in the sense that every character is a unique $\mathbb{Z}$-linear combination of these ψ_κ. [See problem 7, §6.1.]

7. *Kostant's formula for the weight-multiplicities.* Write the weight $\lambda = (l_1, \dots, l_n)$ as $\lambda = l_1\lambda_1 + \cdots + l_n\lambda_n$. Let $\Phi_+ = \{\alpha_{jk} = \lambda_j - \lambda_k \text{ for } j < k\}$. Then the Weyl denominator Δ may be written as

$$\Delta = e^\rho \prod_{\alpha \in \Phi_+} (1 - e^{-\alpha}).$$

(a) Show that

$$\frac{1}{\prod_{\alpha \in \Phi_+} (1-e^{-\alpha})} = \sum_\lambda P(\lambda) e^{-\lambda},$$

where $P(\lambda)$ is the number of ways of writing λ in the form $\lambda = \sum_{\alpha \in \Phi_+} m_\alpha \alpha$ with non-negative integral m_α. The equation is understood as an identity of functions on the domain in $\mathbb{C}^n$ (complex diagonal matrices) where $\mathrm{Re}\alpha > 0$ for all $\alpha \in \Phi_+$ to insure convergence of the series).

(b) Show that the multiplicity of a weight μ in the irreducible character χ_λ is

$$\sum_{s \in W} \mathrm{sgn}(s) P(s(\lambda + \rho) - (\mu + \rho)).$$

[Suggestion: write $\chi_\lambda = \xi_{\lambda+\rho}/\Delta = (e^{-\rho}\xi_{\lambda+\rho})(1/e^{-\rho}\Delta)$. Formally multiply out the product of the two factors in parentheses, using the definition of $\xi_{\lambda+\rho}$ and the above formula for $1/e^{-\rho}\Delta$; collect the coefficients of e^{μ}. Explain why this formal manipulation gives the desired result.] The formula is due to Kostant (1959).

8. *Steinberg's formula for the multiplicities in tensor products.* Show that the multiplicity of χ_λ in $\chi_{\lambda'}\chi_{\lambda''}$ is given by

$$\sum_{s', s'' \in W} \mathrm{sgn}(s's'') P(s'(\lambda' + \rho) + s''(\lambda'' + \rho) - \lambda - 2\rho).$$

[Suggestion: write $\chi_{\lambda'}\chi_{\lambda''} = \sum_\lambda m_\lambda \chi_\lambda$ in the form

$$(e^{-\rho})\xi_{\lambda'+\rho}\xi_{\lambda''+\ rho}(1/e^{-\rho}\Delta) = \sum_\lambda m_\lambda \xi_{\lambda+\rho}.$$

Formally multiply out the left side as in problem 7. Fix a dominant λ and compare the coefficients of $e^{\lambda+\rho}$: on the right this coefficient is m_λ. The formula is due to Steinberg (1961).

9. *The Brauer–Weyl formula for the multiplicities in tensor products.* For any weight λ denote by λ_+ the dominant weight W-conjugate to λ. For regular λ (i.e. $l_j \neq l_k$ for $j \neq k$) set $\mathrm{sgn}(\lambda) = \mathrm{sgn}(s)$ where $s \in W$ is the unique element with $s\lambda = \lambda_+$; for *singular* λ set $\mathrm{sgn}(\lambda) = 0$. Show that

$$\chi_{\lambda'}\chi_{\lambda''} = \sum_\mu m_{\lambda'}(\mu)\,\mathrm{sgn}(\mu + \lambda'' + \rho)\chi_{(\mu+\lambda''+\rho)_+ - \rho},$$

where $m_{\lambda'}(\mu)$ is the multiplicity of the weight μ in $\chi_{\lambda'}$. Deduce that the multiplicity of χ_λ in $\chi_{\lambda'}\chi_{\lambda''}$ is

$$\sum_{s \in W} \mathrm{sgn}(s) m_{\lambda'}(s(\lambda + \rho) - (\lambda'' + \rho)).$$

[Suggestion: justify the following calculation:

$$\begin{aligned}
\chi_{\lambda'}\xi_{\lambda''+\rho} &= \sum_{\mu,s} m_{\lambda'}(\mu)\,\mathrm{sgn}(s)\,e^{\mu+s(\lambda''+\rho)} \\
&= \sum_{\mu,s} m_{\lambda'}(\mu)\,\mathrm{sgn}(s)\,e^{s(\mu+\lambda''+\rho)} \\
&= \sum_{\mu} m_{\lambda'}(\mu)\,\mathrm{sgn}(\mu+\lambda''+\rho)\sum_{s}\mathrm{sgn}(s)\,e^{s(\mu+\lambda''+\rho)_+} \\
&= \sum_{\mu} m_{\lambda'}(\mu)\,\mathrm{sgn}(\mu+\lambda''+\rho)\,\xi_{(\mu+\lambda''+\rho)_+}
\end{aligned}$$

Complete the proof.]

The corresponding formula for the orthogonal group goes back to Brauer's dissertation of 1925; for the classical groups it is discussed in Weyl (1939). According to a note there, both Brauer and Weyl were aware that the formula holds for any semisimple group. The formula is sometimes (inappropriately) ascribed to Klimyk.

10. Assume $\mu+\lambda''$ is dominant for every weight μ in $\chi_{\lambda'}$. Show that the multiplicity of $\chi_{\mu+\lambda''}$ in $\chi_{\lambda'}\chi_{\lambda''}$ equals the multiplicity $m_{\lambda'}(\mu)$ of the weight μ in $\chi_{\lambda'}$

11. (a) Show that the multiplicity of χ_λ in $\chi_{\lambda'}\bar{\chi}_{\lambda''}$ is given by

$$\sum_{s\in W}\mathrm{sgn}(s)m_\lambda(s(\lambda''+\rho)-(\lambda'+\rho)).$$

$\bar{\chi}$ is the contragredient character of χ (= complex conjugate of χ). [Suggestion: show first that

$$(\text{multiplicity of } \chi_\lambda \text{ in } \chi_{\lambda'}\bar{\chi}_{\lambda''}) = (\text{multiplicity of } \chi_{\lambda'} \text{ in } \chi_\lambda\bar{\chi}_{\lambda''}).$$

Then use the Brauer–Weyl formula.]

(b) Deduce from (a) that the multiplicity of the trivial character $\chi\equiv 1$ in $\chi_{\lambda'}\bar{\chi}_{\lambda''}$ is one if $\lambda'=\lambda''$ and zero otherwise. Explain why this is obvious in any case.

6.5 Representations of Lie algebras

Let $\mathfrak{g}$ be a real Lie algebra. A *representation* π of $\mathfrak{g}$ on a real or complex vector space V associates to each $X\in\mathfrak{g}$ a linear transformation $\pi(X)$ of V so that

$$\begin{aligned}
\pi(\alpha X+\beta Y) &= \alpha\pi(X)+\beta\pi(Y), \\
\pi([X,Y]) &= [\pi(X),\pi(Y)],
\end{aligned}$$

for all $X, Y \in \mathsf{g}$ and $\alpha, \beta \in \mathbb{R}$, i.e $\pi : \mathsf{g} \to \mathsf{gl}(V)$ is a homomorphism of *real* Lie algebras.

As with groups, we frequently call the pair V, π a representation of g. Unless explicitly stated otherwise we assume that V is *complex* and *finite-dimensional.* Nevertheless, we require the condition $\pi(\alpha X + \beta Y) = \alpha\pi(X) + \beta\pi(Y)$ only for real α, β, even if g is a complex Lie algebra. If g is complex and this condition holds for complex α, β as well, then we call the representation π of g *holomorphic.* (Calling it 'complex' would be too confusing).

Every representation V, π of a linear group G gives rise to a representation of its Lie algebra g, also denoted V, π: the representation $\pi : \mathsf{g} \to \mathsf{gl}(V)$ is the Lie map of $\pi : G \to \mathrm{GL}(V)$. Explicitly this means that for $X \in \mathsf{g}$,

$$\pi(X)v = \frac{d}{d\tau}\pi(\exp \tau X)v\bigg|_{\tau=0},$$

for all $v \in V$. It will be clear from the context whether π is to be considered a representation of G or of g. A representation V, π of a complex group G is holomorphic, in the sense that $\pi : G \to \mathrm{GL}(V)$ is a holomorphic map, if and only if the corresponding representation π of its Lie algebra is holomorphic. This follows from the formula $\pi(\exp X) = \exp \pi(X)$. For example, the representation $\pi(a) = a$ of $\mathrm{GL}(n, \mathbb{C})$ is holomorphic, the representation $\pi(a) = \bar{a}$ is not; and correspondingly for the representations $\pi(X) = X$ and $\pi(X) = \bar{X}$ of $\mathsf{gl}(n, \mathbb{C})$.

Example 1. Let F be the real vector space of all complex-valued polynomial functions $f(x) = f(\xi_1, \ldots, \xi_n)$ of degree $\leq d$ on $\mathbb{R}^n$ (some d some n). Define a representation π of $\mathrm{GL}(n, \mathbb{R})$ on F by the formula

$$\pi(a)f(x) = f(a^{-1}x).$$

$\mathrm{GL}(n, \mathbb{R})$ acts on $\mathbb{R}^n$ by linear transformations as usual:

$$ax = \sum_{ij} a_{ij}\xi_j e_i \quad \text{if } x = \sum_j \xi_j e_j.$$

The corresponding representation of $\mathsf{gl}(n, \mathbb{R})$ on F is given by

$$\pi(X)f(x) = df_x(-Xx) = \sum_{ij} -X_{ij}\xi_j \frac{\partial f}{\partial \xi_i}.$$

To verify this, calculate:

$$\begin{aligned}\pi(X)f(x) &= \frac{d}{d\tau}\pi(\exp\tau X)f(x)\bigg|_{\tau=0}\\ &= \frac{d}{d\tau}f(\exp(-\tau X)x)\bigg|_{\tau=0}\\ &= df_x(-Xx),\end{aligned}$$

which is the required formula.

In his example one can also take for F the space of C^∞ functions on $\mathbb{R}^n$, except that this space is not finite-dimensional. One may also replace functions on $\mathbb{R}^n$ by functions on $\mathbb{C}^n$ and the group $\mathrm{GL}(n, \mathbb{R})$ by $\mathrm{GL}(n, \mathbb{C})$. In order that the formula

$$\pi(X)f(x) = \sum_{ij} -X_{ij}\xi_j \frac{\partial f}{\partial \xi_j}$$

for $\pi(X)f(x)$ remain true one must then require f to be holomorphic, not just C^∞.

'*Invariant, reducible, irreducible,* g*-map, equivalent*' are defined for representations of Lie algebras as for groups. A representation V, π of a group G was said to be unitary if V has a positive definite, Hermitian inner product such that $(\pi(a)x, \pi(a)y) = (x, y)$ for all $a \in G$, $x, y \in V$, i.e. $\pi(a) \in \mathrm{U}(V) =$ the unitary group of $V, (,)$. The corresponding condition for representation V, π of a Lie algebra g to be *unitary* is that $(\pi(X)x, y) = -(x, \pi(X)y)$ for all $X \in \mathsf{g}$, $x, y \in V$, i.e. $\pi(X) \in \mathsf{u}(V)$.

Proposition 2. *Let V, π and W, ρ be representations of a connected linear group $G, A : V \to W$ a linear map.*

(a) $U \subset V$ is G-invariant if and only if it is g-invariant.

(b) $A : V \to W$ is a G-map if and only if it is a g-map.

(c) V, π is G-irreducible if and only it is g-irreducible.

(d) $V, \pi, (,)$ is G-unitary if and only it is g unitary.

(e) π and ρ are G-equivalent if and only if they are g-equivalent.

Proof. This is immediate from the formula $\pi(\exp X) = \exp \pi(X)$, since a connected G is generated by $\exp X$. QED

If V, π and W, ρ are representations of two linear groups G and H, respectively, the representation $V \otimes W$, $\pi \otimes \rho$ of $G \times H$ was defined by $\pi \otimes \rho(a, b) = \pi(a) \otimes \rho(b)$. Setting $a = \exp(\tau X)$ and $b = \exp(\tau Y)$ and differentiating at $\tau = 0$, one gets for the corresponding representation of the Lie algebra $\mathsf{g} \times \mathsf{h}$ the formula

$$\pi \otimes \rho(X, Y) = \pi(X) \otimes 1 + 1 \otimes \rho(Y).$$

This formula is used to define the tensor product $\pi \otimes \rho$ for any representations π and ρ of Lie algebras g and h. When $\mathsf{h} = \mathsf{g}$ we also get a representation $\pi \otimes \rho$ of g by setting

$$\pi \otimes \rho(X) = \pi(X) \otimes 1 + 1 \otimes \rho(X).$$

The relations between the representations of a group and its Lie algebra have an important consequence for $\mathrm{GL}(n, \mathbb{C})$, known as *Weyl's Unitarian trick.* (The term '*Unitarian Trick*' is due to Weyl himself.)

Theorem 3 (Weyl's Unitarian Trick). *Restriction to K gives a one-to-one correspondence between the holomorphic representations of G and the representations of K. This correspondence preserves irreducibility and equivalence.*

Amplification. (a) Let V, π a holomorphic representation of G, W a subspace of V. The following are equivalent:

(i) W is G-invariant.

(ii) W is K-invariant.

(b) Let $A : V \to W$ be a linear map between the spaces of two holomorphic representations V, π and W, ρ of G. The following are equivalent:

(i) A is a G-map.

(ii) A is K-map.

(c) Every representation π of K extends uniquely to a holomorphic representation of G (on the same space)

Proof. The crucial observation is that

$$\mathsf{g} = \mathsf{k} \oplus i\mathsf{k} \ \text{[direct sum of vector spaces]}.$$

From this equation the assertions (a)–(c) are clear when the groups G and K are replaced by their Lie algebras g and k. For (a) and (b) we may immediately pass from the representations of the Lie algebras to the representations of the groups, since G and K are connected. This is not the case for (c), because quite generally not every homomorphism between matrix Lie algebras comes from a homomorphism of the corresponding connected matrix groups. Instead we argue as follows.

Uniqueness in (c) is clear from $\mathsf{g} = \mathsf{k} + i\mathsf{k}$. For existence it suffices to consider an irreducible representation π of K, since any representation is a direct sum of irreducibles under K. We know from the proof of the Peter–Weyl theorem that π may be realized on a space V of restrictions to K of polynomial functions $f(a, \bar{a})$ in the matrix coefficients of a and $\bar{a}$, K operating by the left regular representation. Since $f(a, \bar{a}) = f(a, (a^{-1})^t)$ on K, these functions on K extend (necessarily uniquely) to holomorphic functions on G. Considered as a space of functions on G, V is g-stable (again because $\mathsf{g} = \mathsf{k} + i\mathsf{k}$) and provides the desired extension of π to G. QED

We note an important consequence.

Theorem 4 (Theorem of Complete Reducibility). *Every holomorphic representation of* $\mathrm{GL}(n, C)$ *is a direct sum of irreducible representations.*

Historical comment. The complete reducibility of the representations of complex semisimple Lie algebras was taken for granted by E. Cartan, in 1913, and therein lies the incompleteness of his arguments pointed out by Weyl in 1925. For the classical Lie algebras, however, the gap is not serious in the sense that the complete reducibility of the particular representations that Cartan uses to construct the irreducibles may be ascertained without difficulty.

Example 5. Representations of SL(2, ℂ). SL(2, ℂ) acts on functions on $\mathbb{C}^2$ by

$$a \cdot f(z) = f(a^{-1}z).$$

Correspondingly, sl(2, ℂ) acts on C^∞ functions on $\mathbb{C}^2$ by

$$X \cdot f(z) = \frac{d}{d\tau} f(\exp(-\tau X)z)\bigg|_{\tau=0} = df_z(-Xz).$$

If one writes $z = (\xi, \eta)$, the basis H, X_+, X_- of sl(2, ℂ) defined by

$$H = \begin{bmatrix} 1 & 0 \\ 0 & -1 \end{bmatrix}, \quad X_+ = \begin{bmatrix} 0 & 1 \\ 0 & 0 \end{bmatrix}, \quad X_- = \begin{bmatrix} 0 & 0 \\ 1 & 0 \end{bmatrix}$$

is represented by the operators

$$H \to -\xi\frac{\partial}{\partial \xi} + \eta\frac{\partial}{\partial \eta}, \quad X_+ \to -\eta\frac{\partial}{\partial \xi}, \quad X_- \to -\xi\frac{\partial}{\partial \eta}.$$

Let V_l, $l = 0, 1, 2, \ldots$, be the space of homogeneous holomorphic polynomials of degree $l \in \xi, \eta$, considered as functions on $\mathbb{C}^2$. V_l is invariant under the action of SL(2, ℂ) described above and therefore gives a representation π_l of SL(2, ℂ) and of its Lie algebra sl(2, ℂ). A basis for V_l is given by the $l + 1$ monomials

$$\eta^l, \xi\eta^{l-1}, \ldots, \xi^{l-1}\eta, \xi^l.$$

The operators H, X_+, X_- act according to the formulas:

$$\begin{aligned} H \cdot \xi^k\eta^{l-k} &= (l - 2k)\xi^k\eta^{l-k} \\ X_+ \cdot \xi^k\eta^{l-k} &= -k\xi^{k-1}\eta^{l-(k-1)} \\ X_- \cdot \xi^k\eta^{l-k} &= (-l + k)\xi^{k-1}\eta^{l-(k-1)}. \end{aligned}$$

These operations may be described as follows. The *weights* of the representation V_l, π_l are

$$-l, -l + 2, \ldots, l - 2, l,$$

i.e. these are the eigenvalues of the operator H. The corresponding *weight spaces* (eigenspaces of H) are one-dimensional. The operator X_+ shifts the *weight spaces* (listed in the order shown) one step to the right, the right-most weight space being annihilated. The operator X_- shifts the weight spaces one step to the left, the left-most weight space being annihilated. One sees from this description that V_l is generated by any non-zero weight vector, hence is irreducible (since any non-zero invariant subspace, being H-invariant, contains a weight vector).

A diagonal matrix $(\epsilon, \epsilon^{-1})$ in SL(2, ℂ) operates on the weight-spaces by scalars with eigenvalues

$$\epsilon^{-l}, \epsilon^{-l+2}, \ldots, \epsilon^{l-2}, \epsilon^l,$$

hence its trace is

$$\epsilon^{-l} + \epsilon^{-l+2} + \cdots + \epsilon^{l-2} + \epsilon^l = \frac{\epsilon^{(l+1)} - \epsilon^{-(l+1)}}{\epsilon - \epsilon^{-1}}.$$

In particular, the restriction of the representation V_l, π_l to SU(2) has character

$$\chi_l = \frac{\epsilon^{(l+1)} - \epsilon^{-(l+1)}}{\epsilon - \epsilon^{-1}}.$$

This agrees with restriction of the irreducible character of U(2) as given by Weyl's formula,

$$\chi_\lambda = \frac{\sum_s \operatorname{sgn}(s)\epsilon^{s(\lambda+\rho)}}{\Delta},$$

with $\lambda = (l+k, k)$ for any $k \in \mathbb{Z}$. (For U(2), $e^\lambda = \epsilon_1^{l_1}\epsilon_2^{l_2}$.)

The representation V_l, π_l of SL(2, $\mathbb{C}$) extends in a natural way to a representation of GL(2, $\mathbb{C}$), in which a scalar matrix ζ operates by the scalar ζ^{-l}. From the above discussion one sees that the restriction of this representation to U(2) is the irreducible representation with character χ_λ of highest weight $\lambda = (0, -l)$. The irreducible representation of U(2) with highest weight $\lambda = (k, -l+k)$, $k \in \mathbb{Z}$, is realized by $(\det^k) \otimes \pi_l$, which acts on the same space $V_l \approx \mathbb{C} \otimes V_l$ by

$$\pi_\lambda(a) = (\det a)^k \pi_l(a).$$

The same formula defines a holomorphic representation of GL(2, $\mathbb{C}$), necessarily irreducible, and it follows from the unitarian trick that every irreducible holomorphic representation of GL(2, $\mathbb{C}$) is of this type.

Problems for §6.5

1. Show that every finite-dimensional, irreducible, holomorphic representation V, π of $\mathfrak{sl}(2, \mathbb{C})$ is equivalent to a representation V_l, π_l as in Example 5. [Suggestion: start with an eigenvector v for $\pi(H)$ with $\pi(X_+)v = 0$. To see that such a vector exists, consider $\pi(X_+)^k v$, $k = 0, 1, \ldots$, for *some* eigenvector for $\pi(H)$. It is not necessary to know *a priori* that V decomposes into eigenspaces for $\pi(H)$.]

2. (a) Show that a differentiable function $f(x) = f(\xi_1, \xi_2, \xi_3)$ on $\mathbb{R}^3$ is invariant under the rotation group SO(3), i.e.

$$f(ax) = f(x) \quad \text{for all } a \in \mathrm{SO}(3) \text{ and all } \quad x \in \mathbb{R}^3,$$

   if and only if

$$\xi_i \frac{\partial f}{\partial \xi_j} - \xi_j \frac{\partial f}{\partial \xi_i} = 0,$$

   for all $i \neq j$ $(i, j = 1, 2, 3)$.

   (b) Discuss generalizations of part (a).

3. Let $\mathfrak{g}$ be a *complex* Lie algebra, $\pi : \mathfrak{g} \to \mathfrak{gl}(E)$ a representation of $\mathfrak{g}$ ($\mathbb{R}$-linear homomorphism).

(a) Show that π can be uniquely written as $\pi = \pi^+ + \pi^-$ where π^+ is $\mathbb{C}$-linear $[\pi^+(iX) = i\pi^+(X)]$ and π^- is $\mathbb{C}$-anti-linear $[\pi^-(iX) = -i\pi^-(X)]$, namely

$$\pi^{\pm}(X) = \tfrac{1}{2}\left(\pi(X) \pm i\pi(i^{-1}X)\right).$$

(b) Show that $(X, Y) \to \pi^+(X) + \pi^-(Y)$ is a holomorphic representation of $\mathsf{g} \times \mathsf{g}$, also denoted π.

(c) Show that π is irreducible as representation of g if and only if it is irreducible as representation of $\mathsf{g} \times \mathsf{g}$.

(d) Show that every irreducible representation of $\mathrm{GL}(n, \mathbb{C})$ is of the form $V^+ \otimes V^-$, $\pi^+ \otimes \pi^-$ where V^+, π^+ and V^-, π^- are, respectively, holomorphic and anti-holomorphic irreducible representations of $\mathrm{GL}(n, \mathbb{C})$. [Suggestion: Use Proposition 10, §6.3.]

4. Let g be a *real* Lie algebra. The complexified vector space $\mathsf{g}_{\mathbb{C}} = \mathsf{g} \oplus i\mathsf{g}$ [appendix to §1.1] is in the obvious way a complex Lie algebra. Show that there is a natural one-to-one correspondence between ($\mathbb{R}$-linear) representations of g and holomorphic ($\mathbb{C}$-linear) representations of $\mathsf{g}_{\mathbb{C}}$. This correspondence preserves equivalence and irreducibility.

5. Suppose g is a complex Lie algebra, but considered a real Lie algebra. Show that its complexification $\mathsf{g}_{\mathbb{C}}$ is isomorphic with $\mathsf{g} \times \mathsf{g}$. [Warning: the two different multiplications by i in $\mathsf{g}_{\mathbb{C}}$ tend to be confusing. Distinguish them in notation, e.g. write elements of $\mathsf{g}_{\mathbb{C}}$ as $X + jY$. Suggestion: consider $\frac{1}{2}(X \pm jiY)$.]

   Comment: Problems 4 and 5 redo problem 3.

6. *Exponentiation of representations.* Every finite-dimensional representation of the Lie algebra of a linear group *exponentiates* to a representation of some linear covering group. Example:

   (a) All finite-dimensional representations of $\mathsf{su}(2)$ exponentiate to $\mathrm{SU}(2)$. Which finite-dimensional, irreducible representations of $\mathsf{su}(2) \approx \mathsf{so}(3)$ exponentiate to representations of $\mathrm{SO}(3)$? [Equivalently (why?): which of the representations π_l of $\mathrm{SU}(2)$ descend to $\mathrm{SO}(3)$ through the covering map $\mathrm{SU}(2) \to \mathrm{SO}(3)$? See problem 1.]

   Not every *infinite-dimensional* representation of a Lie algebra exponentiates. Two examples:

   (b) (d/dx) does not exponentiate on $C^\infty(0,1)$, but does exponentiate on $C^\infty(\mathbb{R})$ in some sense. In what sense?

   (c) $\mathsf{sl}(2, \mathbb{C})$ acts on the space $R(\mathbb{C}^2)$ of holomorphic *polynomial* functions $p(z) = p(\xi, \eta)$ as in Example 5. It acts on the dual space $R(\mathbb{C}^2)^*$ by the contragredient representation: $(X \cdot \varphi)(f) = -\varphi(X \cdot f)$. Let V be the subspace of $R(\mathbb{C}^2)^*$ of linear functionals φ of the form $\varphi(f) = Df(1,0)$ where D is a constant coefficient differential operator.

Show that the functionals

$$\left.\frac{\partial^k}{\partial \xi^k}\right|_{(1,0)}, \quad k = 0, 1, 2, \ldots,$$

form a basis for V consisting of weight vectors (eigenvectors for H) with weights (eigenvalues) $l = 0, 1, 2, \ldots$. Conclude that this representation of $\mathsf{sl}(2, \mathbb{C})$ does *not* exponentiate to $\mathrm{SL}(2, \mathbb{C})$. Worse than that: its restriction to $\mathsf{sl}(2, \mathbb{R})$ does not exponentiate to $\mathrm{SL}(2, \mathbb{R})$ or to any covering group of $\mathrm{SL}(2, \mathbb{R})$. [Suggestion: consider how the element

$$s = \begin{bmatrix} 0 & -1 \\ 1 & 0 \end{bmatrix}$$

would have to act on V, or any element in a covering group that acts like $\mathrm{Ad}(s)$ on $\mathsf{sl}(2, \mathbb{R})$. V is what is called a *Verma module* for $\mathsf{sl}(2, \mathbb{C})$.]

7. Let $\mathsf{hw}(3, \mathbb{R})$ be the three-dimensional real Lie algebra with basis P, Q, Z and relations

$$[P, Q] = Z, [Z, P] = [Z, Q] = 0.$$

($\mathsf{hw}(3, \mathbb{R})$ is called the three-dimensional *Heisenberg–Weyl Lie algebra.*)

(a) Show that $\mathsf{hw}(3, \mathbb{R})$ is isomorphic with the Lie algebra of the *Heisenberg–Weyl group* $\mathrm{HW}(3, \mathbb{R})$ of upper triangular 3×3 matrices with 1's on the diagonal.

(b) Show that for all α, $\beta \in \mathbb{C}$, the rule

$$P \to \alpha \frac{d}{dx}, \quad Q \to \beta x, \quad Z \to \alpha\beta 1$$

defines a representation $\pi_{\alpha\beta}$ on the *infinite-dimensional* space $C^\infty(\mathbb{R})$. (x denotes here the operator $f(x) \to x f(x)$.)

(c) For $\alpha \in \mathbb{R}$, find a representation of the group $\mathrm{HW}(3, \mathbb{R})$ that in some sense corresponds to the representation $\pi_{\alpha\beta}$ of $\mathsf{hw}(3, \mathbb{R})$. Explain in what sense. Why $\alpha \in \mathbb{R}$? Discuss modifications so as to obtain a representation of $\mathrm{HW}(3, \mathbb{R})$ for *all* $\alpha, \beta \in \mathbb{C}$.

(d) Assume $\alpha\beta \neq 0$ and $\alpha \in \mathbb{R}$. Show that $\pi_{\alpha\beta} \sim \pi_{\alpha'\beta'}$ if and only if $\alpha\beta = \alpha'\beta'$.

8. Notation as in problem 7.

(a) Show that the space $F = \{$functions of the form $p(x)e^{-x^2} \mid p(x)$ a complex-valued polynomial$\}$ is stable under $\pi_{\alpha\beta}(\mathsf{hw}(3, \mathbb{R}))$. How about $\pi_{\alpha\beta}(\mathrm{HW}(3, \mathbb{R}))$?

(b) Show that F is irreducible under $\pi_{\alpha\beta}(\mathsf{hw}(3, \mathbb{R}))$ if and only if $\alpha\beta \neq 0$. Describe the invariant subspaces of F in case $\alpha = 0$ or $\beta = 0$. [Suggestion: to deal with the case $\beta = 0$, show first that every function in F can be uniquely written in the form $p(d/dx)e^{-x^2}$ where p is a polynomial.]

(c) Determine for which values of α, β the representation of $\mathsf{hw}(3, \mathbb{R})$ on a function of the form $p(x)e^{-x^2}$ is unitary for the inner product

$$(f, g) = \int_{\mathbb{R}} f(x)\overline{g(x)}\, dx.$$

Comment. In quantum mechanics the relation

$$[P, Q] = -ih1$$

among two operators P, Q (usually on a Hilbert space), with h as Planck's constant, is called *Heisenberg's Commutation relation* or *canonical commutation relation.* Its interpretation in terms of group theory is due to Weyl (1931).

9. Let $\mathrm{SL}(2, \mathbb{C})$ operate on $\mathbb{C}^2 = \{$*row vectors* $\zeta = [x, y]\}$ by matrix multiplication *on the right*, and correspondingly on functions on $\mathbb{C}^2$ by

$$a \cdot \varphi(\zeta) = \varphi(\zeta a).$$

Show that the map $\varphi \to f$ defined by $f(\xi, \eta) = \varphi(-\eta, \xi)$ turns this operation into

$$a \cdot f(z) = f(a^{-1}z),$$

where

$$z = \begin{bmatrix} \xi \\ \eta \end{bmatrix}$$

and $\mathrm{SL}(2, \mathbb{R})$ acts on column vectors by matrix multiplication on the left.

10. Notation as in Example 5.

(a) Define a bilinear form $\langle f, g \rangle$ on V_l by requiring that

$$\langle \eta^j \xi^{l-j}, \eta^{l-k} \xi^k \rangle = \frac{(-1)^{l-k}}{k!(l-k)!} \delta_{jk}, \quad j, k = 0, 1, \ldots, l.$$

Show that this bilinear from is non-degenerate and invariant under $\mathrm{SL}(2, \mathbb{C})$. Determine when this form is symmetric, when skew-symmetric. It may be useful to note that

$$\frac{1}{l!}(x\eta - y\xi)^l = \sum_{k=0}^{l} \frac{(-1)^{l-k}}{k!(l-k)!} x^k y^{l-k} \eta^k \xi^{l-k}.$$

(b) Identify V_l and V_l^* by means of the form $\langle f, g \rangle$. Show that the linear transformation of V_l defined by $f(\xi, \eta) \to f(-\eta, \xi)$ establishes an equivalence of π_l with its contragredient $\check{\pi}_l$.

11. Notation as in Example 5. Define a positive definite Hermitian inner product on V_l by requiring that the elements

$$\frac{1}{\sqrt{!(l-k)!}}\xi^k\eta^{l-k}, \quad k = 0, 1, \ldots, l,$$

form an orthonormal basis. Show that π_l is then unitary *as representation* of SU(2). One can see that

$$\frac{1}{l!}(\bar{x}\xi + \bar{y}\eta)^l = \sum_{k=0}^{l} \frac{1}{k!(l-k)!}\bar{x}^k\bar{y}^{l-k}\xi^k\eta^{l-k}.$$

12. (a) Let $X \in \mathfrak{gl}(n, \mathbb{C})$ be a nilpotent $n \times n$ matrix. Show that there is a triple of matrices $\{H, X_+, X_-\}$ with $X = X_+$ which span a subalgebra of $\mathfrak{gl}(n, \mathbb{C})$ isomorphic with $\mathfrak{sl}(2, \mathbb{C})$ by the map

$$\begin{bmatrix} 1 & 0 \\ 0 & -1 \end{bmatrix} \to H, \quad \begin{bmatrix} 0 & 1 \\ 0 & 0 \end{bmatrix} \to X_+, \quad \begin{bmatrix} 0 & 0 \\ 1 & 0 \end{bmatrix} \to X_-.$$

[Suggestion: write $X = \sum X_i$ as a sum of Jordan blocks. Remember how the X_+ in Example 5 operates in V_l.]

(b) Let V, π be the representation of $\mathfrak{sl}(2, \mathbb{C})$ on $V = \mathbb{C}^n$ defined in part (a). Show that this representation decomposes as

$$V \approx \sum_l \mathbb{C}^{m(l)} \otimes V_l, \quad \pi \approx \sum_l 1 \otimes \pi_l.$$

(c) Show that (as a vector space) the centralizer $\mathfrak{z}(X)$ of X in $\mathfrak{gl}(n, \mathbb{C})$ is isomorphic with

$$\sum_{k,l} \mathrm{Hom}_{\mathbb{C}}(\mathbb{C}^{m(k)}, \mathbb{C}^{m(l)}) \otimes \mathrm{Hom}_X(V_k, V_l),$$

where $m(l)$ is the number of Jordan blocks of X of size $d_l = \dim V_l = 2[l/2] + 1$. Deduce that

$$\dim \mathfrak{z}(X) = \sum_l d_l m(l)^2 + 2\sum_{k<l} d_k m(k)m(l).$$

6.6 The Borel–Weil theorem for $\mathrm{GL}(n, \mathbb{C})$

Weyl's character formula provides a formula for the irreducible characters of U(n), but does not provide a realization of the representations themselves. Several methods for constructing such realizations are known starting with Cartan's algebraic classification of the irreducible representations. In this section we discuss a method, which historically came much later, due to Borel and Weil (1954).

Throughout this section $G = \mathrm{GL}(n, \mathbb{C})$, $K = \mathrm{U}(n)$, $\mathfrak{g} = \mathfrak{gl}(n, \mathbb{C})$, and $\mathfrak{k} = \mathfrak{u}(n)$. H denotes the diagonal subgroup of G. We use the same notation for elements of H as for elements of the diagonal subgroup T of K: $h \in H$ is written as $h = (\epsilon_1, \ldots, \epsilon_n)$, $\epsilon_j \in \mathbb{C}^\times$. Its Lie algebra is $\mathfrak{h} = \{$diagonal subalgebra of $\mathfrak{g}\}$; $H \in \mathfrak{h}$ is typically written as $H = (\lambda_1, \ldots, \lambda_n)$. The diagonal entries ϵ_k of $h \in H$ are considered functions on H, the diagonal entries λ_k of $H \in \mathfrak{h}$ functions on $\mathfrak{h}$. Thus $\epsilon_k(h) = e^{\lambda_k(H)}$ if $h = \exp H$. The λ_k form a basis of $\mathfrak{h}^*$. The linear functional $\lambda = l_1\lambda_1 + \cdots + l_n\lambda_n$ is also written as an n-tuple $\lambda = (l_1, \ldots, l_n)$. When all l_k are integral, then e^λ *lifts* from t to T (as explained in Example 3, §6.2):

$$e^\lambda = \epsilon_1^{l_1} \cdots \epsilon_n^{l_n}.$$

Then $e^\lambda : H \to \mathbb{C}^\times$ is a holomorphic homomorphism and every holomorphic homomorphism $H \to \mathbb{C}^\times$ is of this form. As for T, these *integral* λ or e^λ are called *weights* of H.

Any holomorphic representation V, π of H, in particular any holomorphic representation of G restricted to H, may be decomposed into *weight spaces* V_λ,

$$V = \bigoplus_\lambda V_\lambda,$$

so that $h \in H$ acts by e^λ on V_λ. (One may first decompose under the unitary subgroup T of H: a unitarian trick.) On H, the character $\chi(a) = \operatorname{tr} \pi(a)$ of a representation V, π of G is given by

$$\chi = \sum_\lambda m_\lambda e^\lambda,$$

where $m_\lambda = \dim V_\lambda$ is the *multiplicity* of the weight λ in π (or in χ).

H, or any subgroup of G conjugate to it, is called a *Cartan subgroup*[1] of G. Similarly, $\mathfrak{h}$, or any subalgebra conjugate to it, is called a *Cartan subalgebra* of $\mathfrak{g}$.

Example 1 (Adjoint representation and roots). The weights of the adjoint representation of G on $\mathfrak{g}$ are $\alpha = \alpha_{jk}$ defined by

$$\alpha_{jk} = \lambda_j - \lambda_k.$$

The non-zero weights among these ($j \neq k$) occur with multiplicity one and are called the *roots* of $\mathfrak{g}$ (or of the pair $\mathfrak{g}, \mathfrak{h}$). To the root $\alpha = \alpha_{jk}$ there corresponds the *root vector* $E_\alpha = E_{jk}$, the matrix with jk entry 1 and 0 otherwise. It spans the one-dimensional *root space* $\mathfrak{g}_\alpha = \mathbb{C}E_\alpha$.

The roots $\alpha_{ij} = \lambda_j - \lambda_k$ with $j < k$ are called the *positive roots*, their negatives the *negative roots*. The set of all positive roots is denoted Φ_+. As for K, a weight $\lambda = (l_1, \ldots, l_n)$, $l_j \in \mathbb{Z}$, is *dominant weight* if $l_1 \geq \cdots \geq l_n$.

---

[1] §3.2 is not required.

The root vectors E_α corresponding to the positive roots span the Lie subalgebra of strictly upper triangular matrices in g, denoted n:

$$n = \sum_{\alpha \in \Phi_+} \mathbb{C} E_\alpha.$$

Together with h, they span the subalgebra of all upper triangular matrices in g, denoted b:

$$\mathsf{b} = \mathsf{h} + \mathsf{n} \quad \text{[direct sum of vector spaces]}.$$

The corresponding groups are denoted N and B; N is a normal subgroup of B, and

$$B = HN$$

in the sense that every $b \in B$ is *uniquely* of the form $b = hn$ with $h \in H$ and $n \in N$. Thus the natural map $H \to B/N$ is bijective. By the inverse of this map, every weight λ gives a homomorphism $B \to \mathbb{C}^\times$, also denoted e^λ:

$$e^\lambda(b) = e^\lambda(h) \qquad \text{if } b = hn \quad \text{with } h \in H,\ n \in N.$$

Using the correspondence between holomorphic representations of G and representations of K we shall show that the 'Theorem of the highest weight' may be reformulated as follows.

Theorem 2 (Theorem of the Highest Weight, second version). *An irreducible holomorphic representation V, π of G has a joint eigenvector v for B, unique up to scalar multiples:*

$$\pi(b)v = e^\lambda(b)v \quad \text{for all } b \in B.$$

$v = v_\lambda$ is the highest weight vector of π as representation of K, λ the highest weight.

Proof. Suppose $v \in V$ is a weight-vector for H with weight λ:

$$\pi(h)v = e^\lambda(h)v$$

for all $h \in H$. For any root vector $E_\alpha \in \mathsf{g}$,

$$\begin{aligned} \pi(h)\pi(E_\alpha)v &= \pi(hE_\alpha h^{-1})\pi(h)v \\ &= e^\alpha(h)\pi(E_\alpha)e^\lambda(h)v \\ &= e^{\lambda+\alpha}(h)\pi(E_\alpha)v \end{aligned}$$

so $\pi(E_\alpha)v$ has weight $\lambda + \alpha$. If α is a positive root, then $\lambda + \alpha$ is higher than λ (in the lexicographic ordering of the weights). If λ is the highest weight of π, and v the weight vector for λ, then necessarily $\pi(E_\alpha)v = 0$ for all positive roots α. Therefore,

$$\pi(Y)v = 0 \quad \text{for all } Y \in \mathsf{n}$$

and of course

$$\pi(H)v = \lambda(H)v \quad \text{for all } H \in \mathsf{h}.$$

Writing a general element $b \in B = HN$ as $b = \exp(H)\exp(Y)$ with $H \in \mathsf{h}$ and $Y \in \mathsf{n}$ (as is possible), one finds that indeed

$$bv = e^{\lambda}(b)v$$

for all $b \in B$.

Conversely, suppose λ is a weight and $0 \neq v \in V$ satisfies the above equation for all $b \in B$. It will follow from the next lemma that λ is then indeed the highest weight and v the unique weight vector with this weight, which will complete the proof of the theorem. QED

Lemma 3. *V is spanned by vectors of the form*

$$\pi(E_{-\alpha_k}) \cdots \pi(E_{-\alpha_1})v, \tag{1}$$

where the α_j are positive roots. Furthermore, if non-zero, (1) is a weight vector with weight $\lambda - \alpha_1 - \cdots - \alpha_k$, of π and all such weight vectors are linear combinations of these.

Proof. By irreducibility it suffices to show that the subspace spanned by vectors of the form (1) is g-invariant. For this it suffices that this subspace is invariant under $\pi(E_\beta)$ for all positive roots β. Consider a vector of the form

$$\pi(E_{-\alpha_k}) \ldots \pi(E_\beta)\pi(E_{-\alpha_j}) \ldots \pi(E_{-\alpha_1})v. \tag{2}$$

We have

$$\pi(E_\beta)\pi(E_{-\alpha_j}) = \pi(E_{-\alpha_j})\pi(E_\beta) + \pi[E_\beta, E_{-\alpha_j}]. \tag{3}$$

$[E_\beta, E_{-\alpha_j}]$ is zero or a root-vector belonging to $\beta - \alpha_j$. In any case, substituting (3) into (2) one gets a linear combination of vectors of the same type as in (2), but with smaller j. This process continues until a $\pi(E_\beta)$ appears to the right of all $\pi(E_\alpha)$, but then the vector (2) is 0, because $\pi(E_\beta)v = 0$ for a positive root β. This proves the lemma. QED

Theorem 4 (Borel–Weil Theorem). *Let λ be a dominant weight. Let*

$$F_\lambda = \{f : G \to \mathbb{C} | f \text{ is holomorphic and } f(ab) = e^{\lambda}(b)f(a) \text{ for all } b \in B\}.$$

The representation of K on F_λ by left translation is the irreducible representation of highest weight λ.

Proof. Let F denote the space of all holomorphic functions on G. By restriction to K, F is $(K \times K)$-isomorphic to a dense subspace of $L^2(K)$. For $f \in F$, expand f into an L^2-convergent Fourier series on K according to Theorem 8′, §6.2:

$$f = \sum_\mu f_\mu,$$

where

$$f_\mu(a) = \frac{\deg \chi_\mu}{|K|} \int_K \bar{\chi}_\mu(c) f(c^{-1}a)\, dc \tag{4}$$

is the component of f transforming according to $\pi_\mu \otimes \check{\pi}_\mu$ under $K \times K$. (See Proposition 9, §6.3.) V^μ, π_μ is the irreducible representation of K with highest weight μ, χ_μ its character. Note that these f_μ are still holomorphic functions on G. The representation of $K \times K$ on these f_μ is equivalent to $V^\mu \otimes (V^\mu)^*$, $\pi_\mu \otimes \check{\pi}_\mu$. A holomorphic function $f \in F$ satisfies

$$f(ab) = e^\lambda(b) f(a) \quad \text{for all } b \in B$$

for a given λ if and only if all of its components f_μ do so (formula (4)). By the theorem of the highest weight, this implies that $f_\mu = 0$ for $\mu \neq \lambda$ and $f = f_\lambda$ is of the form $a \to \langle \pi_\lambda(a)v, v_{-\lambda}\rangle$ for a unique $v \in V^\lambda$. Here $v_{-\lambda}$ is the (up to scalars unique) non-zero vector in $(V^\lambda)^*$ of weight $-\lambda$; we used the isomorphism (c) $\to$ (a) of Lemma 9, §6.2. Conversely, every such function $a \to \langle \pi_\lambda(a)v, v_{-\lambda}\rangle$ is in F_λ, by the amplification (c) of Theorem 3, §6.5. This gives the desired isomorphism $V^\lambda \to F_\lambda, v \to f$. QED

Problems for §6.6

1. Let T be the unitary diagonal subgroup of $G = \mathrm{GL}(n, \mathbb{C})$. Show that for every $f \in F_\lambda$,

$$\int_T f(tat^{-1})\, dt = f(1) f_\lambda(a),$$

where f_λ is the (suitably normalized) highest weight vector in F_λ. The invariant integral over T is normalized so that $\int_T dt = 1$. Explain the normalization of f_λ.

2. (a) Let μ_k, $k = 1, 2, \ldots, n$, be the weight defined by

$$\mu_k = \lambda_1 + \cdots + \lambda_k.$$

Show that every weight λ can be uniquely written in the form $\lambda = m_1\mu_1 + \cdots + m_n\mu_n$ with $m_k \in \mathbb{Z}$ and that λ is dominant if and only if $m_1 \geq 0, \ldots, m_{n-1} \geq 0$ (m_n is arbitrary).

(b) Show that the highest weight vector f_λ in F_λ is given by the formula

$$f_\lambda(a) = (\det{}_1(a^{-1}))^{m_1} \ldots (\det{}_n(a^{-1}))^{m_n},$$

where $\det_k(a) = \det_{ij \leq k}(a_{ij})$ for $a \in M_n(\mathbb{C})$ and the exponents m_k are defined by $\lambda = m_1\mu_1 + \cdots + m_n\mu_n$.

3. Let V, π be a holomorphic representation of $G = \mathrm{GL}(n, \mathbb{C})$, V^N the subspace of V fixed by the unipotent upper triangular group N. Show that V^N is H-stable and that the multiplicity of the irreducible representation π^λ of G in V, π equals the multiplicity of the weight λ of H in $V^N, \pi|_H$.

4. (a) Let V', π' and V'', π'' be two holomorphic representations of $G = \mathrm{GL}(n, \mathbb{C})$. Show that the multiplicity of the irreducible representation V^λ, π^λ in $V' \otimes V''$, $\pi' \otimes \pi''$ equals the dimension of the space of N-maps $(V')^* \to V''$ satisfying $\pi''(h) \circ \varphi = e^\lambda(h)\varphi \circ \check{\pi}'(h)$. [Suggestion: problem 3 above and problem 4 , §6.1.]

   (b) Let $V^{\lambda'}, \pi^{\lambda'}$ and $V^{\lambda''}, \pi^{\lambda''}$ be the irreducible representations of G with the indicated highest weights. Use part (a) to show that the only possible λ for which π^λ can occur in $\pi^{\lambda'} \otimes \pi^{\lambda''}$ are of the form $\lambda = \mu + \lambda''$ where λ is a weight in $\pi^{\lambda'}$. [Suggestion: for a non-zero N-map φ: $(V^{\lambda'})^* \to V^{\lambda''}$, choose a weight ν on $(V^{\lambda'})^*$ so that $\varphi(v_\nu) \neq 0$ for some vector v_ν of weight ν, while no higher weight admits such a vector. Consider $\varphi(v_\nu)$. Generally, the weights in V^* are the negatives of the weights in V (why?).]

5. For each root $\alpha = \alpha_{jk}$ let $E_\alpha = E_{jk}$ be the corresponding root vector, and $H_\alpha = E_{jj} - E_{kk}$ the diagonal matrix with jth entry $1, k$th entry -1 and all other entries 0. Show that

$$\begin{bmatrix} 1 & 0 \\ 0 & -1 \end{bmatrix} \to H_\alpha, \quad \begin{bmatrix} 0 & 1 \\ 0 & 0 \end{bmatrix} \to E_\alpha, \quad \begin{bmatrix} 0 & 0 \\ 1 & 0 \end{bmatrix} \to E_{-\alpha}$$

   defines an isomorphism from $\mathfrak{sl}(2, \mathbb{C})$ to the subalgebra of $\mathfrak{sl}(n, \mathbb{C})$ spanned by $H_\alpha, E_\alpha, E_{-\alpha}$.

6.7 Representations of the classical groups

We now extend the representation theory of $\mathrm{U}(n)$ and $\mathrm{GL}(n, \mathbb{C})$ developed in §§6.4–6.6 to all compact or complex classical groups using the results of §§3.1–3.3. We shall of course not repeat those arguments that carry over without change.

As in §3.2, G denotes a complex classical group, $\mathrm{SL}(E)$, $\mathrm{SO}(E)$, or $\mathrm{Sp}(E)$, H the Cartan subgroup of G defined there, $K = G \cap \mathrm{U}(E)$ the unitary subgroup of G, $T = K \cap H$ the Cartan subgroup of K. Recall that K is a *real form* of G:

$$\mathfrak{g} = \mathfrak{k} + i\mathfrak{k} \tag{1}$$

a direct sum of real vector spaces. Just as for $\mathrm{GL}(n, \mathbb{C})$, this simple observation has important consequences:

Theorem 1 (Weyl's Unitarian Trick). *Restriction to K gives a one-to-one correspondence between the holomorphic representations of G and the representations of K. This correspondence preserves irreducibility and equivalence.*

Theorem 2 (Theorem of Complete Reducibility). *Every holomorphic representation of G is a direct sum of irreducible representations.*

The proofs are the same as those for the case $G = \mathrm{GL}(n, \mathbb{C})$, $K = \mathrm{U}(n)$ in § 6.6.

Theorem 4, §3.2 says that intersection with T provides a one-to-one correspondence between K-conjugacy classes in K and W-orbits in T. Weyl's Integration Formula (which we state here for class functions only) gives the expression for the integral of a function of the K-conjugacy classes in K when considered as a function of the W-conjugacy classes in T:

Theorem 3 (Weyl's Integration Formula for Class Functions). *For every class-function f on K,*

$$\int_K f = \frac{1}{|W|}\int_T f\Delta\bar{\Delta}, \tag{2}$$

where

$$\Delta = \prod_{\alpha\in\Phi_+}(e^{\alpha/2} - e^{-\alpha/2})$$

Explanation. The invariant integrals are normalized so that $\int_K dk = 1$ and $\int_T dt = 1$. The formula is understood in the sense that the integral on the left exists if and only if the integral on the right exists, and if so the integrals are equal. The product defining the *Weyl denominator* Δ is over the set Φ_+ of positive roots. The function Δ as it stands is a function on t and may not lift to a function on T: while e^α is always a well-defined weight of H (in virtue of its definition), its square root $e^{\alpha/2}$ is generally double valued. But the product $\Delta\bar{\Delta}$ is a well-defined function on T, because

$$\Delta\bar{\Delta} = \prod_{\alpha\in\Phi}(e^\alpha - 1).$$

We shall return to this point later.

The proof that Weyl's Integration Formula holds up to a constant factor is exactly the same as in the case $K = \mathrm{U}(n)$ (Theorem 6 of §5.3). To prove that the constant factor is 1 we used the formula

$$\Delta = \sum_{s\in W}\operatorname{sgn}(s)\, e^{s\rho}, \tag{3}$$

there a form of Vandermonde's determinant identity. The same formula, appropriately interpreted, is valid in general. To explain the meaning of this formula in general, define $\operatorname{sgn}(s)$ to be the determinant of $s \in W$ as a linear transformation of h, t or L, the real span of the roots. (The determinants are the same, $(-1)^l$, if s can be written as a product of l reflections.) The element $\rho \in L$ appearing in the formula (3) is defined by

$$\rho = \tfrac{1}{2}\sum_{\alpha\in\Phi_+}\alpha.$$

Its explicit expression as an n-tuple is listed in Table 6.1.

It is apparent from this table that *ρ is strictly dominant.* In type A_{n-1} there is the usual freedom of adding an n-tuple $(l,\ldots,l)$ to ρ. In type B_n ($K = \mathrm{SO}(2n+1)$) ρ is not T-integral, i.e. e^ρ does not lift to a well-defined

Table 6.1 The half-sum $\rho = 1/2 \sum_{\alpha \in \Phi_+} \alpha$

Type	ρ
A_{n-1}	$(n-1, \ldots, 1, 0)$
B_n	$(n - \frac{1}{2}, \ldots, \frac{3}{2}, \frac{1}{2})$
C_n	$(n, \ldots, 2, 1)$
D_n	$(n-1, \ldots, 1, 0)$

function on T, but is double-valued on T. While ρ need not be T-integral, $s\rho - \rho$ always *is*, as in fact

$$s\rho - \rho = \sum_{\alpha} \alpha, \quad \text{sum over } \alpha \in s\Phi_+ \cap -\Phi_+ \tag{4}$$

indeed the roots in $s\Phi_+ \cap \Phi_+$ cancel in $s\rho - \rho$, while those in $s\Phi_+ \cap -\Phi_+$ occur twice to produce the given expression.

The above definitions give a meaning to the formula (3). We now restate it formally as

Theorem 4 (Weyl's Denominator Formula). $\Delta = \sum_{s \in W} \operatorname{sgn}(s) e^{s\rho}$

The proof requires some further definitions and lemmas. Make W operate on functions of $\mathfrak{t}$ by setting

$$sf(h) = f(s^{-1} \cdot h)(s \cdot h = shs^{-1}).$$

For $f = e^{\lambda}$ this amounts to $se^{\lambda} = e^{s\lambda}$. If $sf = f$ for all $s \in W$, then f is called *W-symmetric* if $sf = \operatorname{sgn}(s)f$ for all $s \in W$, then f is called *W-antisymmetric*. Functions on T are frequently considered functions on $\mathfrak{t}$ by composition with $\exp : \mathfrak{t} \to T$, and the terminology '$W$-symmetric' and '$W$-antisymmetric' is applied to functions on T as well. We also recall that e^{λ}, $\lambda \in L$, lifts in this way from $\mathfrak{t}$ to T precisely when λ is *T-integral*, i.e. the components $(l_1, \ldots, l_n)$ of λ are integers.

The first step in the proof of Theorem 4 is

Lemma 5. *Δ is W-antisymmetric.*

Proof. Since W is generated by the s_α it suffices to prove that $s_\alpha \Delta = -\Delta$. The roots β for which $s_\alpha \beta \neq \beta$ occur in pairs β, $s_\alpha \beta$. For a root $\beta \in \Phi_+$ of this kind, set $\beta' = \pm s_\alpha \beta$, the sign being chosen so that $\beta' \in \Phi_+$ as well. If $\beta' \neq \beta$ the pair $\beta, \beta' \in \Phi_+$ contributes the factor

$$(e^{\beta/2} - e^{-\beta/2})(e^{\beta'/2} - e^{-\beta'/2}).$$

If $s_\alpha \beta = +\beta'$ then s_α interchanges these two factors; if $s_\alpha \beta = -\beta'$ then s_α interchanges these two factors and multiplies each by -1. In either case the product of the two remains unchanged. For a root $\beta \in \Phi_+$ with $s_\alpha \beta = \beta$ the factor $e^{\beta/2} - e^{-\beta/2}$ remains unchanged by s_β. Thus the only change comes from the factor $e^{\alpha/2} - e^{-\alpha/2}$, which contributes a minus sign. As a result the whole product Δ changes sign under the influence of s_α. QED

Lemma 6. *Let φ be a function on T of the form*

$$\varphi = \sum_\lambda n_\lambda e^{\lambda+\rho}, \tag{5}$$

finite sum over T-integral weights λ with integral coefficients $n_\lambda \in \mathbb{Z}$. Assume there is $\alpha \in \Phi$ so that $\varphi = 0$ when $e^\alpha = 1$. Then one can write

$$\varphi = (e^\alpha - 1)\psi,$$

where ψ is a function of the same type as φ.

Proof. Choose a basis $\mu_1, \ldots, \mu_r$ ($r = \dim L$) for L consisting of weights of the form $\lambda + \rho$ with λ T-integral so that every such $\lambda + \rho$ is an integral linear combination of the μ_j and $\mu_1 = \alpha/m$, some positive integer m. (It suffices to arrange that the λ_j and ρ be integral linear combinations of the μ_j, and this is possible with $\mu_1 = \alpha/m$, as one sees from the explicit formulas for the α; in fact one can take $m = 1$ or 2.) Set $\tau_j = e^{\mu_j}$. Equation (5) says that φ may be written as a polynomial $f(\tau_1, \ldots, \tau_r)$ in $(\tau_j)^{\pm 1}$ with integral coefficients. The assumption is that

$$f(\tau_1, \ldots, \tau_r) = 0 \quad \text{for } \tau_1^m = 1 \tag{6}$$

and we have to show that then

$$f(\tau_1, \ldots, \tau_r) = (\tau_1^m - 1)g(\tau_1, \ldots, \tau_r) \tag{7}$$

for a $g(\tau_1, \ldots, \tau_r)$ of the same type.

The relation (5) holds identically for $\tau_j \neq 0$ (all j) when the τ_j are interpreted as independent complex variables: by assumption it holds as function on t, therefore as function on $\mathfrak{h} = \mathfrak{t} + i\mathfrak{t}$ (by holomorphicity), therefore as function on H, therefore for $\tau_j \neq 0$ (all j). Furthermore, we may multiply $f(\tau_1, \ldots, \tau_r)$ by any monomial in the τ_j in order to prove (7). In particular we may assume that $f(\tau_1, \ldots, \tau_r)$ is a polynomial in the τ_j only (no inverses). Then it is clear that (7) holds for a polynomial $g(\tau_1, \ldots, \tau_r)$ whose coefficients are necessarily integral when those of f are. QED

Lemma 7. *Let φ be a W-antisymmetric function on T of the form*

$$\varphi = \sum_\lambda n_\lambda e^{\lambda+\rho},$$

finite sum over T-integral weights λ with integral coefficients $n_\lambda \in \mathbb{Z}$. Then one can write

$$\varphi = \Bigl(\prod_{\alpha \in \Phi_+} (e^\alpha - 1)\Bigr)\psi,$$

where ψ is a function of the same type as φ.

Proof. Let $\alpha \in \Phi_+$. Since φ is W-antisymmetric, $\varphi(s_\alpha \cdot t) = -\varphi(t)$ for all $t \in T$. Since $s_\alpha \cdot t = t$ when $e^\alpha(t) = 1$ this equation shows that $\varphi = 0$ when $e^\alpha = 1$. Thus one may apply Lemma 6 to write $\varphi = (e^\alpha - 1)\psi$. ψ is of the same type as φ and satisfies $\psi = 0$ when $e^\beta = 1$ for $\beta \neq \alpha$ in Φ_+. One may apply Lemma 6 again until one arrives at the stated equation. QED

Lemma 8. *Let ξ be a W-antisymmetric function on $\mathfrak{t}$ of the form*

$$\xi = \sum_\lambda n_\lambda e^{\lambda+\rho},$$

finite sum over T-integral weights λ with integral coefficients $n_\lambda \in \mathbb{Z}$. Then one can write

$$\xi = \Delta\chi,$$

where χ is a W-symmetric function of the form

$$\chi = \sum_\lambda m_\lambda e^\lambda,$$

finite sum over T-integral weights λ with integral coefficients $m_\lambda \in \mathbb{Z}$.

Proof. One has

$$\frac{\xi}{\Delta} = \frac{e^{-\rho}\xi}{e^{-\rho}\Delta}.$$

The numerator $e^{-\rho}\xi$ is of the form required in Lemma 7; the denominator $e^{-\rho}\Delta$ is

$$e^{-\rho}\Delta = \prod_{\alpha\in\Phi_+}(1 - e^{-\alpha}).$$

The assertion therefore follows from Lemma 7. QED

Proof of Theorem 4. By Lemma 8,

$$\sum_{s\in W} \operatorname{sgn}(s) e^{s\rho} = \Delta\chi$$

for some χ as in that lemma. Among the exponents $s\rho$ on the left side ρ is the unique highest one. For Δ one finds

$$\Delta = \prod_{\alpha\in\Phi_+}(e^{\alpha/2} - e^{-\alpha/2}) = e^\rho + \cdots$$

the dots indicating terms with lower exponents. As a quotient of two W-antisymmetric functions, χ is W-symmetric, and therefore contains the dominant W-conjugate of any of its terms. Suppose λ is a highest exponent among those in χ, m_λ its coefficient. Then $\lambda + \rho$ is a highest exponent among those in $\Delta\chi$, hence equals ρ. This leaves $\lambda = 0$ and $m_\lambda = 1$ as the only possibility, i.e. $\chi = 1$. QED

We now return to the question of the normalization of the integrals in (2). We assume the integrals are normalized so that $\int_K dk = 1$ and $\int_T dt = 1$. We know that then (2) holds up to a constant factor. To see that the factor is 1 it suffices to show that

$$\frac{1}{|W|}\int_T \Delta\bar{\Delta} = 1.$$

From Schur's orthogonality relations (or by direct evaluation) we know that for any two weights λ, μ,

$$\int_T e^{\lambda}e^{-\mu} = \delta_{\lambda\mu}.$$

If one applies this relation to the $|W|$ weights in $e^{\rho}\Delta$ one gets the desired relation $\int_T \Delta\bar{\Delta} = |W|$. (The factor e^{ρ} ensures that the exponents are integral and is harmless here since $\Delta\bar{\Delta} = e^{\rho}\Delta\overline{e^{\rho}\Delta}$.)

Finally we get to

Theorem 9 (Weyl's Character Formula). *As functions on T, the irreducible characters of K are given by the formula*

$$\chi_\lambda = \frac{1}{\Delta}\sum_{s\in W} \operatorname{sgn}(s)\, e^{s(\lambda+\rho)} \tag{8}$$

with λ T-integral and dominant.

Comment. (a) The formula may also be written as

$$\chi_\lambda = \frac{\sum_{s\in W} \operatorname{sgn}(s)\, e^{s(\lambda+\rho)-\rho}}{\prod_{\alpha\in\Phi_+}(1-e^{\alpha})}. \tag{9}$$

In this expression both numerator and denominator lift to T, which need not be the case in (8) (in type B).

(b) From Lemma 8 we know that the denominator in the quotient in (8) and (9) in fact divides the numerator, i.e. the quotient can be written in the form

$$\sum_{\mu} m_\mu e^{\mu},$$

as is required of a character.

Proof. The proof is now essentially the same as for $K = \mathrm{U}(n)$; there are three preliminary observations.

1. Let χ be a character of G. Set $\xi = \Delta\chi$ as function on $\mathfrak{t}$. Then ξ is antisymmetric and of the form

$$\xi = \sum_{\lambda} n_\lambda e^{\lambda+\rho}, \tag{10}$$

finite sum over T-integral weights λ with coefficients $n_\lambda \in \mathbb{Z}$.

This is immediate from the expressions

$$\chi = \sum_\lambda m_\lambda e^\lambda, \tag{11}$$

$$\Delta = e^\rho \prod_{\alpha\in\Phi_+} (1 - e^{-\alpha}). \tag{12}$$

2. For any $\kappa \in L$ set

$$\xi_\kappa = \sum_{s\in W} \operatorname{sgn}(s) e^{s\kappa}.$$

Any antisymmetric function ξ on $\mathfrak{t}$ of the form (10) can be written as

$$\xi = \sum_\lambda n_\lambda \xi_{\lambda+\rho}, \tag{13}$$

finite sum over *dominant* T-integral weights λ with the same coefficients $n_\lambda \in \mathbb{Z}$.

If λ is dominant, then $\lambda + \rho$ is regular, because ρ is strictly dominant. The $s(\lambda+\rho)$ are then distinct and the functions $e^{s(\lambda+\rho)}$ linearly independent, because $e^{s(\lambda+\rho)-\rho}$ are orthogonal functions on T. ξ, being odd, contains $\operatorname{sgn}(s)e^{s(\lambda+\rho)}$ with the same coefficient n_λ as $\lambda + \rho$. One may take for λ the unique dominant weight in its W-orbit (Proposition 2(c), §3.3) and combine the terms $\operatorname{sgn}(s)e^{s(\lambda+\rho)}$ to the sum $\xi_{\lambda+\rho}$ with the coefficient n_λ.

3. For dominant T-integral weights λ, μ,

$$\frac{1}{|W|}\int_T \xi_{\lambda+\rho}\bar{\xi}_{\mu+\rho} = \begin{cases} 1 & \text{if } \lambda = \mu, \\ 0 & \text{otherwise.} \end{cases} \tag{14}$$

The integrand is the sum over $s, r \in W$ of terms $\operatorname{sgn}(sr)e^{s(\lambda+\rho)-r(\mu+\rho)}$. The exponents are T-integral (because $s\rho - r\rho = s(\rho - s^{-1}r\rho)$ is a sum of roots in view of (4)); the integral is 0 unless the exponent is 0, which happens precisely when $s = r$.

Now let χ be an *irreducible* character. Write

$$\Delta\chi = \sum_\lambda n_\lambda \xi_{\lambda+\rho}$$

as in (13). The equation

$$\int_K \chi\bar{\chi} = 1$$

becomes

$$\sum_\lambda (n_\lambda)^2 = 1$$

in view of Weyl's integration formula (2) and eqn (14). Therefore there is a single λ with $n_\lambda = \pm 1$ and all other $n_\mu = 0$. Thus

$$\Delta\chi = \pm\xi_{\lambda+\rho}. \tag{15}$$

Let μ be a highest exponent in χ. Then $\mu+\rho$ is a highest exponent in $\Delta\chi$ (by formula (12) or (8)). This exponent $\mu+\rho$ must be the highest exponent $\lambda+\rho$ in $\pm\xi_{\lambda+\rho}$, and the multiplicity m_λ of e^λ in χ must equal the coefficient ±1 of $e^{\lambda+\rho}$ in $\pm\xi_{\lambda+\rho}$. Therefore $\mu=\lambda$ and the sign in (15) is $+$, i.e. $\Delta\chi=\xi_{\lambda+\rho}$, which is Weyl's formula.

It remains to show that for *every* dominant integral λ, $\xi_{\lambda+\rho}=\Delta\chi$ for some irreducible character χ. Assume, on the contrary, that $\xi_{\mu+\rho}$ is *not* of this form (μ dominant integral). Then

$$\int_T \xi_{\mu+\rho}\bar{\xi}_{\lambda+\rho}=0,$$

for all λ for which $\xi_{\lambda+\rho}$ is of the form $\xi_{\lambda+\rho}=\Delta\chi$. In view of Weyl's integration formula (2) this implies that for all irreducible χ

$$\int_K \left(\frac{\xi_{\mu+\rho}}{\Delta}\right)\bar{\chi}=0. \tag{16}$$

($\xi_{\mu+\rho}/\Delta$ is a W-symmetric function on T, which may be considered a class-function on K; it is analytic because of Lemma 8, but this is not even needed since the existence of the integral over K follows from the existence of the corresponding integral over T.) Because of the completeness of the irreducible characters (Peter–Weyl theorem), (16) implies that $\xi_{\mu+\rho}/\Delta=0$, which is absurd. QED

Note, incidentally, that for the trivial character $\chi\equiv1$ Weyl's formula reads $1=(1/\Delta)\sum_{s\in W}\operatorname{sgn}(s)e^{s\rho}$; we thus recover Weyl's Denominator Formula.

A first consequence of Weyl's character formula is

Theorem 10 (Weyl's Dimension Formula). *The degree of the irreducible representation with highest weight λ is*

$$\frac{\prod_{\alpha\in\Phi_+}(\lambda+\rho,\alpha)}{\prod_{\alpha\in\Phi_+}(\rho,\alpha)}.$$

Proof. The degree in question is $\chi_\lambda(1)$. Since the Weyl denominator Δ vanishes on singular elements of T, $\chi_\lambda(1)$ must be computed as the limit of $\chi_\lambda(t)$ as $t\to1$ from within the regular set in T, or from within the regular set in H, since χ_λ is a continuous function on H. We shall take $t=\exp(\tau\rho)$, ρ considered as an element of h: we identify h with h^* by means of the $\mathbb{C}$-bilinear extension of the inner product on $L=i\mathsf{t}^*$ to a bilinear form on h^*. The numerator of Weyl's Formula becomes:

$$\begin{aligned}\xi_{\lambda+\rho}(\exp(\tau\rho)) &= \sum_{s\in W}\operatorname{sgn}(s)e^{\tau\tau(s(\lambda+\rho),\rho)}\\ &= \Delta(\exp\tau(\lambda+\rho)) \quad \text{(by Theorem 4)}\\ &= \prod_{\alpha\in\Phi_+}(\tau(\lambda+\rho,\alpha)+\cdots),\end{aligned}$$

Table 6.2 Determinantal formulas for characters

Type	K	χ_λ	$\deg \chi_\lambda$
A_{n-1}	$\mathrm{SU}(n)$	$\dfrac{\lvert\epsilon^{k_1},\ldots,\epsilon^{k_n}\rvert}{\lvert\epsilon^{n-1},\ldots,\epsilon^{0}\rvert}$	$\dfrac{\prod_{i<j}(k_i-k_j)}{\prod_{i<j}(j-i)}$
B_n	$\mathrm{SO}(2n+1)$	$\dfrac{\lvert\epsilon^{k_1}-\epsilon^{-k_1},\ldots,\epsilon^{k_n}-\epsilon^{-k_n}\rvert}{\lvert\epsilon^{n-\frac{1}{2}}-\epsilon^{-n+\frac{1}{2}},\ldots,\epsilon^{\frac{1}{2}}-\epsilon^{-\frac{1}{2}}\rvert}$	$\dfrac{\prod_i k_i\prod_{i<j}(k_i\pm k_j)}{\prod_i(i-\frac{1}{2})\prod_{i<j}((j-\frac{1}{2})\pm(i-\frac{1}{2}))}$
C_n	$\mathrm{Sp}(n)$	$\dfrac{\lvert\epsilon^{k_1}-\epsilon^{-k_1},\ldots,\epsilon^{k_n}-\epsilon^{-k_n}\rvert}{\lvert\epsilon^{n}-\epsilon^{-n},\ldots,\epsilon^{1}-\epsilon^{-1}\rvert}$	$\dfrac{\prod_i k_i\prod_{i<j}(k_i\pm k_j)}{\prod_i i\prod_{i<j}(j\pm i)}$
D_n	$\mathrm{SO}(2n)$	$\dfrac{\lvert\epsilon^{k_1}+\epsilon^{-k_1},\ldots,\epsilon^{k_n}+\epsilon^{-k_n}\rvert\lvert\epsilon^{k_1}-\epsilon^{-k_1},\ldots,\epsilon^{k_n}-\epsilon^{-k_n}\rvert}{\lvert\epsilon^{n-1}+\epsilon^{-n+1},\ldots,\epsilon^{0}+\epsilon^{-0}\rvert\lvert\epsilon^{n-1}-\epsilon^{-n+1},\ldots,\epsilon^{0}-\epsilon^{-0}\rvert}$	$\dfrac{\prod_{i<j}(k_i\pm k_j)}{\prod_{i<j}((j-1)\pm(i-1))}$

Notation

A_{n-1}	$\mathrm{SU}(n)$	$k_1>\cdots>k_n,\ k_j\in\mathbb{Z}$
B_n	$\mathrm{SO}(2n+1)$	$k_1>\cdots>k_n>0\ k_j\in\frac{1}{2}+\mathbb{Z}$
C_n	$\mathrm{Sp}(n)$	$k_1>\cdots>k_n>0,\ k_j\in\mathbb{Z};$
D_n	$\mathrm{SO}(2n)$	$k_1>\cdots>\lvert k_{n-1}\rvert>k_n,\ k_j\in\mathbb{Z}$

as one sees by expanding the factors $e^{\alpha/2} - e^{-\alpha/2}$ in Δ. The dots indicate higher terms in τ. Similarly, for the denominator:

$$\Delta(\exp(\tau\rho)) = \prod_{\alpha\in\Phi_+} (\tau(\rho,\alpha) + \cdots).$$

From these formulas one sees that the quotient approaches the given expression as $\tau \to 0$. QED

The irreducible characters of the classical groups may be written as quotients of determinants, a notation often used by Weyl himself. We list these *determinantal formulas* along with the formulas for the degrees, in Table 6.2.

Explanation. $|\epsilon^{k_1}, \ldots, \epsilon^{k_n}|$ is the determinant of the matrix $(\epsilon_i^{k_j})$ whose ith rows are obtained from the one written down by replacing ϵ by ϵ_i. The other determinants are interpreted analogously. The n-tuple $(k_1, \ldots, k_n)$ represents the strictly dominant element $\kappa = \lambda + \rho$ of L. It has integral entries, except for $\mathrm{SO}(2n+1)$ where they are half-integral ($k_j \in (1/2) + \mathbb{Z}$); they are subject to the conditions listed under the heading '*Notation*' in the Table 6.2. In type A_{n-1} two n-tuples differing by an n-tuple of the form $(l, \ldots, l)$ are identified, as usual. The indices i and j run from 1 to n in all cases. The double sign in $\prod_{i\leq j}(k_i \pm k_j)$ indicates that both terms occur as factors.

An immediate consequence of Weyl's Character Formula and Proposition 2(c), §3.3 is the uniqueness of the highest weight, already noted during the proof of the formula.

Theorem 11 (Theorem of the Highest Weight). *Every irreducible representation of K has a unique highest weight; it occurs with multiplicity one and characterizes the representation up to equivalence. These highest weights are exactly the dominant weights.*

As for $\mathrm{GL}(n, \mathbb{C}) - \mathrm{U}(n)$ there is a reformulation of the Theorem of the highest weight leading to the Borel–Weil Theorem.

Let B denote the connected subgroup of G with Lie algebra $\mathsf{b} = \mathsf{h} + \mathsf{n}$, where

$$\mathsf{n} = \sum_{\alpha\in\Phi_+} \mathbb{C}E_\alpha$$

is spanned by the positive root vectors. B, or any subgroup of G conjugate to it, is called a *Borel subgroup* of G. n is a Lie algebra, and in fact an ideal of b, a consequence of the relations (6) and (7) of §3.2. Let N be the connected subgroup of G with Lie algebra n. We require the following lemma.

Lemma 12. *N is a normal subgroup of B. $B = HN$ and every $b \in B$ is uniquely of the form $b = hn$, $h \in H, n \in N$.*

Comment. Thus B is a *semidirect product*, as defined before problem 7, §2.5.

Proof. This could be verified using the explicit description of E_α; but it is simpler to proceed as follows. N is normal in B because n is an ideal of b. It follows that HN is a subgroup of B, which must be all of B, as its Lie algebra is $\mathfrak{h}+\mathfrak{n}=\mathfrak{b}$. It remains to prove the uniqueness property of the decomposition $B=HN$, which amounts to $H\cap N=1$.

Recall that E denotes the complex vector space on which $G\subset \mathrm{GL}(E)$ acts. Let

$$E=\sum_{\lambda}(E)_{\lambda}$$

be the decomposition of E into weight spaces $(E)_\lambda$ for H. For $x_\lambda \in (E)_\lambda$, $\alpha\in\Phi$, and $h\in H$,

$$h(E_\alpha x_\lambda)=(hE_\alpha h^{-1})hx_\lambda = e^{\lambda+\alpha}(h)E_\alpha x_\lambda,$$

so that $E_\alpha(E)_\lambda \subset (E)_{\lambda+\alpha}$. Choose an ordered basis for E compatible with the decomposition $E=\sum_\lambda (E)_\lambda$ so that for $\alpha\in\Phi_+$ the basis vectors in $(E)_{\alpha+\lambda}$ precede those in $(E)_\lambda$. The E_α are then strictly upper triangular. Thus $N=\exp(\mathfrak{n})$ is unipotent triangular while H is diagonal. The relation $H\cap N=1$ is now evident. QED

Since $B/N\approx H$, a homomorphism $e^\lambda : H\to \mathbb{C}^\times$ may be considered a homomorphism $e^\lambda : B\to\mathbb{C}^\times$:

$$e^\lambda(hn)=e^\lambda(h) \quad \text{for } h\in H,\ n\in N.$$

As in the case $G=\mathrm{GL}(n,\mathbb{C})$ one now gets:

Theorem 13 (Theorem of the Highest Weight, second version). *An irreducible holomorphic representation π of G has a joint eigenvector v for B, unique up to scalar multiples:*

$$\pi(b)v=e^\lambda(b)v \quad \textit{for all } b\in B.$$

$v=v_\lambda$ is the highest weight vector of π as representation of K, λ the highest weight.

Theorem 14 (Borel–Weil Theorem). *Let λ be a dominant weight. Let*

$$F_\lambda=\{f: G\to\mathbb{C} \mid f \textit{ is holomorphic and } f(ab)=e^\lambda(b)f(a) \textit{ for all } b\in B\}.$$

The representation of K on F_λ by left translation is the irreducible representation of highest weight λ.

The proofs of these theorems given in §6.6 for $G=\mathrm{GL}(n,\mathbb{C})$ apply word for word to any complex classical group G.

Problems for §6.7

1. Verify Table 6.1.

2. Prove Theorem 4 type-by-type with the help of Vandermonde's determinant identity.

3. Describe the representations of K of minimal degree for each of the four types A–D. (Ignore the trivial representation.)

4. Verify the determinantal formulas for the irreducible characters in Table 6.2.

5. Verify the degree formulas for the irreducible characters in Table 6.2.

6. Show that a representation all of whose weights are W-conjugate is irreducible. Give some examples of such representations.

7. For each root α let E_α be a corresponding root vector, and $H_\alpha = [E_\alpha, E_{-\alpha}]$. Show that E_α and $E_{-\alpha}$ may be normalized so that

$$\begin{bmatrix}1 & 0\\ 0 & -1\end{bmatrix} \to H_\alpha, \quad \begin{bmatrix}0 & 1\\ 0 & 0\end{bmatrix} \to E_\alpha, \quad \begin{bmatrix}0 & 0\\ 1 & 0\end{bmatrix} \to E_{-\alpha}$$

defines an isomorphism from $\mathsf{sl}(2, \mathbb{C})$ to the subalgebra of $\mathsf{sl}(n, \mathbb{C})$ spanned by $H_\alpha, E_\alpha, E_{-\alpha}$ [Problem 5, §6.6.]

8. *Integrality property of weights.* Let V, π be a holomorphic representation of $\mathfrak{g}$. Let λ be a weight of π. Show that

$$2\frac{(\lambda,\alpha)}{(\alpha,\alpha)} \in \mathbb{Z}$$

for every root $\alpha \in \Phi$. [Suggestion: recall that $s_\alpha\lambda = \lambda - 2((\lambda,\alpha)/(\alpha,\alpha))\alpha$; use problem 16 to reduce to a problem concerning $\mathsf{sl}(2, \mathbb{C})$ or $\mathsf{su}(2)$; then use Example 11, §2.6 and Example 5, §6.5.]

Problems 9–15 extend results from $\mathrm{GL}(n, \mathbb{C})$ and $\mathrm{U}(n)$ to any complex classical group G and its real form K. The extensions are routine. If the problems have previously been worked out for $\mathrm{GL}(n, \mathbb{C})$ and $\mathrm{U}(n)$ it will suffice to indicate the modifications needed, if any. It is mainly for reference that the results are listed here once more. The notation is as explained in this section.

9. *Kostant's formula for the weight-multiplicities.* (a) Show that

$$\frac{1}{\prod_{\alpha\epsilon\Phi_+}(1 - e^{-\alpha})} = \sum_\lambda P(\lambda)e^{-\lambda},$$

where $P(\lambda)$ is the number of ways of writing λ in the form $\lambda = \sum_{\alpha\epsilon\Phi_+} m_\alpha\alpha$ with non-negative integral m_α.

(b) Show that the multiplicity of a weight μ in the irreducible character χ_λ is $\sum_{s\in W} \mathrm{sgn}(s)P(s(\lambda+\rho) - (\mu+\rho))$. [Problem 7, §6.4.]

10. *Steinberg's formula for the multiplicities in tensor products.* Show that the multiplicity of χ_λ in $\chi_{\lambda'}\chi_{\lambda''}$ is given by

$$\sum_{s',s''\in W} sgn(s's'')P(s'(\lambda'+\rho) + s''(\lambda''+\rho) - \lambda - 2\rho)$$

[problem 8, §6.4].

11. *The Brauer–Weyl formula for the multiplicities in tensor products.* For any weight λ denote by λ_+ the dominant weight W-conjugate to λ. For regular λ (i.e. $l_j \neq l_k$ for $j \neq k$) set $\operatorname{sgn}(\lambda) = \operatorname{sgn}(s)$ where $s \in W$ is the unique element with $s\lambda = \lambda_+$; for singular λ set $\operatorname{sgn}(\lambda) = 0$. Show that

$$\chi_{\lambda'}\chi_{\lambda''} = \sum_{\mu} m_{\lambda'}(\mu)\operatorname{sgn}(\mu + \lambda'' + \rho)\chi_{(\mu+\lambda''+\rho)_+-\rho},$$

where $m_{\lambda'}(\mu)$ is the multiplicity of the weight μ in $\chi_{\lambda'}$ Deduce that the multiplicity of χ_λ in $\chi_{\lambda'}\chi_{\lambda''}$ is

$$\sum_{s\in W} \operatorname{sgn}(s)m_{\lambda'}(s(\lambda+\rho) - (\lambda''+\rho))$$

(problem 9, §6.4).

12. Assume $\mu + \lambda''$ is dominant for every weight μ in $\chi_{\lambda'}$. Show that the multiplicity of $\chi_{\mu+\lambda''}$ in $\chi_{\lambda'}\chi_{\lambda''}$ equals the multiplicity $m_{\lambda'}(\mu)$ of the weight μ in $\chi_{\lambda'}$ (problem 10, §6.4).

13. (a) Show that the multiplicity of χ_λ in $\chi_{\lambda'}\bar{\chi}_{\lambda''}$ is given by

$$\sum_{s\in W} \operatorname{sgn}(s)m_{\lambda}(s(\lambda''+\rho) - (\lambda'+\rho)).n$$

$\bar{\chi}$ is the contragredient character of χ (= complex conjugate of χ).

(b) Deduce from (a) that the multiplicity of the trivial character $\chi \equiv 1$ in $\chi_{\lambda'}\bar{\chi}_{\lambda''}$ is one if $\lambda' = \lambda''$ and zero otherwise. Explain why this is obvious in any case (problem 11 of §6.4).

14. Show that for every $f \in F_\lambda$,

$$\int_T f(tat^{-1})\,dt = f(1)f_\lambda(a),$$

where f_λ is the (suitably normalized) highest weight vector in F_λ. (problem 1, §6.6).

15. Let V, π be a holomorphic representation of G, V^N the subspace of V fixed by N. Show that V^N is H-stable and that the multiplicity of the irreducible representation π^λ of G in V, π equals the multiplicity of the weight λ of H in V^N, $\pi|_H$. [Notation as in this section. See also problem 3, §6.6.]

16. (a) Let V', π' and V'', π'' be two holomorphic representations of G. Show that the multiplicity of the irreducible representation V^λ, π^λ in $V' \otimes V''$, $\pi' \otimes \pi''$ equals the dimension of the space of N-maps $(V')^* \to V''$ satisfying $\pi''(h) \circ \varphi = e^\lambda(h)\varphi \circ \check{\pi}'(h)$.

(b) Let $V^{\lambda'}$, $\pi^{\lambda'}$ and $V^{\lambda''}, \pi^{\lambda''}$ be the irreducible representations of G with the indicated highest weights. Use part (a) to show that the only possible λ for which π^λ can occur in $\pi^{\lambda'} \otimes \pi^{\lambda''}$ are of the form $\lambda = \mu + \lambda''$ where λ is a weight in $\pi^{\lambda'}$. [Compare problem 15. Also problem 4, §6.6.]

Appendix: Analytic Functions and Inverse Function Theorem

We start by reviewing some facts from calculus, even though this material is assumed known. Let $f : \mathbb{R}^n \cdots \rightarrow \mathbb{R}^m$ be a map defined on an open subset of $\mathbb{R}^n$. (The broken arrow denotes a partially defined map; its domain of definition is assumed to be open.) f *is differentiable at a point* x of its domain if there is a linear map $L : \mathbb{R}^n \rightarrow \mathbb{R}^m$ so that

$$f(x+h) = f(x) + L(h) + o(h),$$

meaning

$$\lim_{h \to 0} \frac{\|f(x+h) - f(x) - L(h)\|}{\|h\|} = 0.$$

If so, the linear map L is called the *differential* (or *derivative*) of f at x, denoted df_x. Relative to the standard bases in $\mathbb{R}^n$ and $\mathbb{R}$, this linear map is represented by the matrix $(\partial f_i / \partial \xi_j)$ of the partials of the component functions f_i of f with respect to the standard coordinates ξ_j, but this fact is never used here to calculate differentials. Instead, we shall frequently calculate df_x using

$$df_x(v) = \left. \frac{df(x(\tau))}{d\tau} \right|_{\tau=0},$$

where $x(\tau)$ is any differentiable curve in $\mathbb{R}^n$ with $x(0) = x$ and $x'(0) = v$. This is a special case of the *Chain Rule*. The general *Chain Rule* says that

$$d(g \circ f)_x = (dg)_{f(x)} \circ df_x$$

provided the composite $g \circ f$ makes sense and the functions f and g are differentiable at the points in question. f is simply called *differentiable* if it is differentiable at every point of its domain; f is said to be of *class* C^k if its partials up to order k exist and are continuous. C^0 is understood to mean "continuous", C^∞ to mean "partials of all orders", and C^ω to mean "(real) analytic", that is, convergent Taylor series. C^1 implies "differentiable".

The most important fact for us will be the *Inverse Function Theorem*: If $f : \mathbb{R}^n \to \mathbb{R}^n$ is a C^k map, $1 \leq k \leq \infty, \omega$, with an invertible differential at a point $c \in \mathbb{R}^n$, then f is itself *locally* C^k invertible at c. This means in detail that f gives a one-to-one correspondence $U \to V$ between an open neighbourhood U of c and an open neighbourhood V of $f(c)$ whose inverse $V \to U$ is again of class C^k. As a consequence, if df_x is invertible for all x in an open set U, then $f(U)$ is open. If, in addition, f is one-to-one on all of U, then its inverse, defined on $f(U)$, is again of class C^k.

We shall use the expression "f is a C^k *bijection* between open subsets of $\mathbb{R}^n$" to mean that both f and its inverse are of class C^k on the open subsets in question. A C^ω bijection is also said to be *bi-analytic*. Without qualification, "*analytic*" will always be understood to mean "real analytic". For "complex analytic" the term *holomorphic* will also be used.

The definition of "differentiable" and "C^k" carries over from $\mathbb{R}^n$ to arbitrary (finite dimensional) real vector spaces in an evident manner. The same is true of the Inverse Function Theorem.

The Inverse Function Theorem stated above in full generality will actually only be used only in the analytic case $k = \omega$, except for the proof of Theorem 3 of §2.2, where the case $k = 1$ is used. In fact, if desired, the "$k = 1$" case could also be replaced by the "$k = \omega$" case by replacing "C^1" by "C^ω" in the definition of "tangent space" $\mathbf{g}$ at the beginning of §2.2.

It may be appropriate to mention at this point some matters of principle concerning this "C^1" vs. "C^ω" issue. The natural category of mappings to use in connection with Lie groups is C^ω. The reason is that Lie groups are in a natural way C^ω manifolds (§4.1), and their homomorphisms are analytic maps, even if initially only required to be C^1. (Even C^0 would suffice, but that is a deeper fact). For linear groups, this will become clear in §2.5, for general Lie groups in §4.3. And it is in order to see this phenomenon that we start with C^1 in §2.2, especially as this can be done with virtually no extra effort.

In view of the importance of analytic functions for the theory of Lie groups, we shall now develop the basic facts concerning such functions *ab initio*, leading up to the analytic Inverse Function Theorem. But this topic is strictly auxiliary, and the reader of some experience with analytic functions (e.g., from complex analysis), or who feels comfortable taking the basic facts for granted (e.g., in view of experience with differentiable functions and Taylor series) may certainly omit the remainder of this appendix. In the same vein, it should be pointed out that all that is to follow for real analytic functions applies equally to complex analytic functions, and could conversely be deduced from the complex case, using the observation that a real analytic function extends (locally) to a complex analytic function, defined by the same power series in a neighbourhood of each point of its domain. After these remarks we begin the formalities.

A function $f : \mathbb{R}^n \cdots \to \mathbb{R}^m$ defined on an open subset U of $\mathbb{R}^n$ is *analytic* (in U) if, in a neighbourhood of every $c \in U$, f admits a representation as a norm-convergent (defined below) power series

$$f(x) = \sum_k a_k (x - c)^k.$$

The sum is over all n-tuples $k = (k_1, \ldots, k_n)$ of integers ≥ 0, the coefficients a_k come from $\mathbb{R}^m$, and the powers are defined by

$$h^k = \eta_1^{k_1} \ldots \eta_n^{k_n}$$

for $h = (\eta_1, \ldots, \eta_n)$. The series will often be written as

$$f(c+h) = \sum_k a_k h^k.$$

To discuss the convergence of such power series, we use the following notation: $|h| = (|\eta_1|, \ldots, |\eta_n|)$, $r = (\rho_1, \ldots, \rho_n)$ and $|h| \leq r$ means $|\eta_1| \leq \rho_1, \ldots, |\eta_n| \leq \rho_n$; similarly for $\leq$ replaced by $<$. It is understood throughout that $r > 0$ in this sense. For an n-tuple of indices $k = (k_1, \ldots, k_n)$, we set

$$|k| = k_1 + \ldots + k_n$$

(which must be distinguished from the notation $|h|$ for $h \in \mathbb{R}^n$).

A simple but basic fact about convergence of power series is:

Lemma A.1 (Abel's Lemma). (a) *If $\sum_k \|a_k\| r^k < \infty$, then there is a constant M so that $\|a_k\| r^k \leq M$.*

(b) *If there is a constant M so that $\|a_k\| r^k \leq M$, then for all $0 < s < r$ there is a constant C so that $\sum_k \|a_k\| s^k \leq C$.*

Proof. (a) This is clear since in fact $\|a_k\| r^k \to 0$.

(b) Suppose $s < r$. Then,

$$\sum_k \|a_k\| s^k = \sum_k (\|a_k\| r^k) \frac{s^k}{r^k} \leq M \sum_k \frac{s^k}{r^k} = M \prod_{i=1}^n \left(\frac{1}{1 - \sigma_i/\rho_i} \right).$$

QED

The condition in (a), $\sum_k \|a_k\| r^k < \infty$, is meant when one says the series $\sum_k a_k h^k$ is *norm-convergent* (or *converges in norm*) for $|h| \leq r$. According to (b), a series that converges in norm for $|h| < r$ (i.e. $|h| \leq s$, all $s < r$) converges *uniformly* for $|h| \leq s < r$ and therefore, represents a continuous function on $|h| \leq r$. Thus, an analytic function is continuous on its domain. We shall prove shortly that it is in fact C^∞. But first we need:

Proposition A.2. *Suppose $f(x) = \sum_k a_k (x-c)^k$ is norm-convergent for $|x - c| < r$. Then $f(x)$ is analytic on $|x - c| < r$.*

Proof. We may as well assume $c = 0$. Thus, we assume $\sum_k \|a_k\| s^k < \infty$ for $0 < s < r$ and have to show that for any b with $|b| < r$, there is a representation of $f(b+h)$ as a power series in h for $|h| < t$, some $t > 0$. We start with the binomial expansion

$$(b+h)^k = \sum_{j \leq k} \binom{k}{j} b^{k-j} h^j.$$

The *binomial coefficients* $\binom{k}{j}$ are defined by this equation. They are explicitly given by the formula

$$\binom{k}{j} = \frac{k!}{j!(k-j)!},$$

where we use the notation

$$k! = k_1! \cdots k_n!.$$

For $|b+h| < r$, we have

$$f(b+h) = \sum_k a_k \left(\sum_{j \le k} \binom{k}{j} b^{k-j} h^j \right). \tag{A.1}$$

When $|b| \le s < r$, this series is norm-convergent for $|h| < r - s$. The proposition will follow if we show that the series may be re-arranged as a power series in h, which is permissible provided

$$\sum_k \sum_{j \le k} \|a_k\| \binom{k}{j} |b|^{k-j} |h|^j < \infty. \tag{A.2}$$

For fixed k, one has

$$\sum_{j \le k} \|a_k\| \binom{k}{j} |b|^{k-j} |h|^j = \|a_k\| (|b| + |h|)^k$$

so that (A.2) holds for $|b| + |h| < r$, that is $|h| < r - |b|$. QED

We remark that the formula (A.1) may be written as

$$f(b+h) = \sum_j \frac{1}{j!} D^j f(b) h^j,$$

where $D^j f(b)$ is obtained by formally differentiating the power series $f(x) = \sum_k a_k (x-c)^k$ and evaluating at $x = b$:

$$D^j \sum_k a_k (x-c)^k = \sum_{k \ge j} a_k \frac{k!}{(k-j)!} (x-c)^{k-j}.$$

We shall now show that this formal differentiation gives the derivatives of the function f in the usual sense.

Proposition A.3. *An analytic function f is differentiable, in fact C^∞. Its power series $\sum_k a_k (x-c)^k$, convergent in norm for $|x-c| < r$, may differentiated term-by-term; the resulting power series converges again in norm for $|x-c| < r$ and represents the corresponding (partial) derivative of f. Furthermore, the power series representation of f at c is unique and is given by* Taylor's Formula

$$f(x) = \sum_j \frac{1}{j!} D^j f(c) (x-c)^j.$$

Proof. The formula (A.5) for $f(c+h)$, valid for $|c|+|h| < r$, shows that $f(c+h) = f(c) + L(h) + o(h)$, where $L(h)$ is the part of the series that is of first order in h and $o(h)$ denotes here the remaining series, whose terms are all of order ≥ 2 in h, say

$$o(h) = \sum_{|k|\geq 2} c_k h^k.$$

This series is norm-convergent on $|h| < s$ for any $s < r - |c|$. Choose such an $s > 0$. From $\|c_k\| s^k \leq M$ one gets that $\|c_k\| \leq M/s^k$, hence

$$\|o(h)\| \leq \sum_{|k|\geq 2} \text{const.}\ \frac{|h|^k}{s^k} \leq M\|h\|^2 \sum_{|k|\geq 0} \frac{|h|^k}{s^k} \leq \text{const.}\|h\|^2.$$

It follows that f is differentiable at c with differential $df_c(h) = L(h)$. Since

$$L(h) = \sum_{|k|=1} a_k D^k f(c) h^k,$$

the first partials $D^k f(c)$, $|k| = 1$, of f at c are indeed obtained by differentiating the series for $f(x)$ term-by-term, as in (A.5), and evaluating at c. Since we already know that the series $D^k f(x)$ in powers of $(x - c)$ converges again for $|x - c| < r$ if the series for $f(x)$ does, the argument may now be applied with f replaced by $D^k f$, to prove the proposition as stated. QED

The next item concerns composition of analytic maps.

Proposition A.4. *Let $f : \mathbb{R}^n \cdots \to \mathbb{R}^m$ and $g : \mathbb{R}^m \cdots \to \mathbb{R}^l$ be analytic maps. Then $g \circ f$ is analytic where defined.*

Proof. We may assume $f(0) = 0$, $g(0) = 0$, and have to show that $g \circ f$ has a power series expansion around 0. Write

$$f(x) = \sum_k a_k x^k, \quad g(y) = \sum_j b_j y^j.$$

Choose $s > 0$ so that the series for $g(y)$ converges in norm for $|y| \leq s$, and then $r > 0$ so that the series for $f(x)$ converges in norm for $|x| \leq r$ and in addition

$$\sum_k |a_k| r^k < s$$

(componentwise inequality). Then for $|h| < r$,

$$g \circ f(x) = \sum b_j \left(\sum_k a_k h^k \right)^j. \tag{A.3}$$

We must show that the right side may be rewritten as a norm-convergent power series in h. This will be the case if

$$\sum |b_j| \left(\sum_k |a|_k r^k \right)^j < \infty \quad \text{(componentwise)}. \tag{A.4}$$

For when one multiplies out the series in (A.3) and (A.4), one gets a series of terms of the form $b_j a_k^p h^q$ and $|b_j||a_k^p||r|^q$, respectively, corresponding terms have the *same* non-negative, integral coefficients in both series and satisfy

$$b_j a_k^p h^q \leq |b_j||a_k^p||r|^q.$$

Thus, under the assumption (A.4) the series (A.3), when multiplied out, converges in norm and may therefore be re-arranged at will, in particular it may be written both in the form (A.3) and as a power series in h. The validity of (A.4) is a consequence of the choice of r, which guarantees that

$$\sum |b_j| \left(\sum_k |a_k| r^k \right)^j < \sum_j |b_j| s^j < \infty.$$

QED

Remark A.5. When one rewrites (A.3) as a power series in h, there are actually only finitely many terms which contribute to the coefficient of a given power h^l: the terms with $|j| > |l|$ make no contribution, as all non-zero terms $a_k h^k$ of the inner series have $|k| \geq 1$ (in view of the condition $f(0) = 0$).

We now come to what is the main point for us:

Theorem A.6 (Inverse Function Theorem). *Let $f : \mathbb{R}^n \cdots \to \mathbb{R}^n$ be an analytic map, $c \in \mathbb{R}^n$ a point in the domain of f so that df_c is invertible. Then there is a neighbourhood U of c and a neighbourhood V of $f(c)$ so that $f : U \to V$ is bijective and the inverse map $g : V \to U$ is analytic on V.*

Proof. We may assume $c = 0$ and $f(c) = 0$. We show that there is a unique power series $g(y)$, norm-convergent in a neighbourhood of 0, so that $g(0) = 0$ and $f \circ g(y) \equiv y$. g has then also an invertible differential at 0 (because $df_0 \circ dg_0 = 1$), so we may apply the same result with f replaced by g to conclude that $g \circ f(x) \equiv x$ as well. Replacing f by $f \circ (df_0)^{-1}$, we may assume $df_0 = 1$, in addition to $f(0) = 0$. We can then write $f(x)$ in the form

$$f(x) = x - \sum_{|k| \geq 2} a_k x^k = x - F(x). \tag{A.5}$$

(The reason for the minus sign will become clear presently.) We have to find a convergent power series $g(y) = \sum_{|j| \geq 1} b_j y^j$ so that $x = g(y)$ satisfies

$$y = x - F(x) \tag{A.6}$$

identically. We write this equation as

$$x = y + F(x)$$

or explicitly with $x = g(y)$:

$$\sum_j b_j y^j = y + \sum_k a_k \left(\sum_i b_i y^i \right)^k . \tag{A.7}$$

There are no terms of degree 0, and comparing terms of degree 1 one finds that $\sum_{|j|=1} b_j y^j = y$, so that these b_j with $|j| = 1$ are the standard basis vectors of $\mathbb{R}^n$. Generally, b_j is a linear combination with *positive* integral (binomial) coefficients of terms $a_k b_i^l$ with k and i strictly less than j. Successively substituting for the previous b_i we find that

$$b_j = P_j(a_k), \tag{A.8}$$

where the $P_j(a_k)$ are linear combinations of powers a_k^l, $k < j$, with positive integral coefficients independent of F. This observation is the basis of Cauchy's *method of majorants*, which we now employ.

Suppose we can find a power series $\tilde{f}$ of the same sort as f, written as $\tilde{f}(x) = x - \tilde{F}(x)$, with $\tilde{F}(x) = \sum_{|k|>2} \tilde{a}_x x^k$, for which the corresponding $\tilde{g}(y) = \sum_j \tilde{b}_j y^j$ converges in norm for $|y| < s$. Suppose further that

$$|a_k| < \tilde{a}_k \tag{A.9}$$

for all k. Then, the coefficients b_j of $g(y)$ satisfy

$$|b_j| = |P_j(a_k)| \leq P_j(|a_k|) \leq P_j(\tilde{a}_k) = \tilde{b}_j,$$

from which it follows that $g(x) = \sum_j b_j y^j$ converges for $|y| < s$ as well.

Such an $\tilde{f}(x) = x - \tilde{F}(x)$ may be constructed explicitly as follows. In the scalar case, $n = 1$, we write $\tilde{\varphi}(x) = x - \tilde{\Phi}(x)$ for $\tilde{f}$ and take for $\tilde{F}$ the series

$$\tilde{\Phi}(x) = M \sum_{k \geq 2} (mx)^k = \frac{M(mx)^2}{1 - mx}$$

with $M, m > 0$ to be determined. Abel's Lemma shows that (A.9) is then satisfied for appropriate choice of M and m. Furthermore, the solution of

$$y = x - \frac{M(mx)^2}{1 - mx}$$

that equals 0 when $y = 0$ is

$$x = \tilde{\psi}(y) = \frac{(1+m)y - \sqrt{(1+my)^2 - 4(m + Mm^2)y}}{2(m + Mm^2)}. \tag{A.10}$$

This function may be expanded into a convergent power series about $y = 0$, and this gives the desired inverse function for $f(x)$ in the scalar case $n = 1$. For general n, take $\tilde{f}(x) = x - \tilde{F}(x)$ with

$$\tilde{F}(x) = \tilde{\Phi}(\lambda(x))e,$$

where $\lambda(x) = \sum_i \xi_i$ and $e = 1/n(1, 1, \ldots, 1)$. From the definition of $\tilde{\Phi}$ one sees that the coefficients $\tilde{a}_k$ satisfy (A.9) provided M and m are chosen suitably. The function $\tilde{f}$ is now

$$\tilde{f}(x) = x - \tilde{\Phi}(\lambda(x))e,$$

and its inverse $\tilde{g}$ is explicitly given by

$$\tilde{g}(y) = y + \{-\lambda(y) + \tilde{\psi}(\lambda(y))\}e.$$

To check this, substitute (using the linearity of $\lambda(x)$ in x and $\lambda(e) = 1$):

$$\begin{aligned}
\tilde{f}(\tilde{g}(y)) &= y + \{-\lambda(y) + \tilde{\psi}(\lambda(y))\}e - \tilde{\Phi}\left(\lambda(y) + \{-\lambda(y) + \tilde{\psi}(\lambda(y))\}\right)e \\
&= y - \lambda(y)e + \{\tilde{\psi}(\lambda(y)) - \tilde{\Phi}(\tilde{\psi}(\lambda(y)))\}e \\
&= y - \lambda(y)e + \tilde{\varphi}(\tilde{\psi}(\lambda(y)))e \\
&= y.
\end{aligned}$$

This finishes the proof of the theorem. QED

Reference. The material of this appendix is classical; the presentation follows Serre (1965).

References

(Not meant as a guide to the literature; only publications actually cited are listed.)

Artin, E. (1957). *Geometric Algebra*, Interscience Publishers, New York.

Baker, H. F. (1905). Alternants and continuous groups, *Proc. London Math. Soc.*, **3**, 24–47.

Borel, A. and Weil, A. (1954). Représentations linéaires et espaces homogènes Kählerians de groupes de Lie compacts. Séminaire Bourbaki (Exposé de J.-P. Serre.).

Bourbaki, N. (1960). *Eléments de Mathématiques, Groupes et Algèbres de Lie*, Fascicule XXVI, Chapitre 1, Hermann, Paris.

Campbell, J. E. (1897). On a law of combination of operators bearing on the theory of continuous transformation groups, *Proc. London Math. Soc.*, **28**, 381–90.

Campbell, J. E. (1898). On the law of combination of operators (second paper), *Ibid.*, **39**, 14–32.

Cartan, E. (1913). Sur les groupes projectifs qui ne laissent invariante aucune multiplicité plane, *Bull. Soc. Math. France*, **41**, 53–96.

Cartan, E. (1927). La géométrie des groupes simples, *Annali di Mat.*, **4**, 209–56.

Cartan, H. (1963). *Elementary Theory of Analytic Functions of One or Several Complex Variables*, Hermann, Paris, and Addison Wesley, Reading, Mass.

Duistermaat, J. and Kolk, J. (1988). *Lie Groups*, Springer Verlag, New York.

Dynkin, E. (1947). Calculation of the coefficients of the Campbell–Hausdorff formula, *Dokl. Akad. Nauk.*, **57**, 323–6 (in Russian).

Euler, L. (1775). Formula generales pro translatione quacunce corporum rigidorum. Novi Comment. *Acad. Sci. Petropolitanae*, **20**.

Freudenthal, H. (1941). Die Topologie der Lieschen Gruppen als Algebraisches Phänomen. I, *Annals of Math.*, **42**, 1051–74.

Freudenthal, H. and de Vries, H. (1969). *Linear Lie Groups*, Academic Press, New York.

Frobenius, G. (1896). *Über Gruppencharaktere*, Sitzungsber. Preuss. Akad., 985–1021.

Frobenius, G. (1896). *Über die Primfactoren der Gruppendeterminante*, Sitzungsber. Preuss. Akad., 1343–82.

Gutkin, E. (1986). Geometry and combinatorics of groups generated by reflections, *L'Enseignement Mathématique*, **32**, 95–110.

Hausdorff, F. (1906). Die symbolische Exponentialformel in der Gruppentheorie, *Leibziger Ber.*, **58**, 19–48.

Helgason, S. (1978). *Differential Geometry, Lie Groups, and Symmetric Spaces*, Academic Press, New York.

Herstein, I. N. (1964). *Topics in Algebra*, Blaisdell Publishing Co., Waltham.

Hilbert, D. and Cohn-Vossen, S. (1952). *Geometry and the Imagination*, Chelsea Publishing Co., New York.

Hoffman, K. and Kunze, R. (1961). *Linear Algebra*, Prentice Hall, Englewood Cliffs.

Humphreys, J. (1972) *Introduction to Lie Algebras and Representation Theory*, Springer Verlag, New York.

Hurwitz, A. (1897). *Nachr. Gött. Ges. Wissensch.*, 71.

Jacobi, C. G. J. (1827). Euleri formulae de transformatione coordinatarum, *Crelle J.*, **2**, 188–9.

Jacobi, C. G. J. (1841). De functionibus alternantibus earumque divisione per productum e differantiis elementorum conflatum, *Crelle J.*, **22**, 360–71.

Jacobi, C. G. J. (1868). Über diejenigen Probleme der Mechanik in welchen eine Kräftefunction existirt und über die Theorie der Störungen. *Ges. Werke V*, 217–395.

Jacobson, N. (1979). *Lie Algebras*, Dover Publications, Inc., (first published 1966).

Kronecker, L. (1884). *Näherungsweise ganzzahlige Auflösung linearer Gleichungen*, Monatsber. Preuss. Akad., 1179–93.

Lang, S. (1969). *Analysis*, Addison-Wesley, Inc., Reading, Mass.

Lie, S. *Theorie der Transformationsgruppen I, II, III.* Unter Mitwirkung von Prof. Dr. F. Engel, Teubner, Leipzig, 1888, 1890, 1893.

Marsden, J. (1974). *Elementary Classical Analysis*, W. H. Freeman and Co., San Francisco.

Massey, W. S. (1967). *Algebraic Topology: An Introduction*, Springer Verlag, New York.

Nelson, E. (1967). *Tensor Analysis*, Princeton University Press and The University of Tokyo Press, Princeton.

Poincaré, H. (1952–54). *Oeures complètes*, Gauthier-Villars, Paris.

Poincaré, H. (1899). *Sur les groupes continus.* Cambridge Philos. Trans., **18** 220–55, = Oevre, t. 3, 173–212.

Schur, F. (1891). Zur Theorie der endlichen Transformationsgruppen, *Abh. Math. Sem. Univ. Hamburg*, **4**, 15–32.

Schur, I. (1924). Neue Anwendungen der Integralrechnung auf Probleme der Invariantentheorie, I, II, III, *Preuss. Akad.*, 189, 297, 346.

Serre, J.-P. (1965). *Lie Algebras and Lie Groups*, W. A. Benjamin, Inc., New York.

Serre, J.-P. (1966). *Algèbres de Lie semi-simples complexes*, W. A. Benjamin, Inc., New York.

Spivak, M. (1979). *Differential Geometry*, vol. I, Publish or Perish, Inc., Berkeley.

Steinberg, R. (1961). A general Clebsch–Gordan theorem, *Bull. Amer. Math. Soc.*, **67**, 406–7.

Weyl, H. (1950). *The Theory of Groups and Quantum Mechanics*, Dover, New York, orig. edition 1931.

Weyl, H. (1939). *The Classical Groups*, Princeton University Press, Princeton.

Weyl, H. Theorie der Darstellung kontinuierlicher halb-einfacher Gruppen durch lineare Transformationen, I, II, III, *Math. Zeitschrift* (1925) 271–309, (1926) 328–76, 377–95, 789–91.

Index